Staphylococcal Infections (Host and Pathogenic Factors)

Staphylococcal Infections (Host and Pathogenic Factors)

Editor

Rajan P. Adhikari

MDPI • Basel • Beijing • Wuhan • Barcelona • Belgrade • Manchester • Tokyo • Cluj • Tianjin

Editor
Rajan P. Adhikari
Bacterial Therapeutics
Integrated Biotherapeutics inc.
Rockville
United States

Editorial Office
MDPI
St. Alban-Anlage 66
4052 Basel, Switzerland

This is a reprint of articles from the Special Issue published online in the open access journal *Microorganisms* (ISSN 2076-2607) (available at: www.mdpi.com/journal/microorganisms/special_issues/Staphylococcal_Infections).

For citation purposes, cite each article independently as indicated on the article page online and as indicated below:

LastName, A.A.; LastName, B.B.; LastName, C.C. Article Title. *Journal Name* **Year**, *Volume Number*, Page Range.

ISBN 978-3-0365-1418-5 (Hbk)
ISBN 978-3-0365-1417-8 (PDF)

Cover image courtesy of ABHIRA

Contents

About the Editor

Rajan P. Adhikari

Dr. Adhikari has 22+ years of experience as a bacteriologist and molecular biologist. He completed his PhD at the University of Otago, New Zealand, where he characterized the newly emerged cMRSA. He spent a year of his PhD research at the University of Zurich in the laboratory of Prof. Brigitte Berger-Bachi. He is a recipient of the NZODA scholarship, New Zealand. After completion of his PhD, he joined Dr. Richard P. Novick's laboratory in New York University as a post-doctoral fellow where he worked on the molecular pathogenesis of S. aureus for five years. For the past twelve years, he has been working with bacterial vaccines and therapeutic areas at IBT. His research focuses on the vaccine and therapeutic development against Gram-positive and -negative pathogens. He is an expert in bacterial pathogenesis, small molecules, the microbiome, and phage therapy. He has successfully secured and completed multiple NIAID/NIH-funded grants as a principal investigator in vaccine and therapeutics areas.

microorganisms

Editorial

Staphylococcal Infections: Host and Pathogenic Factors

Rajan P. Adhikari

Integrated Biotherapeutics Inc., Rockville, MD 20878, USA; rajan@IntegratedBiotherapeutics.com

check for updates

Citation: Adhikari, R.P. Staphylococcal Infections: Host and Pathogenic Factors. *Microorganisms* **2021**, *9*, 1080. https://doi.org/10.3390/microorganisms9051080

Received: 20 April 2021
Accepted: 21 April 2021
Published: 18 May 2021

Publisher's Note: MDPI stays neutral with regard to jurisdictional claims in published maps and institutional affiliations.

In 1880, the Scottish surgeon Sir Alexander Ogston first described staphylococci in pus from a surgical abscess in a knee joint: "The masses looked like bunches of grapes" [1]. In 1884, the German physician Friedrich Julius Rosenbach differentiated the staphylococci by the color of their colonies: *S. aureus* (from the Latin *aurum*, gold) [2]. For another 20 years, very little was known on the pathophysiology of this bug. Based on a PubMed search record, the first Staphylococcal paper was published in 1900 on a case report [3]. More and more scientists gradually engaged to study diseases caused by this bacterium. There were some 10 publications recorded during the period 1900–1910, which translate in average to one publication/year. Research on this bug exploded in the 20th century, which is reflected in a recent PubMed search. It yields 47,974 publications records when searched using the keyword "Staphylococcus" in the title. During the period 2010–2020, the average publication was 2000 articles/year. This record makes Staphylococcus the single most researched bacterium based on this publication track record.

Over time, numerous Staphylococcus species were discovered, consisting of more than 45 staphylococcal species and 24 subspecies classified using molecular methods [4]. These various species of Staphylococcus are clinically important as 30% of the healthy human population is colonized with various Staphylococcus spp. Some strains are opportunistic pathogens and can cause a minor infection to life-threatening diseases. Pathogenicity of these different strains depends on several virulence factors: Level of protein expression as well as the robustness of the regulatory networks expressing these virulence factors. These factors consist of numerous toxins, enterotoxins (some of which act as superantigens), enzymes, and proteins with other functions (cytoplasmic, extracellular, and surface) that are tightly regulated by two-components (TC), transcriptional and translational regulators, as well as quorum-sensing (QS) regulatory networks [5]. This Special Issue is dedicated to the studies and recent advancements in our understanding of staphylococcal virulence mechanisms that enable *Staphylococcus* spp. either to successfully establish themselves as a colonizer or to overcome the host's defense system to cause infection.

Fourteen wonderful papers are included in this issue with a wide spectrum of Staphylococcal research. A vaccine paper by Dr. Anderson from the Pfizer Vaccine group entitled "Performance of a four-Antigen *S. aureus* vaccine in preclinical models of Invasive diseases [6]" clearly reflects a critical problem faced by all vaccine companies struggling to demonstrate that these vaccines are clinically efficacious so that they can be approved by regulatory agencies: A lack of correlation between preclinical efficacies with human clinical trials [6]. The SA4Ag vaccine described in this paper clearly demonstrates the significant decrease in the organ bacterial loads in a deep tissue infection, a bacteremia, a pyelonephritis mouse model, as well as a complete protection of endocarditis in a rat model, which is still not enough to provide significant protection in human surgery-associated invasive *S. aureus* infection.

This is not a single case. Most of the anti-staphylococcal vaccine and therapeutics that failed in clinical trials 2 and 3 have similar stories. Merck V710, an *S. aureus* iron-regulated surface determinant B (IsdB) vaccine, provided a significant protection in different animal models (both by active and passive immunization) [7–11], whereas it failed in a blinded randomized trial [12]. Anti-toxin engineered mAb successfully neutralized six major *S. aureus* toxins in an in vitro study, as well as exhibited great efficacies in various animal models [13–16], but failed in a human clinical trial (https://clinicaltrials.gov/ct2/show/NCT02940626 (accessed on 19 April 2021)) due to the lack of enough efficacy. In recent years, these failed efficacy studies sparked a clear debate among the Staphylococcal vaccinologists and therapeutic scientists into two schools of thought: One believes in surface protein and other believes in extracellular toxins and proteins as better targets. A great review by Millar et al. [17] has published evidence that extracellular toxins such as pore-forming toxins (PFT) and superantigens targeted therapeutics/vaccine are more likely to provide better protection over the approaches to induce antibodies to facilitate opsonophagocytosis [17]. Numerous recent published papers focused on targeting secreted toxins and virulence factors [18–23] as vaccine targets. Again, in animal models, great protection and efficacy were reported in this approach. On the other side, many scientists believe that the surface proteins, capsule, and cell wall structure such as the Wall teichoic acid (WTA) [24–28] and lipoteichoic acid (LTA) are better targets which are well characterized. In addition, the important virulence factors helping bacterial adhesion and invasions are an important target for future vaccine and therapeutics development. Equally convincing data are available to support this hypothesis (in different animal models) [11,29–33]. Not only the vaccine, but also different therapeutics options for anti-staphylococcal infections failed to provide enough protection in the human clinical trial even though their preclinical data in animal models were impressive [12,13,28]. I completely agree and hope that all the scientists working in these areas agree with the author [6], in that a clear animal efficacy model is needed for every intended vaccine and therapeutic testing, which correlates with the human clinical outcome before going into these expensive clinical trials to save resources, human subjects, as well as time.

A paper by Dr. Holtfreter, "Discovery of *S. aureus* Adhesion Inhibitors by Automated Imaging and Their Characterization in a Mouse Model of Persistent Nasal Colonization" describes a novel automated high throughput screening method that can quantify the bacterial adhesion in human epithelial cells [34]. Since adhesion inhibitors/blockers/neutralizers interfere with the entry of pathogens into the cell, it will be a highly effective treatment for the first line of defense [35–38]. Many studies have generated high-quality data in favor of arresting pathogens in these stages [39–44].

A paper by Dr. Bischoff's group from Saarland University explained the role of histidine-containing phosphocarrier protein HPr (encoded by *ptsH*) in carbon catabolite repression (CCR) and infectivity [45]. CCR has been established as a connector for metabolome to the virulence factors [46–53]. Though the impact of HPr on CCR is well studies in other Gram positive bacteria, it was largely unknown in *S. aureus* This paper clearly demonstrates that the inactivation of *ptsH* alters the transcription of genes involved in the TCA cycle as well as alpha hemolysin, a well-characterized virulence factor in *S.aureus*. A significant reduction in biofilm production was reported in the *ptsH* mutant under static and flow conditions, which correlates in a reduction in CFU/catheter fragment in a *S. aureus*-based murine foreign body infection model. Putting these data together, there is a clear potential of *ptsH* to be a target gene for vaccine and therapeutics. *srt*A is an another gene function reported in this Special Issue by Dr. Becker's group [54]. Though well characterized in *S. aureus* [55–59], very little is known about the role of this gene in another species: *S. lugdunensis*. This paper demonstrates that though there is no significant decrease in adherence and invasion in a human cell line, the mutant *srt*A exhibited a decrease in biofilm production, as well as affected the transcription of two different adhesins genes (Fbl and vWbF). Since *srt*A in different pathogens is considered as a therapeutic target [60,61], the characterization of this locus in new species is critical to understanding the pathogen.

Two papers included in this issue are focused on the characterization of *S. aureus* biofilm inhibitors. Dr. Reigada reported the "Combined Effect of Naturally-Derived Biofilm Inhibitors and Differentiated HL-60 Cells in the Prevention of *S. aureus* Biofilm Formation" [62] and the other by Dr. Jimi from Fukuoka University on "The Effects of Silver Sulfadiazine on Methicillin-Resistant *S. aureus* Biofilms" [63]. Staphylococcal biofilm is still a huge health burden throughout the world as 80% of the infection in medical device associated infection (usually hospital acquired) results in biofilm. In the former paper [62], three naturally derived biofilm inhibitors: Dehydroabietic acid (DHA) 1 and 2 and the third one flavan derivative (FLA1) were tested along with differentiated HL-60 cells in implantable titanium devices and low-density polyethylene endotracheal tubes. Out of the three tested inhibitors, DHA1 exhibited the optimal anti-biofilm profile in coculture. These compounds can have a great potential for anti-biofilm drugs in medical devices. In a later paper [63], anti-biofilm activities of the existing drug silver sulfadiazine is reported.

Dr. Månsson in her paper "Methicillin-Resistant *S. epidermidis* (MRSE) Lineages in the Nasal and Skin Microbiota of Patients Planned for Arthroplasty Surgery" [64] reports a clinical study where 45% of patients were colonized with MRSE, among them 15% were with multidrug-resistant strains. A previously reported lineage associated with prosthetic joint infections was among the isolates. This type of screening prior to hospitalization and proper antimicrobial prophylaxis helps minimize cross infections. Dr. Dekker reported the molecular characterization of the *S. aureus* isolates from chronically infected wounds in rural Ghana [65]. Twenty-eight isolates were characterized by whole genome sequencing and the resistance profile was analyzed to determine the population structure of the isolates in the rural part.

The bacterial strain-specific model development in different animals is very critical for efficacy testing of the vaccine, therapeutics, as well as an antimicrobial development pipeline. An epidemiological study for the predominant clones and animal model development based on these existing clones are vital for the success of any drug and biologics. In a very elaborate study, Dr. Zhang compares "Community-Associated MRSA Strain USA300 from Other MRSA Strains in A Murine Skin Infection Model" [66]. In a dermatopathology readout, USA300 induced dermonecrosis with extensive open [67] lesions, increased neutrophil recruitment, and increased the production of cytokines associated with disease severity when compared with USA400 and M92 (colonizing control) strains. This study is highly relevant as USA300 is still one of the leading causes of community and hospital-acquired infections in the USA, as well as many countries around the world [68–73].

A paper by Dr. Ehrhardt's group from Jena University reported a significant impact on the regulation of pro-inflammatory factors contributing to a synergistic effect on cells' intrinsic innate response when a human *S. aureus* small colony variants colonizer is subsequently infected with influenza virus [74].

Dr. Johler's group from the University of Zurich reported the possibility of post-transcriptional modifications in SEC expression under lactic acid stress conditions, in some strains of *S. aureus*, based on the difference in the level of sec mRNA with protein [75]. Since enterotoxins are a major cause of staphylococcal food poisoning in humans [76,77], the pH-dependent regulations (transcription and translations) of these genes under different food storage conditions are of interest in terms of food safety as well as food shelf life.

A paper by Dr. Mishra, Dr. Bayer, and Dr. Rose's group from UCLA on "Cell Membrane Adaptations Mediate β-Lactam-Induced Resensitization of Daptomycin-Resistant (DAP-R) *Staphylococcus aureus* In Vitro" reported the multiple mechanisms involving the resensitization of DAP-R back to sensitive strains by prolonged exposure to cloxacillin [78]. The reported mechanisms involved the accumulation of multiple point mutations in the *mpr*F gene, resulting in overall changes in cell membrane composition and function [79]. This study is highly significant during the era in which most antibiotics are ineffective due to the emergence of drug resistance and a very limited discovery pipeline because of its high cost. Understanding these resistance mechanisms will help to repurpose these old drugs which are relatively cost-effective [80–82].

There are two review papers included in this issue. One is "Human *mecC*-Carrying MRSA: Clinical Implications and Risk Factors" [83]. This relatively new *mecC* type first reported in 2011 has now been reported in animals as well as in humans. This review covers epidemiological data for *mecC* carrying MRSA strains including the resistance profile, and virulence factors associated with different clonal complexes. Another is, "No Change, No

Life? What We Know about Phase Variation in *Staphylococcus aureus*" [84] by Dr. Gor. In this review based on the relatively new topic, the author discusses a different aspect of gene switching and phase variation in *S. aureus*. Since, heterogeneity and phase variations are common phenomena in *S. aureus* [85,86] and can be clearly visible in terms of an antibiotic heterogeneous population, small colony variants, and persister colonies [74,87–92], this review shed the light on the possible mechanism of these genes switching.

Overall, this issue has an impressive participation of scientists throughout the world. We would like to thank all the authors, contributors, and reviewers for their valuable time and their important contribution to this Special Issue.

Funding: The APC was funded by AI136143 from National Institute of Allergey and Infectious Diseases.

Conflicts of Interest: This article is an opinion of the author and does not reflect the company's interest.

References

1. Classics in infectious diseases. On abscesses. Alexander Ogston (1844–1929). *Rev. Infect. Dis.* **1984**, *6*, 122–128.
2. Rosenbach, A. *Mikro-Organismen bei den Wund-Infections-Krankheiten desMenschen*; Wiesbaden, J.F., Ed.; Bergmann: Wiesbaden, Germany, 1884; p. 18.
3. Berg, A.A., VI. A Case of Acute Osteomyelitis of the Femur, with General Systemic *Staphylococcus aureus* Infection, Terminating in Recovery. *Ann. Surg.* **1900**, *31*, 332–339. [CrossRef] [PubMed]
4. GiovanniGherardi, G.B. *VincenzoSavini, Staphylococcal Taxonomy*; Academic Press: Cambridge, MA, USA, 2018; Volume 1.
5. Novick, R.P. Autoinduction and signal transduction in the regulation of staphylococcal virulence. *Mol. Microbiol.* **2003**, *48*, 1429–1449. [CrossRef]
6. Scully, I.L.; Timofeyeva, Y.; Illenberger, A.; Lu, P.; Liberator, P.A.; Jansen, K.U.; Anderson, A.S. Performance of a Four-Antigen Staphylococcus aureus Vaccine in Preclinical Models of Invasive *S. aureus* Disease. *Microorganisms* **2021**, *9*, 177. [CrossRef] [PubMed]
7. Joshi, A.; Pancari, G.; Cope, L.; Bowman, E.P.; Cua, D.; Proctor, R.A.; McNeely, T. Immunization with *Staphylococcus aureus* iron regulated surface determinant B (IsdB) confers protection via Th17/IL17 pathway in a murine sepsis model. *Hum. Vaccin. Immunother.* **2012**, *8*, 336–346. [CrossRef] [PubMed]
8. Ebert, T.; Smith, S.; Pancari, G.; Clark, D.; Hampton, R.; Secore, S.; Towne, V.; Fan, H.; Wang, X.; Wu, X.; et al. A fully human monoclonal antibody to *Staphylococcus aureus* iron regulated surface determinant B (IsdB) with functional activity in vitro and in vivo. *Hum. Antibodies* **2010**, *19*, 113–128. [CrossRef] [PubMed]
9. Brown, M.; Kowalski, R.; Zorman, J.; Wang, X.; Towne, V.; Zhao, Q.; Secore, S.; Finnefrock, A.C.; Ebert, T.; Pancari, G.; et al. Selection and characterization of murine monoclonal antibodies to *Staphylococcus aureus* iron-regulated surface determinant B with functional activity in vitro and in vivo. *Clin. Vaccine Immunol.* **2009**, *16*, 1095–1104. [CrossRef]
10. Raedler, M.D.; Heyne, S.; Wagner, E.; Shalkowski, S.K.; Secore, S.; Anderson, A.S.; Cook, J.; Cope, L.; McNeely, T.; Retzlaff, M.; et al. Serologic assay to quantify human immunoglobulin G antibodies to the *Staphylococcus aureus* iron surface determinant B antigen. *Clin. Vaccine Immunol.* **2009**, *16*, 739–748. [CrossRef]
11. Kuklin, N.A.; Clark, D.J.; Secore, S.; Cook, J.; Cope, L.D.; McNeely, T.; Noble, L.; Brown, M.J.; Zorman, J.K.; Wang, X.M.; et al. A novel *Staphylococcus aureus* vaccine: Iron surface determinant B induces rapid antibody responses in rhesus macaques and specific increased survival in a murine *S. aureus* sepsis model. *Infect. Immun.* **2006**, *74*, 2215–2223. [CrossRef]
12. McNeely, T.B.; Shah, N.A.; Fridman, A.; Joshi, A.; Hartzel, J.S.; Keshari, R.S.; Lupu, F.; DiNubile, M.J. Mortality among recipients of the Merck V710 *Staphylococcus aureus* vaccine after postoperative *S. aureus* infections: An analysis of possible contributing host factors. *Hum. Vaccin. Immunother.* **2014**, *10*, 3513–3516. [CrossRef]
13. Stulik, L.; Rouha, H.; Labrousse, D.; Visram, Z.C.; Badarau, A.; Maierhofer, B.; Gross, K.; Weber, S.; Kramaric, M.D.; Glojnaric, I.; et al. Preventing lung pathology and mortality in rabbit *Staphylococcus aureus* pneumonia models with cytotoxin-neutralizing monoclonal IgGs penetrating the epithelial lining fluid. *Sci. Rep.* **2019**, *9*, 5339. [CrossRef] [PubMed]
14. Rouha, H.; Weber, S.; Janesch, P.; Maierhofer, B.; Gross, K.; Dolezilkova, I.; Mirkina, I.; Visram, Z.C.; Malafa, S.; Stulik, L.; et al. Disarming *Staphylococcus aureus* from destroying human cells by simultaneously neutralizing six cytotoxins with two human monoclonal antibodies. *Virulence* **2018**, *9*, 231–247. [CrossRef] [PubMed]
15. Rouha, H.; Badarau, A.; Visram, Z.C.; Battles, M.B.; Prinz, B.; Magyarics, Z.; Nagy, G.; Mirkina, I.; Stulik, L.; Zerbs, M.; et al. Five birds, one stone: Neutralization of alpha-hemolysin and 4 bi-component leukocidins of *Staphylococcus aureus* with a single human monoclonal antibody. *MAbs* **2015**, *7*, 243–254. [CrossRef] [PubMed]
16. Diep, B.A.; Le, V.T.M.; Visram, Z.C.; Rouha, H.; Stulik, L.; Dip, E.C.; Nagy, G.; Nagy, E. Improved Protection in a Rabbit Model of Community-Associated Methicillin-Resistant *Staphylococcus aureus* Necrotizing Pneumonia upon Neutralization of Leukocidins in Addition to Alpha-Hemolysin. *Antimicrob. Agents Chemother.* **2016**, *60*, 6333–6340. [CrossRef]

17. Miller, L.S.; Fowler, V.G., Jr.; Shukla, S.K.; Rose, W.E.; Proctor, R.A. Development of a vaccine against *Staphylococcus aureus* invasive infections: Evidence based on human immunity, genetics and bacterial evasion mechanisms. *FEMS Microbiol. Rev.* **2020**, *44*, 123–153. [CrossRef]

18. Adhikari, R.P.; Karauzum, H.; Sarwar, J.; Abaandou, L.; Mahmoudieh, M.; Boroun, A.R.; Vu, H.; Nguyen, T.; Devi, V.S.; Shulenin, S.; et al. Novel structurally designed vaccine for *S. aureus* alpha-hemolysin: Protection against bacteremia and pneumonia. *PLoS ONE* **2012**, *7*, e38567. [CrossRef]

19. Adhikari, R.P.; Kort, T.; Shulenin, S.; Kanipakala, T.; Ganjbaksh, N.; Roghmann, M.; Holtsberg, F.W.; Aman, M.J. Antibodies to *S. aureus* LukS-PV Attenuated Subunit Vaccine Neutralize a Broad Spectrum of Canonical and Non-Canonical Bicomponent Leukotoxin Pairs. *PLoS ONE* **2015**, *10*, e0137874. [CrossRef] [PubMed]

20. Adhikari, R.P.; Thompson, C.D.; Aman, M.J.; Lee, J.C. Protective efficacy of a novel alpha hemolysin subunit vaccine (AT62) against *Staphylococcus aureus* skin and soft tissue infections. *Vaccine* **2016**, *34*, 6402–6407. [CrossRef]

21. Kailasan, S.; Kort, T.; Mukherjee, I.; Liao, G.C.; Kanipakala, T.; Williston, N.; Ganjbaksh, N.; Venkatasubramaniam, A.; Holtsberg, F.W.; Karauzum, H.; et al. Rational Design of Toxoid Vaccine Candidates for *Staphylococcus aureus* Leukocidin AB (LukAB). *Toxins* **2019**, *11*, 339. [CrossRef] [PubMed]

22. Karauzum, H.; Venkatasubramaniam, A.; Adhikari, R.P.; Kort, T.; Holtsberg, F.W.; Mukherjee, I.; Mednikov, M.; Ortines, R.; Nguyen, N.T.Q.; Doan, T.; et al. IBT-V02: A Multicomponent Toxoid Vaccine Protects Against Primary and Secondary Skin Infections Caused by *Staphylococcus aureus*. *Front. Immunol.* **2021**, *12*, 624310. [CrossRef]

23. Venkatasubramaniam, A.; Liao, G.; Cho, E.; Adhikari, R.P.; Kort, T.; Holtsberg, F.W.; Elsass, K.E.; Kobs, D.J.; Rudge, T.L., Jr.; Kauffman, K.D.; et al. Safety and Immunogenicity of a 4-Component Toxoid-Based *Staphylococcus aureus* Vaccine in Rhesus Macaques. *Front. Immunol.* **2021**, *12*, 621754. [CrossRef]

24. Van Dalen, R.; Molendijk, M.M.; Ali, S.; van Kessel, K.P.M.; Aerts, P.; van Strijp, J.A.G.; de Haas, C.J.C.; Codee, J.; van Sorge, N.M. Do not discard *Staphylococcus aureus* WTA as a vaccine antigen. *Nature* **2019**, *572*, E1–E2. [CrossRef]

25. Fattom, A.I.; Sarwar, J.; Ortiz, A.; Naso, R. A *Staphylococcus aureus* capsular polysaccharide (CP) vaccine and CP-specific antibodies protect mice against bacterial challenge. *Infect. Immun.* **1996**, *64*, 1659–1665. [CrossRef] [PubMed]

26. O'Brien, F.G. Staphylococcus aureus vaccine conjugate—Nabi: Nabi-StaphVAX, StaphVAX. *Drugs R. D.* **2003**, *4*, 383–385.

27. Fattom, A.; Fuller, S.; Propst, M.; Winston, S.; Muenz, L.; He, D.; Naso, R.; Horwith, G. Safety and immunogenicity of a booster dose of *Staphylococcus aureus* types 5 and 8 capsular polysaccharide conjugate vaccine (StaphVAX) in hemodialysis patients. *Vaccine* **2004**, *23*, 656–663. [CrossRef]

28. Fattom, A.I.; Horwith, G.; Fuller, S.; Propst, M.; Naso, R. Development of StaphVAX, a polysaccharide conjugate vaccine against *S. aureus* infection: From the lab bench to phase III clinical trials. *Vaccine* **2004**, *22*, 880–887. [CrossRef] [PubMed]

29. Begier, E.; Seiden, D.J.; Patton, M.; Zito, E.; Severs, J.; Cooper, D.; Eiden, J.; Gruber, W.C.; Jansen, K.U.; Anderson, A.S.; et al. SA4Ag, a 4-antigen *Staphylococcus aureus* vaccine, rapidly induces high levels of bacteria-killing antibodies. *Vaccine* **2017**, *35*, 1132–1139. [CrossRef] [PubMed]

30. Schneewind, O.; Missiakas, D. Sortases, Surface Proteins, and Their Roles in *Staphylococcus aureus* Disease and Vaccine Development. *Microbiol. Spectr.* **2019**, *7*.

31. Zhang, F.; Jun, M.; Ledue, O.; Herd, M.; Malley, R.; Lu, Y. Antibody-mediated protection against *Staphylococcus aureus* dermonecrosis and sepsis by a whole cell vaccine. *Vaccine* **2017**, *35*, 3834–3843. [CrossRef]

32. Stranger-Jones, Y.K.; Bae, T.; Schneewind, O. Vaccine assembly from surface proteins of *Staphylococcus aureus*. *Proc. Natl. Acad. Sci. USA* **2006**, *103*, 16942–16947. [CrossRef]

33. Yu, W.; Yao, D.; Yu, S.; Wang, X.; Li, X.; Wang, M.; Liu, S.; Feng, Z.; Chen, X.; Li, W.; et al. Protective humoral and CD4(+) T cellular immune responses of *Staphylococcus aureus* vaccine MntC in a murine peritonitis model. *Sci. Rep.* **2018**, *8*, 3580. [CrossRef] [PubMed]

34. Fernandes de Oliveira, L.M.; Steindorff, M.; Darisipudi, M.N.; Mrochen, D.M.; Trube, P.; Broker, B.M.; Bronstrup, M.; Tegge, W.; Holtfreter, S. Discovery of *Staphylococcus aureus* Adhesion Inhibitors by Automated Imaging and Their Characterization in a Mouse Model of Persistent Nasal Colonization. *Microorganisms* **2021**, *9*, 631. [CrossRef]

35. Mu, D.; Xiang, H.; Dong, H.; Wang, D.; Wang, T. Isovitexin, a Potential Candidate Inhibitor of Sortase A of *Staphylococcus aureus* USA300. *J. Microbiol. Biotechnol.* **2018**, *28*, 1426–1432. [CrossRef] [PubMed]

36. Wang, G.; Wang, X.; Sun, L.; Gao, Y.; Niu, X.; Wang, H. Novel Inhibitor Discovery of *Staphylococcus aureus* Sortase B and the Mechanism Confirmation via Molecular Modeling. *Molecules* **2018**, *23*, 977. [CrossRef] [PubMed]

37. Hogan, S.; Kasotakis, E.; Maher, S.; Cavanagh, B.; O'Gara, J.P.; Pandit, A.; Keyes, T.E.; Devocelle, M.; O'Neill, E. A novel medical device coating prevents *Staphylococcus aureus* biofilm formation on medical device surfaces. *FEMS Microbiol. Lett.* **2019**, *366*, fnz107. [CrossRef] [PubMed]

38. Elshina, E.; Allen, E.R.; Flaxman, A.; van Diemen, P.M.; Milicic, A.; Rollier, C.S.; Yamaguchi, Y.; Wyllie, D.H. Vaccination with the *Staphylococcus aureus* secreted proteins EapH1 and EapH2 impacts both *S. aureus* carriage and invasive disease. *Vaccine* **2019**, *37*, 502–509. [CrossRef] [PubMed]

39. Li, T.; Huang, M.; Song, Z.; Zhang, H.; Chen, C. Biological characteristics and conjugated antigens of ClfA A-FnBPA and CP5 in *Staphylococcus aureus*. *Can. J. Vet. Res.* **2018**, *82*, 48–54.

40. Liu, Y.; Sul, Y.; Zhang, B.; Su, L.; Jiang, H. Immunological comparison of Efb and ClfA of *Staphylococcus aureus* isolated from bovine. *Sheng Wu Gong Cheng Xue Bao* **2015**, *31*, 1335–1343.

41. Delfani, S.; Mobarez, A.M.; Fooladi, A.A.I.; Amani, J.; Emaneini, M. Protection of mice against *Staphylococcus aureus* infection by a recombinant protein ClfA-IsdB-Hlg as a vaccine candidate. *Med. Microbiol. Immunol.* **2016**, *205*, 47–55. [CrossRef]
42. Veloso, T.R.; Mancini, S.; Giddey, M.; Vouillamoz, J.; Que, Y.; Moreillon, P.; Entenza, J.M. Vaccination against *Staphylococcus aureus* experimental endocarditis using recombinant Lactococcus lactis expressing ClfA or FnbpA. *Vaccine* **2015**, *33*, 3512–3517. [CrossRef]
43. Li, Y.; Li, Z.; Ye, J. Identification of Th cell epitopes on clfA adhesin of *Staphylococcus aureus* and characterization of their role in immunity. *Wei Sheng Wu Xue Bao* **2013**, *53*, 966–975.
44. Brouillette, E.; Lacasse, P.; Shkreta, L.; Belanger, J.; Grondin, G.; Diarra, M.S.; Fournier, S.; Talbot, B.G. DNA immunization against the clumping factor A (ClfA) of *Staphylococcus aureus*. *Vaccine* **2002**, *20*, 2348–2357. [CrossRef]
45. Patzold, L.; Brausch, A.; Bielefeld, E.; Zimmer, L.; Somerville, G.A.; Bischoff, M.; Gaupp, R. Impact of the Histidine-Containing Phosphocarrier Protein HPr on Carbon Metabolism and Virulence in *Staphylococcus aureus*. *Microorganisms* **2021**, *9*, 466. [CrossRef] [PubMed]
46. Crooke, A.K.; Fuller, J.R.; Obrist, M.W.; Tomkovich, S.E.; Vitko, N.P.; Richardson, A.R. CcpA-independent glucose regulation of lactate dehydrogenase 1 in *Staphylococcus aureus*. *PLoS ONE* **2013**, *8*, e54293. [CrossRef] [PubMed]
47. Nuxoll, A.S.; Fuller, J.R.; Obrist, M.W.; Tomkovich, S.E.; Vitko, N.P.; Richardson, A.R. CcpA regulates arginine biosynthesis in *Staphylococcus aureus* through repression of proline catabolism. *PLoS Pathog* **2012**, *8*, e1003033. [CrossRef]
48. Leiba, J.; Hartmann, T.; Cluzel, M.; Cohen-Gonsaud, M.; Delolme, F.; Bischoff, M.; Molle, V. A novel mode of regulation of the *Staphylococcus aureus* catabolite control protein A (CcpA) mediated by Stk1 protein phosphorylation. *J. Biol. Chem.* **2012**, *287*, 43607–43619. [CrossRef]
49. Li, C.; Sun, F.; Cho, H.; Yelavarthi, V.; Sohn, C.; He, C.; Schneewind, O.; Bae, T. CcpA mediates proline auxotrophy and is required for *Staphylococcus aureus* pathogenesis. *J. Bacteriol.* **2010**, *192*, 3883–3892. [CrossRef]
50. Seidl, K.; Muller, S.; Franccois, P.; Kriebitzsch, C.; Schrenzel, J.; Engelmann, S.; Bischoff, M.; Berger-Bachi, B. Effect of a glucose impulse on the CcpA regulon in *Staphylococcus aureus*. *BMC Microbiol.* **2009**, *9*, 95. [CrossRef]
51. Seidl, K.; Bischoff, M.; Berger-Bachi, B. CcpA mediates the catabolite repression of tst in *Staphylococcus aureus*. *Infect. Immun.* **2008**, *76*, 5093–5099. [CrossRef]
52. Seidl, K.; Goerke, C.; Wolz, C.; Mack, D.; Berger-Bachi, B.; Bischoff, M. *Staphylococcus aureus* CcpA affects biofilm formation. *Infect. Immun.* **2008**, *76*, 2044–2050. [CrossRef]
53. Seidl, K.; Stucki, M.; Ruegg, M.; Goerke, C.; Wolz, C.; Harris, L.; Berger-Bachi, B.; Bischoff, M. *Staphylococcus aureus* CcpA affects virulence determinant production and antibiotic resistance. *Antimicrob. Agents Chemother.* **2006**, *50*, 1183–1194. [CrossRef]
54. Hussain, M.; Kohler, C.; Becker, K. Role of SrtA in Pathogenicity of *Staphylococcus lugdunensis*. *Microorganisms* **2020**, *8*, 1975. [CrossRef]
55. Frankel, B.A.; Bentley, M.; Kruger, R.G.; McCafferty, D.G. Vinyl sulfones: Inhibitors of SrtA, a transpeptidase required for cell wall protein anchoring and virulence in *Staphylococcus aureus*. *J. Am. Chem. Soc.* **2004**, *126*, 3404–3405. [CrossRef] [PubMed]
56. Kruger, R.G.; Dostal, P.; McCafferty, D.G. Development of a high-performance liquid chromatography assay and revision of kinetic parameters for the *Staphylococcus aureus* sortase transpeptidase SrtA. *Anal. Biochem.* **2004**, *326*, 42–48. [CrossRef]
57. Kruger, R.G.; Otvos, B.; Frankel, B.A.; Bentley, M.; Dostal, P.; McCafferty, D.G. Analysis of the substrate specificity of the *Staphylococcus aureus* sortase transpeptidase SrtA. *Biochemistry.* **2004**, *43*, 1541–1551. [CrossRef] [PubMed]
58. Weiss, W.J.; Lenoy, E.; Murphy, T.; Tardio, L.; Burgio, P.; Projan, S.J.; Schneewind, O.; Alksne, L. Effect of srtA and srtB gene expression on the virulence of *Staphylococcus aureus* in animal models of infection. *J. Antimicrob. Chemother.* **2004**, *53*, 480–486. [CrossRef] [PubMed]
59. Huang, X.; Aulabaugh, A.; Ding, W.; Kapoor, B.; Alksne, L.; Tabei, K.; Ellestad, G. Kinetic mechanism of *Staphylococcus aureus* sortase SrtA. *Biochemistry* **2003**, *42*, 11307–11315. [CrossRef]
60. Yang, T.; Zhang, T.; Guan, X.; Dong, Z.; Lan, L.; Yang, S.; Yang, C. Tideglusib and Its Analogues as Inhibitors of *Staphylococcus aureus* SrtA. *J. Med. Chem.* **2020**, *63*, 8442–8457. [CrossRef] [PubMed]
61. Selvaraj, C.; Sivakamavalli, J.; Baskaralingam, V.; Singh, S.K. Virtual screening of LPXTG competitive SrtA inhibitors targeting signal transduction mechanism in *Bacillus anthracis*: A combined experimental and theoretical study. *J. Recept. Signal Transduct. Res.* **2014**, *34*, 221–232. [CrossRef]
62. Reigada, I.; Guarch-Perez, C.; Patel, J.Z.; Riool, M.; Savijoki, K.; Yli-Kauhaluoma, J.; Zaat, S.A.J.; Fallarero, A. Combined Effect of Naturally-Derived Biofilm Inhibitors and Differentiated HL-60 Cells in the Prevention of *Staphylococcus aureus* Biofilm Formation. *Microorganisms* **2020**, *8*, 1757. [CrossRef]
63. Ueda, Y.; Miyazaki, M.; Mashima, K.; Takagi, S.; Hara, S.; Kamimura, H.; Jimi, S. The Effects of Silver Sulfadiazine on Methicillin-Resistant *Staphylococcus aureus* Biofilms. *Microorganisms* **2020**, *8*, 1551. [CrossRef] [PubMed]
64. Mansson, E.; Tevell, S.; Nilsdotter-Augustinsson, A.; Johannesen, T.B.; Sundqvist, M.; Stegger, M.; Soderquist, B. Methicillin-Resistant *Staphylococcus epidermidis* Lineages in the Nasal and Skin Microbiota of Patients Planned for Arthroplasty Surgery. *Microorganisms* **2021**, *9*, 265. [CrossRef]
65. Wolters, M.; Frickmann, H.; Christner, M.; Both, A.; Rohde, H.; Oppong, K.; Akenten, C.W.; May, J.; Dekker, D. Molecular Characterization of *Staphylococcus aureus* Isolated from Chronic Infected Wounds in Rural Ghana. *Microorganisms* **2020**, *8*, 2052. [CrossRef]

66. Zhang, J.; Conly, J.; McClure, J.; Wu, K.; Petri, B.; Barber, D.; Elsayed, S.; Armstrong, G.; Zhang, K. A Murine Skin Infection Model Capable of Differentiating the Dermatopathology of Community-Associated MRSA Strain USA300 from Other MRSA Strains. *Microorganisms* **2021**, *9*, 287. [CrossRef] [PubMed]

67. Ikeuchi, K.; Adachi, E.; Sasaki, T.; Suzuki, M.; Lim, L.A.; Saito, M.; Koga, M.; Tsutsumi, T.; Kido, Y.; Uehara, Y.; et al. An Outbreak of USA300 Methicillin-Resistant *Staphylococcus aureus* among People with HIV in Japan. *J. Infect. Dis.* **2021**, *223*, 610–620. [CrossRef] [PubMed]

68. McCaskill, M.L.; Mason, E.O., Jr.; Kaplan, S.L.; Hammerman, W.; Lamberth, L.B.; Hulten, K.G. Increase of the USA300 clone among community-acquired methicillin-susceptible *Staphylococcus aureus* causing invasive infections. *Pediatr. Infect. Dis. J.* **2007**, *26*, 1122–1127. [CrossRef]

69. Ruppitsch, W.; Stoger, A.; Schmid, D.; Fretz, R.; Indra, A.; Allerberger, F.; Witte, W. Occurrence of the USA300 community-acquired *Staphylococcus aureus* clone in Austria. *Eur. Surveill.* **2007**, *12*, E0710251. [CrossRef]

70. Johnson, J.K.; Khoie, T.; Shurland, S.; Kreisel, K.; Stine, O.C.; Roghmann, M. Skin and soft tissue infections caused by methicillin-resistant *Staphylococcus aureus* USA300 clone. *Emerg. Infect. Dis.* **2007**, *13*, 1195–1200. [CrossRef] [PubMed]

71. Sifri, C.D.; Park, J.; Helm, G.A.; Stemper, M.E.; Shukla, S.K. Fatal brain abscess due to community-associated methicillin-resistant *Staphylococcus aureus* strain USA300. *Clin. Infect. Dis.* **2007**, *45*, e113–e117. [CrossRef]

72. Popovich, K.J.; Snitkin, E.; Green, S.J.; Aroutcheva, A.; Hayden, M.K.; Hota, B.; Weinstein, R.A. Genomic Epidemiology of USA300 Methicillin-Resistant *Staphylococcus aureus* in an Urban Community. *Clin. Infect. Dis.* **2016**, *62*, 37–44. [CrossRef] [PubMed]

73. Hota, B.; Lyles, R.; Rim, J.; Popovich, K.J.; Rice, T.; Aroutcheva, A.; Weinstein, R.A. Predictors of clinical virulence in community-onset methicillin-resistant *Staphylococcus aureus* infections: The importance of USA300 and pneumonia. *Clin. Infect. Dis.* **2011**, *53*, 757–765. [CrossRef]

74. Wilden, J.J.; Hrincius, E.R.; Niemann, S.; Boergeling, Y.; Löffler, B.; Ludwig, S.; Ehrhardt, C. Impact of *Staphylococcus aureus* Small Colony Variants on Human Lung Epithelial Cells with Subsequent Influenza Virus Infection. *Microorganisms* **2020**, *8*, 1998. [CrossRef] [PubMed]

75. Etter, D.; Jenni, C.; Tasara, T.; Johler, S. Mild Lactic Acid Stress Causes Strain-Dependent Reduction in SEC Protein Levels. *Microorganisms* **2021**, *9*, 1014. [CrossRef]

76. Ciupescu, L.; Auvray, F.; Nicorescu, I.M.; Meheut, T.; Ciupescu, V.; Lardeux, A.; Tanasuica, R.; Hennekinne, J. Characterization of *Staphylococcus aureus* strains and evidence for the involvement of non-classical enterotoxin genes in food poisoning outbreaks. *FEMS Microbiol. Lett.* **2018**. [CrossRef] [PubMed]

77. Chao, G.; Bao, G.; Cao, Y.; Yan, W.; Wang, Y.; Zhang, X.; Zhou, L.; Wu, Y. Prevalence and diversity of enterotoxin genes with genetic background of *Staphylococcus aureus* isolates from different origins in China. *Int. J. Food Microbiol.* **2015**, *211*, 142–147. [CrossRef]

78. Mishra, N.N.; Bayer, A.S.; Baines, S.L.; Hayes, A.S.; Howden, B.P.; Lapitan, C.K.; Lew, C.; Rose, W.E. Cell Membrane Adaptations Mediate β-Lactam-Induced Resensitization of Daptomycin-Resistant (DAP-R) *Staphylococcus aureus* In Vitro. *Microorganisms* **2021**, *9*, 1028. [CrossRef]

79. Ventola, C.L. The antibiotic resistance crisis: Part 1: Causes and threats. *Pharm. Ther.* **2015**, *40*, 277–283.

80. Van den Driessche, F.; Brackman, G.; Swimberghe, R.; Rigole, P.; Coenye, T. Screening a repurposing library for potentiators of antibiotics against *Staphylococcus aureus* biofilms. *Int. J. Antimicrob. Agents* **2017**, *49*, 315–320. [CrossRef]

81. Bayer, A.S.; Xiong, Y.Q. Redeploying beta-Lactams against *Staphylococcus aureus*: Repurposing with a Purpose. *J. Infect. Dis.* **2017**, *215*, 11–13. [CrossRef]

82. Das, S.; Dasgupta, A.; Chopra, S. Drug repurposing: A new front in the war against *Staphylococcus aureus*. *Future Microbiol.* **2016**, *11*, 1091–1099. [CrossRef]

83. Lozano, C.; Fernandez-Fernandez, R.; Ruiz-Ripa, L.; Gomez, P.; Zarazaga, M.; Torres, C. Human mecC-Carrying MRSA: Clinical Implications and Risk Factors. *Microorganisms* **2020**, *8*, 1615. [CrossRef] [PubMed]

84. Gor, V.; Ohniwa, R.L.; Morikawa, K. No Change, No Life? What We Know about Phase Variation in *Staphylococcus aureus*. *Microorganisms* **2021**, *9*, 244. [CrossRef] [PubMed]

85. Kiem, S.; Oh, W.S.; Peck, K.R.; Lee, N.Y.; Lee, J.; Song, J.; Hwang, E.S.; Kim, E.; Cha, C.Y.; Choe, K. Phase variation of biofilm formation in *Staphylococcus aureus* by IS 256 insertion and its impact on the capacity adhering to polyurethane surface. *J. Korean Med. Sci.* **2004**, *19*, 779–782. [CrossRef]

86. Baselga, R.; Albizu, I.; de la Cruz, M.; del Cacho, E.; Barberan, M.; Amorena, B. Phase variation of slime production in *Staphylococcus aureus*: Implications in colonization and virulence. *Infect. Immun.* **1993**, *61*, 4857–4862. [CrossRef]

87. Manasherob, R.; Mooney, J.A.; Lowenberg, D.W.; Bollyky, P.L.; Amanatullah, D.F. Tolerant Small-colony Variants Form Prior to Resistance within a *Staphylococcus aureus* Biofilm Based on Antibiotic Selective Pressure. *Clin. Orthop. Relat. Res.* **2021**. [CrossRef]

88. Sato, T.; Uno, T.; Kawamura, M.; Fujimura, S. In Vitro Tolerability of Biofilm-Forming Trimethoprim-/Sulfamethoxazole-Resistant Small Colony Variants of *Staphylococcus aureus* against Various Antimicrobial Agents. *Microb. Drug Resist.* **2021**. [CrossRef] [PubMed]

89. Mirani, Z.A.; Urooj, S.; Khan, M.N.; Khan, A.B.; Shaikh, I.A.; Siddiqui, A. An effective weapon against biofilm consortia and small colony variants of MRSA. *Iran J. Basic Med. Sci.* **2020**, *23*, 1494–1498.

90. Stoneham, S.M.; Cantillon, D.M.; Waddell, S.J.; Llewelyn, M.J. Spontaneously Occurring Small-Colony Variants of *Staphylococcus aureus* Show Enhanced Clearance by THP-1 Macrophages. *Front. Microbiol.* **2020**, *11*, 1300. [CrossRef]

91. Tuchscherr, L.; Loffler, B.; Proctor, R.A. Persistence of *Staphylococcus aureus*: Multiple Metabolic Pathways Impact the Expression of Virulence Factors in Small-Colony Variants (SCVs). *Front. Microbiol.* **2020**, *11*, 1028. [CrossRef]
92. Lee, J.; Zilm, P.S.; Kidd, S.P. Novel Research Models for *Staphylococcus aureus* Small Colony Variants (SCV) Development: Co-pathogenesis and Growth Rate. *Front. Microbiol.* **2020**, *11*, 321. [CrossRef]

 MDPI

Article

Performance of a Four-Antigen *Staphylococcus aureus* Vaccine in Preclinical Models of Invasive *S. aureus* Disease

Ingrid L. Scully, Yekaterina Timofeyeva, Arthur Illenberger, Peimin Lu, Paul A. Liberator, Kathrin U. Jansen and Annaliesa S. Anderson *

Pfizer Vaccine Research & Development, Pearl River, NY 10965, USA; ingrid.scully@pfizer.com (I.L.S.); yekaterina.timofeyeva@pfizer.com (Y.T.); arthur.illenberger@pfizer.com (A.I.); peiminlu@msn.com (P.L.); paul.liberator@pfizer.com (P.A.L.); kathrin.jansen@pfizer.com (K.U.J.)
* Correspondence: annaliesa.anderson@pfizer.com; Tel.: +1-845-602-4674

Abstract: A *Staphylococcus aureus* four-antigen vaccine (SA4Ag) was designed for the prevention of invasive disease in surgical patients. The vaccine is composed of capsular polysaccharide type 5 and type 8 CRM_{197} conjugates, a clumping factor A mutant (Y338A-ClfA) and manganese transporter subunit C (MntC). *S. aureus* pathogenicity is characterized by an ability to rapidly adapt to the host environment during infection, which can progress from a local infection to sepsis and invasion of distant organs. To test the protective capacity of the SA4Ag vaccine against progressive disease stages of an invasive *S. aureus* infection, a deep tissue infection mouse model, a bacteremia mouse model, a pyelonephritis model, and a rat model of infectious endocarditis were utilized. SA4Ag vaccination significantly reduced the bacterial burden in deep tissue infection, in bacteremia, and in the pyelonephritis model. Complete prevention of infection was demonstrated in a clinically relevant endocarditis model. Unfortunately, these positive preclinical findings with SA4Ag did not prove the clinical utility of SA4Ag in the prevention of surgery-associated invasive *S. aureus* infection.

Keywords: *Staphylococcus aureus*; invasive disease; surgery-associated infection; sepsis; SA4Ag vaccine; conjugated polysaccharide; ClfA; MntC; protection; animal models

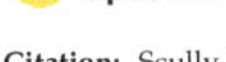 check for updates

Citation: Scully, I.L.; Timofeyeva, Y.; Illenberger, A.; Lu, P.; Liberator, P.A.; Jansen, K.U.; Anderson, A.S. Performance of a Four-Antigen *Staphylococcus aureus* Vaccine in Preclinical Models of Invasive *S. aureus* Disease. *Microorganisms* **2021**, *9*, 177. https://doi.org/10.3390/microorganisms9010177

Received: 8 December 2020
Accepted: 7 January 2021
Published: 15 January 2021

Publisher's Note: MDPI stays neutral with regard to jurisdictional claims in published maps and institutional affiliations.

1. Introduction

Staphylococcus aureus is a Gram-positive bacterium that is carried asymptomatically in the nares of 20–50% of the general population [1]. However, upon a breach of skin or mucosal barriers, it can cause a wide spectrum of diseases, ranging from relatively mild skin infections, such as carbuncles, to life-threatening wound and bloodstream infections [2]. *S. aureus* infections following surgery carry particularly high mortality rates, and survivors require an additional 13–17 days in the hospital, significantly increasing healthcare costs [3]. The burden of *S. aureus* disease is exacerbated by the prevalence of antibiotic-resistant *S. aureus* isolates [4], highlighting the need for an effective prophylactic vaccine.

A consideration in both development and preclinical evaluation of a *S. aureus* vaccine is the organism's ability to rapidly adapt to the host microenvironment [5]. *S. aureus* can enter normally sterile sites through lesions such as those created during surgery or traumatic injury and rapidly deploy an array of pathogenesis mechanisms, rendering *S. aureus* a challenging vaccine target. Consequently, a licensed vaccine against *S. aureus* disease is not yet available, and a clinically validated correlate of protection has not yet been established. Prevention strategies for patients at high risk for invasive *S. aureus* disease, such as surgical patients, include decolonization and antibiotic prophylaxis [6]. These procedures are limited by the variable effectiveness of decolonization and by the development of antibiotic resistance [7–9]. Thus, alternative preventative strategies, such as prophylactic vaccines, would be a welcome addition to the clinician's armamentarium.

With that aim, a four-antigen *S. aureus* vaccine (SA4Ag) was designed for the prevention of invasive *S. aureus* infections following elective surgery. Each vaccine component was carefully selected so that, when combined, the vaccine would contravene major *S. aureus* pathogenesis mechanisms, namely initial adhesion events (ClfA), nutrient acquisition sustaining growth (MntC), and evasion of neutrophil-mediated killing (MntC and capsular polysaccharide (CP) type 5 and type 8 conjugates). The adhesin ClfA enables the attachment of *S. aureus* to human fibrinogen, and antibodies directed against this protein inhibit ClfA-mediated fibrinogen binding [10]. MntC is a highly conserved component of the manganese transporter MntABC that is quickly upregulated in vivo [11] and is expressed during biofilm formation in animal models [12]. Manganese acquisition by *S. aureus* is important for both growth and evasion of neutrophil killing through detoxifying oxygen radicals [11,13]. Finally, the vaccine contains capsular polysaccharide type 5 and type 8 conjugated to cross-reactive material 197 (CRM_{197}), which induce functional antibodies that kill the bacteria following opsonophagocytosis [14,15], the process whereby antibodies coat the bacterium, fix complement, and induce uptake and killing by phagocytes, such as neutrophils. It has become clear that antibodies that bind to a bacterial antigen, such as those measured by an enzyme-linked immunosorbent assay, are not always functional, especially in adults with pre-existing exposure to a pathogen. Thus, demonstration of functional antibody activity, such as through an opsonophagocytic killing assay or fibrinogen binding assay is important as another step in the evaluation of a vaccine candidate, along with in vivo proof of concept studies.

The correlate of protection for *S. aureus* is not yet known, and a vaccine for the prevention of *S. aureus* disease has yet to be commercially licensed. The two candidates previously advanced to the clinic were supported by preclinical serology [16] and/or murine sepsis model studies [17,18]. These studies demonstrated that the vaccine candidates were immunogenic in preclinical models, an important first step in selecting candidates to advance to the clinic. In the case of iron-regulated surface determinant B (IsdB), a single challenge model, the murine sepsis model, was used to support preclinical efficacy [17,18]. The murine sepsis model mimics hematogenous spread through the body, one important aspect of *S. aureus* infection. However, *S. aureus* has notoriously complex pathogenesis mechanisms, which can only be modeled using multiple preclinical models. The development of relevant preclinical models for *S. aureus* is challenging, as *S. aureus* is adapted to the human host environment, and preclinical models have failed to predict clinical efficacy [19,20]. Here, the ability of SA4Ag to protect against invasive disease is demonstrated in three preclinical rodent models of *S. aureus* infection, each mimicking a distinct phase of *S. aureus* infection, namely deep tissue invasion, bacteremia, and distal infection. In the absence of a defined correlate of protection, and due to the limited ability of any single preclinical model to predict *S. aureus* vaccine clinical success, demonstrating efficacy in multiple models of *S. aureus* invasive disease is relevant for advancing a vaccine into clinical trials.

2. Materials and Methods

2.1. Bacterial Strains

The *S. aureus* clinical isolates from Pfizer's internal collection that were used for in vivo analyses are listed in Table 1. These strains represent a diverse set of clinical isolates.

2.2. Animal Studies

All animal work was performed in strict accordance with approved Institutional Animal Care and Use Committee protocols at an Association for Assessment and Accreditation of Laboratory Animal Care, International-accredited facility, under the following Animal Use Protocols: PRL-2011-00105 (approved in 2007), PRL-2011-00249 (approved in 2008), PRL-2011-00102 (approved in 2002) and PRL-2011-00338 (approved in 2002) for mouse studies, and PRL-2011-00440 (approved in 2002) for rat studies. For all animal studies, statistical significance was determined via Student's *t*-test using Welch's correction, and a *p* value of ≤ 0.05 was considered significant (GraphPad Prism). For each model, bacterial challenge dose and immunization schedule were optimized.

Table 1. *S. aureus* Strains Used in These Studies.

Strain Name	Capsule Type	Source	MRSA/MSSA	ST/CC	Comments
CDC3 (USA300)	5	ID	MRSA	ST8/CC8	PVL+
PFESA0241 (USA300)	5	ID	MRSA	ST8/CC8	Mec IV, PVL+ TSST-neg
PFESA0158	5	ID	MSSA	ST28/CC25	
PFESA0186	8	Carriage	MSSA	ST57/CC30	
Reynolds	5	ID	MSSA	ST25/CC25	

ID, Invasive Disease; MRSA, Methicillin-resistant *Staphylococcus aureus*; MSSA, Methicillin-sensitive *Staphylococcus aureus*; PVL+, Panton-Valentine leukocidin-positive; ST/CC, sequence type/clonal complex; TSST-neg, Toxic shock syndrome toxin-negative.

2.2.1. Surgical Site Infection Mouse Model

CD-1 female mice (6–8–week–old, 10–20/group; Charles River Laboratories) were vaccinated subcutaneously at weeks 0, 3, and 6 with 100 µL volume containing 1 µg CP8-CRM$_{197}$ + 1 µg CP5-CRM$_{197}$ + 10 µg Y338A ClfA + 10 µg MntC (SA4Ag, Pfizer, described in [11,14,21]) in QS-21 (Pfizer) or QS-21 alone. Two weeks after the final vaccination, animals underwent surgery, where a small incision was made in the thigh muscle parallel to the femur [22]. A stitch was placed in the deep tissue, then ~300 colony forming units (CFU) of *S. aureus* PFESA0158 in 10 µL phosphate-buffered saline (PBS) was instilled into the surgical site, which was then closed. Two days post-challenge, the mice were euthanized, quadriceps muscles were collected and homogenized, and serial dilutions of tissue homogenate were plated on tryptic soy agar (TSA; Becton Dickinson, Franklin Lakes, NJ, USA) plates (Becton Dickinson) to enumerate bacterial burden.

2.2.2. Murine Bacteremia Model

Groups of 10 female (6–8–week–old) CD-1 mice (Charles River Laboratories, Wilmington, MA, USA) were vaccinated by subcutaneous injection at weeks 0, 3, and 6 with either vehicle or SA4Ag (Pfizer) in 23 µg AlPO$_4$ as adjuvant. On week 8, the animals were challenged by intraperitoneal injection with ~4 × 10^8 CFU of *S. aureus* CDC3 or PFESA0241. Three hours after challenge animals were exsanguinated by cardiac puncture and serial dilutions of blood plated on TSA to enumerate CFU.

2.2.3. Murine Pyelonephritis Model

Groups of 5 female (6–8 week–old) CD-1 mice (Charles River Laboratories) were vaccinated by subcutaneous injection at weeks 0, 3, and 6 with either vehicle or SA4Ag (Pfizer, New York, NY, USA) in 23 µg AlPO$_4$ as adjuvant. On week 8, the animals were challenged by intraperitoneal injection with ~2 × 10^8 CFU of *S. aureus* Reynolds. Two days after challenge, kidneys were harvested and homogenized, and serial dilutions of homogenate plated on TSA to enumerate CFU.

2.2.4. Rat Endocarditis Model

The vaccine was tested in two rat endocarditis models, a standard model [23] and a refined model that more accurately represents clinical disease.

Standard Model

Groups of female Sprague Dawley rats (5–6–week–old, 20/group, Charles River Laboratories) were vaccinated at weeks 0, 3, and 6, with SA4Ag in AlPO$_4$ or with placebo. At week 8, two weeks after the final vaccination, catheter placement surgery was performed to generate sterile valvular vegetations [23]. Due to the complex nature of the model and surgery required, a proportion (<10%) of animals succumbed before infectious challenge due to the surgery and were thus not included in the analysis. Two days after surgery, animals were challenged intravenously with ~4 × 10^6 CFU of *S. aureus* PFESA0158 in 0.1 mL PBS. Two days post-challenge, the rats were euthanized and hearts collected. Bacteria in homogenized heart tissue were enumerated (CFU/mL), and the arithmetic mean and 95% confidence interval (CI) were calculated for each treatment group.

Refined Model

In order to reflect the low challenge inoculum that has been reported in human infections [24], the endocarditis model was refined. Groups of female Sprague Dawley rats (5–6–week–old, 14/group; Charles River Laboratories) were vaccinated at weeks 0, 3, and 6, with SA4Ag in AlPO$_4$ or with AlPO$_4$ alone, and catheter placement surgery was conducted on week 8 as described above. Two days after surgery, animals were challenged intravenously with ~4 × 10^3 CFU of *S. aureus* PFESA0186. Two days post-challenge, the rats were euthanized and hearts and kidneys collected. Tissues were plated and the presence or absence of bacterial CFU scored.

2.3. Assessment of Immune Responses to Vaccination

2.3.1. Opsonophagocytic Activity Assay

Serologic responses to capsular polysaccharides CP5 and CP8 were measured by an opsonophagocytic activity (OPA) assay, as previously described [14]. Briefly, serial dilutions of heat-inactivated immune sera were mixed with either a CP5-expressing or CP8-expressing clinical isolate of *S. aureus* and allowed to opsonize the bacteria. The reaction mixtures were then mixed with baby rabbit complement (Pel-Freez, Rogers, AR, USA) and neutrophil-like HL-60 cells (ATCC® CCL-240™, Manassas, VA, USA). An OPA antibody titer was defined as the reciprocal of the highest serum dilution resulting in 50% reduction of the number of bacterial colonies after incubation for 60 min at 37 °C when compared to the background control from which serum was omitted. Samples without detectable activity at the lowest serum dilution of 100 were assigned OPA titer values of 50.

2.3.2. Competitive Luminex Immunoassay

Competitive Luminex-based immunoassays (cLIA) were used to quantify antigen-specific binding antibodies elicited by the investigational vaccine. The assays monitor the ability of each vaccine component to elicit antibodies that can compete with the binding of antigen-specific monoclonal antibodies (mAbs) that have shown functional activity either in vitro or in vivo. The ClfA mAb prevents binding of live *S. aureus* to fibrinogen [10], while the MntC mAb inhibits manganese uptake [25,26]. The mAbs used for the CP antigens facilitated killing of *S. aureus* as measured by the OPA assay. Spectrally distinct Luminex microspheres were coated individually with each of the antigens and incubated overnight with appropriately diluted serum samples. A mixture of phycoerythrin (PE)-labeled CP5-, CP8-, rmClfA-, and rP305A-specific mouse mAbs is then added to the microsphere/serum mixture, and after incubation, the unbound components are washed off. After reading in a Bio-Plex reader (Bio-Rad, Hercules, CA, USA), signals are converted to Units/mL.

3. Results

3.1. SA4Ag Is Immunogenic in Mice, Rats and Non-Human Primates

SA4Ag is comprised of capsular polysaccharides type 5 and type 8 CRM$_{197}$-conjugates, ClfA, and MntC. SA4Ag was also shown to be able to elicit functional antibody responses in mice, rats and non-human primates, as measured by the OPA assay, which monitors the ability of serum samples to opsonize and induce uptake and killing of target bacteria by a neutrophil-like cell line, or by cLIA, which monitors the ability of serum to compete with mAbs for antigen binding. Importantly, the mAbs used in these assays inhibit the pathological function of the antigens they bind—e.g., fibrinogen binding in the case of ClfA [10] and manganese uptake in the case of MntC [26] (Figure 1). As Figure 1 shows, humans and non-human primates mount responses to the antigens in SA4Ag after a single dose (PD1), even in the absence of adjuvant, while rodents (mice and rats) require multiple immunizations with an adjuvant (e.g., $AlPO_4$) to generate similar magnitude responses. This is likely due to the induction of an anamnestic response to the antigens in SA4Ag in non-human primates and humans, which is absent in the rodent species.

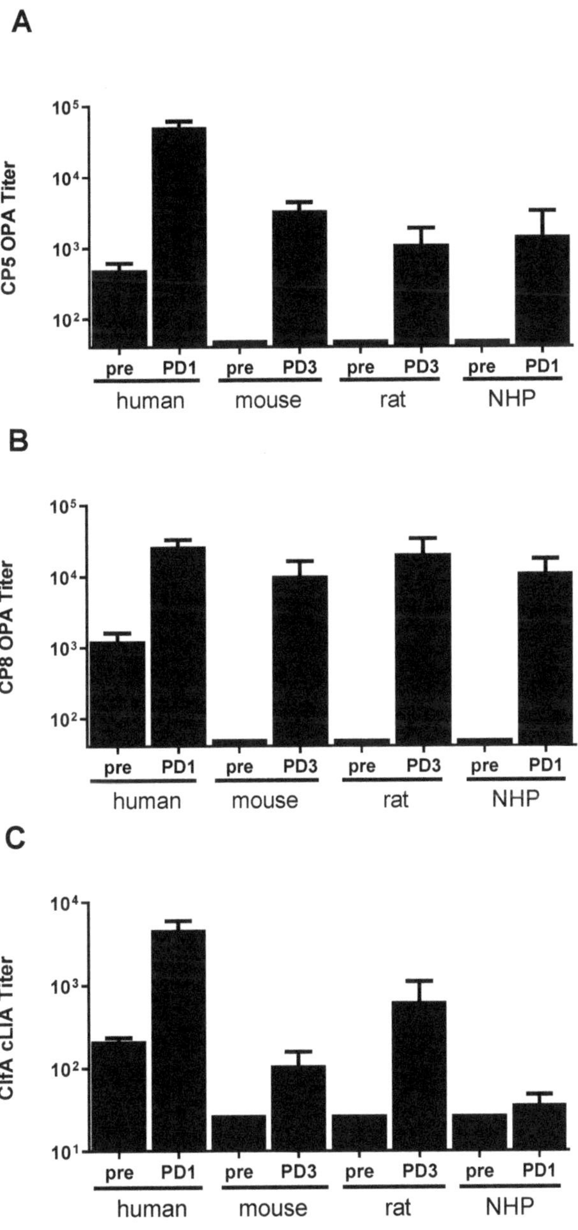

Figure 1. *Cont.*

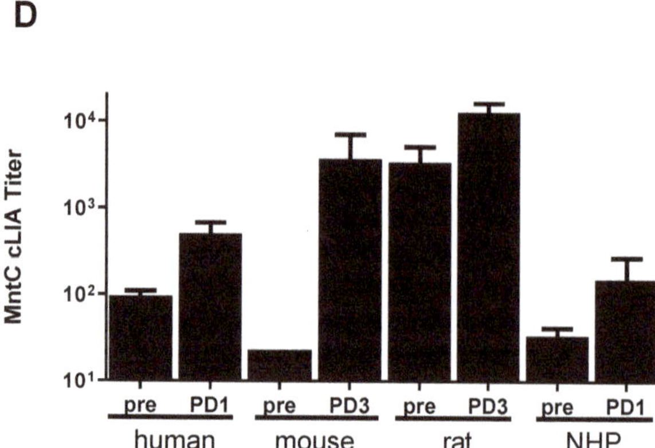

Figure 1. SA4Ag antigens are immunogenic in preclinical species. Immune responses against SA4Ag antigens CP5, CP8, ClfA, and MntC were measured before (pre) and after (PD) immunization. Rodents were immunized three times subcutaneously with SA4Ag + $AlPO_4$ prior to sample collection post-dose 3 (PD3). Non-human primates (NHP) were immunized a single time with SA4Ag without adjuvant. Anti-capsular immune responses were measured by the OPA assay for CP5 (**A**) and CP8 (**B**). Anti-protein immune responses were measured by cLIA for ClfA (**C**) and MntC (**D**). Human responses to a single unadjuvanted dose of SA4Ag are included as a comparator, adapted from [27].

3.2. Immunization with SA4Ag Reduces Bacterial Burden in a Murine Model of Surgical Site Infection

Immunization with SA4Ag was evaluated in a murine model of surgical site infection. Analogous to human deep tissue surgical site infection, *S. aureus* was not introduced systemically but instead a low number of bacterial cells were inoculated into the surgical wound. As shown in Figure 1 and observed by others [28], mice respond relatively poorly to ClfA immunization, in comparison to both rats and humans, even in the presence of $AlPO_4$ adjuvant. To enhance the ClfA response in a model where initial adhesion events are likely important, we considered the addition of an alternate, non–alum-based adjuvant. We limited our selection to adjuvants that are usable in human clinical trials, as some highly reactogenic adjuvants, such as Freund's, overinflate immunogenicity. QS-21, a derivative of the bark of the *Quillaja saponaria* tree, has been used in human clinical trials and is purported to induce a more balanced IgG1/IgG2a response than alum-containing adjuvants in mice [29,30]. Addition of QS-21 to SA4Ag resulted in enhanced ClfA responses (geometric mean titer: 516.4; 95% CI: 222.4–1199.0; unpaired *t*-test with Welch's correction: $p = 0.048$) as compared with $AlPO_4$ (geometric mean titer: 42.8; 95% CI: 17.9–102.1). In the murine surgical site infection model, immunization with SA4Ag adjuvanted with QS-21 reduced bacterial burden (Table 2), illustrating an impact of SA4Ag in this difficult local infection model.

3.3. Immunization with SA4Ag Protects against MRSA Challenge in a Murine Bacteremia Model

S. aureus can disseminate from a local infection via the bloodstream. Therefore, SA4Ag was evaluated for its ability to reduce the bacterial burden in a murine bacteremia model, which mimics very early stages of hematogenous spread. Mice were immunized three times with SA4Ag with $AlPO_4$ and then challenged with either *S. aureus* CDC3 or PFESA0241, both USA300 MRSA isolates. Blood was collected three hours post-challenge and bacteria were enumerated. Immunization with SA4Ag significantly reduced the number of recovered CFU with both USA300 MRSA isolates (Table 3).

Table 2. Immunization with SA4Ag Reduces *S. aureus* Burden in a Murine Surgical Site Infection Model.

Trial	Vaccine	Number of Animals	Mean log CFU/mL (95% CI)	*p* Value vs. Control
1	SA4Ag	20	4.65 (3.90–5.40)	0.0393
	Vehicle	20	5.72 (4.99–6.45)	
2	SA4Ag	15	6.42 (5.07–7.77)	0.0330
	Vehicle	13	7.89 (6.95–8.83)	
3	SA4Ag	17	5.95 (4.60–7.31)	0.2899
	Vehicle	16	6.45 (5.11–7.81)	
Meta-analysis	SA4Ag	48	6.29 (5.57–7.01)	0.0126
	Vehicle	43	7.36 (6.75–7.97)	

Female CD1 mice (*n* = 10–20) were immunized on weeks 0, 3, and 6 with SA4Ag + QS-21 or QS-21 alone. On week 8, animals underwent surgery by placing an incision and stitch into the quadriceps muscle. Approximately 300 colony-forming units (CFU) of *S. aureus* PFESA0158 were instilled into the surgical site. Two days after surgery, tissue was harvested and bacterial burden was enumerated. *p* values were determined by Student's *t*-test.

Table 3. Immunization with SA4Ag Reduces Bacterial Burden in a *S. aureus* Bacteremia Model.

Challenge	Trial	Vaccine	Number of Animals	Mean log CFU/mL (95% CI)	*p* Value vs. Control
S. aureus CDC3 (USA300)	1	SA4Ag	10	4.65 (4.33–4.97)	0.0056
		Vehicle	10	5.35 (4.96–5.73)	
	2	SA4Ag	10	4.49 (4.07–4.90)	0.0147
		Vehicle	10	5.24 (4.76–5.72)	
	3	SA4Ag	10	3.63 (3.25–4.01)	0.0056
		Vehicle	10	4.47 (4.00–4.94)	
	Meta-analysis	SA4Ag	30	4.25 (4.00–4.51)	<0.0001
		Vehicle	30	5.02 (4.75–5.29)	
S. aureus PFESA0241 (USA300)	1	SA4Ag	10	4.17 (3.83–4.51)	0.0491
		Vehicle	10	4.68 (4.25–5.12)	
	2	SA4Ag	10	4.26 (3.93–4.59)	0.0157
		Vehicle	10	4.82 (4.48–5.15)	
	3	SA4Ag	10	4.38 (3.98–4.78)	0.0455
		Vehicle	10	4.88 (4.54–5.22)	
	Meta-analysis	SA4Ag	30	4.27 (4.09–4.45)	0.0002
		vehicle	30	4.79 (4.60–4.98)	

Female CD-1 mice (*n* = 10) were vaccinated at weeks 0, 3, and 6, with SA4Ag or with vehicle alone. Two weeks after the final vaccination animals were challenged with ~2 × 10^8 *S. aureus* CDC3 or PFESA0241. Three hours post-challenge, the mice were euthanized and blood collected. Bacteria in blood were enumerated (colony-forming unit [CFU]/mL), and the log CFU reduction with 95% confidence interval (CI) was calculated compared to the vehicle-treated control. *p* values were determined by Student's *t*-test

3.4. Immunization with SA4Ag Reduces Bacterial Burden in a Murine Pyelonephritis Model

S. aureus can cause infection at sites distant from the initial site of infection, and the kidney is a common end organ for *S. aureus* infection. A murine pyelonephritis model was used to evaluate the ability of SA4Ag to reduce the bacterial burden in the kidney. Mice were immunized with SA4Ag and then challenged with *S. aureus* Reynolds. Two days after challenge, kidneys were harvested and homogenized to enumerate bacterial burden. Immunization with SA4Ag significantly reduced the number of recovered CFU from kidneys (Figure 2).

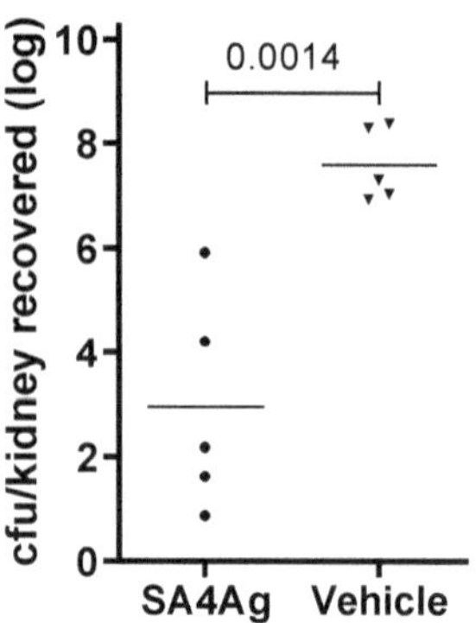

Figure 2. Immunization with SA4Ag reduces bacterial burden in a *S. aureus* pyelonephritis model. Female CD-1 mice (*n* = 5) were vaccinated at weeks 0, 3, and 6, with SA4Ag or with vehicle alone. Two weeks after the final vaccination animals were challenged with ~2 × 10^8 *S. aureus* Reynolds. Two days post-challenge, the mice were euthanized and kidneys collected. Bacteria in kidneys were enumerated (colony-forming unit [CFU]/kidney). *p* value was calculated by Student's *t*-test.

3.5. Immunization with SA4Ag Protects against Both CP5- and CP8-Expressing S. aureus in a Rat Endocarditis Model

S. aureus can cause infection at sites distant from the initial site of infection. SA4Ag was evaluated for its ability to reduce the bacterial burden in a model of foreign body–like infection, i.e., rat endocarditis following catheterization, considered a very stringent preclinical model. A catheter was inserted into the heart via the jugular vein across the aortic valve to create sterile valvular vegetations. Two days after catheter placement, animals were challenged intravenously with *S. aureus*. In two out of three studies, immunization with SA4Ag reduced the bacterial burden of a CP5-expressing *S. aureus* clinical isolate from infected heart tissue (Table 4). Meta-analysis of these three studies demonstrated a significant reduction in recovered CFU (Table 5, *p* = 0.0126).

Table 4. Immunization with SA4Ag Reduces Recovered CFU of a CP5-expressing *S. aureus* Clinical Isolate in a Rat Endocarditis Model.

Trial	Vaccine	Number of Animals	Mean Log CFU/mL (95% CI)	*p* Value vs. Control
1	SA4Ag	16	6.52 (5.20–7.84)	0.0319
	Vehicle	14	7.88 (7.19–8.58)	
2	SA4Ag	15	6.42 (5.07–7.77)	0.0330
	Vehicle	13	7.89 (6.95–8.83)	
3	SA4Ag	17	5.95 (4.60–7.31)	0.2899
	Vehicle	16	6.45 (5.11–7.81)	
Meta-analysis	SA4Ag	48	6.29 (5.57–7.01)	0.0126
	Vehicle	43	7.36 (6.75–7.97)	

Groups of female Sprague Dawley rats (20/group) were vaccinated at weeks 0, 3, and 6, with SA4Ag in $AlPO_4$ or with $AlPO_4$ alone. Two weeks after the final vaccination, catheter placement surgery was performed to generate sterile valvular vegetations. A certain number of animals succumbed due to the surgery and were not included in the analysis. Two days after surgery, animals were challenged intravenously with *S. aureus* PFESA0158. Two days post-challenge, the rats were euthanized and hearts collected. Bacterial burden in heart tissue was enumerated (colony-forming unit [CFU]/mL) and the arithmetic mean and 95% confidence interval (CI) were calculated for each treatment group.

The endocarditis model was next refined to improve clinical relevance by reducing the challenge inoculum. It is believed that *S. aureus*-induced foreign body infections and endocarditis are caused by the seeding of very low numbers of bacteria introduced to the bloodstream either during surgery or as a result of minor tissue injury, such as resulting from a scratch or brushing of teeth [24]. To model human bacterial exposure more closely, the infectious inoculum was reduced to a level just above the threshold for achieving an infection. In this model, *S. aureus* preferentially seeds the damaged valvular tissue and the kidneys; no bacterial burden was detected in the liver, spleen or lungs. In the refined model, the complete absence of detectable infection in both hearts and kidneys following immunization with SA4Ag was evaluated. Rats were immunized subcutaneously three times on weeks 0, 3, and 6 with SA4Ag, and then underwent catheter placement surgery. Two days after catheter placement, rats were challenged intravenously with a CP8-expressing clinical *S. aureus* isolate, PFESA0186. Immunization with SA4Ag reduced the number of animals with detectable *S. aureus* infection in the hearts and kidneys in two separate experiments. Meta-analysis of the two experiments showed a significant reduction in the number of infected animals upon immunization, representing a vaccine efficacy of 88.7% (Table 5).

Table 5. Immunization with SA4Ag Reduces the Number of *S. aureus* Infections in a Rat Endocarditis Model of Infection.

Experiment	Vaccine	Number of Infected Animals	Number of Uninfected Animals	*p* Value vs. Control
1	SA4Ag	0	11	0.0983
	Vehicle	4	9	
2	SA4Ag	1	11	0.0730
	Vehicle	6	7	
Meta-analysis	SA4Ag	1	22	<0.0001
	Vehicle	10	16	

Female Sprague Dawley rats ($n = 11$–14) were vaccinated at weeks 0, 3, and 6, with SA4Ag in $AlPO_4$ or with $AlPO_4$ alone. Two weeks after the final vaccination, catheter placement surgery was performed to generate sterile valvular vegetations. A certain number of animals succumb due to the surgery and were not included in the analysis. Two days after surgery, animals were challenged intravenously with ~4 × 10³ colony-forming unit (CFU) *S. aureus* PFESA0186, a CP8-expressing clinical isolate. Two days post-challenge, the rats were euthanized, and tissue bacterial burden was quantitated in hearts and kidneys.

4. Discussion

A four-antigen *S. aureus* vaccine (SA4Ag) was designed as a candidate to prevent invasive *S. aureus* disease in postsurgical populations. Therefore, SA4Ag was evaluated for the ability to protect against invasive *S. aureus* disease in a series of preclinical models that mimic aspects of human postsurgical infection. Individual components of SA4Ag have previously been shown to elicit robust antibody responses and efficacy in rodents [11,14,21,31], but efficacy of the combined SA4Ag vaccine in multiple preclinical models has not previously been reported. SA4Ag was found to significantly reduce or abrogate infection in murine models of surgical site infection, bacteremia, and in two iterations of a rat endocarditis model. These models mimic the progression of *S. aureus* postsurgical invasive disease from a local deep tissue infection, through hematogenous spread, and dissemination to distant tissue sites, such as the heart valves and kidneys. Together, the responses seen in these models demonstrated that vaccination with SA4Ag elicits an immune response that restrains multiple stages of invasive preclinical *S. aureus* infection and suggested that similar positive outcomes could be achieved in the clinic. Importantly, in the refined rat endocarditis model, a complete absence of infection was observed in 96% of vaccinated animals, reflecting a vaccine efficacy of 88.7%. This is the first time to our knowledge that a sterilizing immune response has been demonstrated after administration of an *S. aureus* vaccine in a preclinical model, and supported moving the SA4Ag into clinical development.

S. aureus disease is challenging to effectively model in animals due to the organism's adaptation to the human host environment and the lack of a defined correlate of protection for invasive *S. aureus* disease. As demonstrated in Figure 1, rodents, which are commonly used in preclinical models of *S. aureus* pathogenesis, required three doses of SA4Ag with an adjuvant to achieve immune responses similar to those seen in non-human primates and humans after a single unadjuvanted dose of SA4Ag. It is possible, therefore, that the response of repeatedly immunized naïve animals does not fully recapitulate the maturation of an immune response originally elicited through natural exposure. The response to ClfA may serve as a case in point. Naïve mice respond poorly to ClfA, even after multiple immunizations in the presence of adjuvant, while humans and non-human primates respond well after a single dose. Interestingly, even though humans respond well to a single dose of ClfA-containing SA4Ag, implying ClfA is eliciting an anamnestic response, a functional anti-ClfA response, which can block the binding of ClfA to fibrinogen, is only observed after immunization [10]. This highlights the need to match the right vaccine with antigen presentation in the correct format.

Invasive *S. aureus* infection can proceed from a local inoculation site, through dissemination in the bloodstream, to distant tissues. A given vaccine candidate may show efficacy in one or more models, but it is the combination of vaccine candidates and evaluation in relevant models of infection that improves the robustness of the conclusions drawn from the preclinical program. Based in part on the preclinical data included here, the *S. aureus* vaccine was advanced into clinical testing. Prior to the clinical assessment of SA4Ag, two other *S. aureus* vaccines also failed to demonstrate efficacy in the clinic, despite having supportive preclinical data. Both vaccines had a single antigenic target, either capsular polysaccharide or IsdB. The StaphVax vaccine, comprised of capsular polysaccharides type 5 and type 8 conjugated to Pseudomonas exotoxin A, did not meet its Phase III clinical endpoint. It has been suggested that this may have resulted from quality variations of the conjugate used in the trial, as opposed to a failure of the mechanism of action [32]. Thus, *S. aureus* capsular polysaccharides had remained an attractive vaccine candidate but potentially not sufficient to protect against invasive *S. aureus* disease. A second single-antigen *S. aureus* vaccine, V710, contained the iron uptake component, IsdB, and also did not meet its primary efficacy endpoint (prevention of serious postoperative *S. aureus* infections following cardiothoracic surgery) in a randomized Phase 2b/3 trial. Additionally, V710 was associated with increased mortality among patients who developed *S. aureus* infections [33]. IsdB did protect mice in preclinical lethal challenge models [17,18]. However, *S. aureus* has multiple redundant iron acquisition systems and prefers human hemoglobin as an iron source, which may have limited the predictive value of the models used to evaluate this antigen [34]. In preclinical models, the protection seen with IsdB was improved with incorporation of additional *S. aureus* antigens [35]. The recognition that a single antigen approach may not be sufficient to protect against *S. aureus*-mediated disease spurred the development of several multi-antigen approaches in addition to SA4Ag [36–38].

In an early Phase 1/2 clinical study (NCT01364571), SA4Ag elicited robust functional immune responses and showed an acceptable safety profile [27]. However, a large randomized, placebo-controlled Phase 2b study (STRIVE; NCT02388165) in adult patients undergoing elective open posterior spinal fusion surgery [39], the results of a pre-planned interim analysis indicated low statistical probability to meet the pre-defined primary efficacy objective—prevention of invasive *S. aureus* disease (bloodstream and surgical site infections), leading to the decision to terminate the study and discontinue the clinical development [40]. This failure indicates that despite protection seen in the current preclinical studies, the animal models that were used had limitations that did not enable prediction of clinical efficacy of the vaccine. One such potential shortcoming is the measure of protection. For example, the absence of detectable infection was used as the criterion of induced protection for the refined endocarditis model. A second shortcoming is that the infection models all evaluated early time points before any animals could have succumbed to infection. Examination at later time points might have determined any impact of the vaccine on survival, and changes in pathology could be identified (e.g., wound size and induration, paralysis or some other signs of morbidity, or histological changes). Examination of the immune responses elicited by the vaccine in more detail (innate vs. adaptive) might also have provided more insight into the mechanism(s) of protection.

Since there is no human correlate of protection against *S. aureus* infection, establishing reliable animal models for preclinical evaluation of *S. aureus* vaccine candidates remains very challenging [19,20]. This is further supported by the observation that while non-human primates required a single dose of SA4Ag without adjuvant to generate functional immune responses, rodents required multiple doses of SA4Ag with adjuvant to achieve similar levels of immune response. Therefore, current preclinical efficacy models play an important yet circumscribed role in illustrating the protective potential of a *S. aureus* vaccine candidate. Animal models are useful in this context for demonstrating proof of principle or proof of mechanism, but ultimately clinical efficacy data, obtained using carefully selected vaccine formulations and target populations, is required until a reliable correlate of protection is identified [19,20]. Identification of a relevant correlate of protection for invasive *S. aureus* disease and development of preclinical models that accurately reflect these human correlates is needed to improve effective preclinical evaluation of *S. aureus* vaccine candidates.

5. Conclusions

Although a *S. aureus* vaccine candidate, SA4Ag, showed promise in the various preclinical models presented, it ultimately did not show clinical efficacy. SA4Ag is the latest in a series of *S. aureus* vaccines that showed protection in preclinical models, which did not translate into positive clinical outcomes. This highlights the need to identify relevant correlates of protection for invasive *S. aureus* infection, and to develop appropriate preclinical models that better predict clinical efficacy.

Author Contributions: Conceptualization, K.U.J., A.S.A., P.A.L., I.L.S.; methodology, Y.T., P.L., A.I.; formal analysis, I.L.S., P.A.L., A.S.A.; writing—original draft preparation, I.L.S., A.S.A.; writing—review and editing, I.L.S., Y.T., A.I., P.L., P.A.L., K.U.J., A.S.A. All authors have read and agreed to the published version of the manuscript.

Funding: The research was funded by Pfizer Inc.

Institutional Review Board Statement: All animal studies were performed in an AAALAC-accredited facility under IACUC-approved protocols.

Informed Consent Statement: Not applicable. This study did not involve humans; human data is reproduced from a cited publication.

Data Availability Statement: The data presented in this study are available on request from the corresponding author.

Acknowledgments: The authors thank the following Pfizer colleagues: excellent and dedicated comparative medicine colleagues who assisted in the procedures reported in this manuscript; Danka Pavliakova and Hong Chen for the provision of cLIA data; and Nataliya Kushnir for editorial assistance.

Conflicts of Interest: I.L.S., Y.T., A.I., P.A.L., K.U.J. and A.S.A. are employees of Pfizer Inc. and may hold stock or stock options. P.L. was an employee of Pfizer at the time the work was done.

References

1. Kluytmans, J.; van Belkum, A.; Verbrughn, H. Nasal carriage of *Staphylococcus aureus*: Epidemiology, underlying mechanisms, and associated risks. *Clin. Microbiol. Rev.* **1997**, *10*, 505–520. [CrossRef]
2. Lowy, F.D. *Staphylococcus aureus* infections. *N. Engl. J. Med.* **1998**, *339*, 520–532. [CrossRef] [PubMed]
3. Noskin, G.A.; Rubin, R.J.; Schentag, J.J.; Kluytmans, J.; Hedblom, E.C.; Jacobson, C.; Smulders, M.; Gemmen, E.; Bharmal, M. National trends in *Staphylococcus aureus* infection rates: Impact on economic burden and mortality over a 6-year period (1998–2003). *Clin. Infect. Dis.* **2007**, *45*, 1132–1140. [CrossRef] [PubMed]
4. Carrel, M.; Perencevich, E.N.; David, M.Z. USA300 Methicillin-Resistant *Staphylococcus aureus*, United States, 2000–2013. *Emerg. Infect. Dis.* **2015**, *21*, 1973–1980. [CrossRef] [PubMed]
5. Malachowa, N.; Whitney, A.R.; Kobayashi, S.D.; Sturdevant, D.E.; Kennedy, A.D.; Braughton, K.R.; Shabb, D.W.; Diep, B.A.; Chambers, H.F.; Otto, M.; et al. Global changes in *Staphylococcus aureus* gene expression in human blood. *PLoS ONE* **2011**, *6*, e18617. [CrossRef]
6. Bode, L.; Kluytmans, J.; Wertheim, H.; Bogaers, D.; Vandenbroucke-Grauls, C.; Roosendaal, R.; Troelstra, A.; Box, A.; Voss, A.; Van Der Tweel, I.; et al. Preventing surgical-site infections in nasal carriers of *Staphylococcus aureus*. *N. Engl. J. Med.* **2010**, *362*, 9–17. [CrossRef]

7. Kalmeijer, M.D.; Coertjens, H.; Van Nieuwland-Bollen, P.M.; Bogaers-Hofman, D.; De Baere, G.A.J.; Stuurman, A.; Van Belkum, A.; Kluytmans, J.A.J.W. Surgical site infections in orthopedic surgery: The effect of mupirocin nasal ointment in a double-blind, randomized, placebo-controlled study. *Clin. Infect. Dis.* **2002**, *35*, 353–358. [CrossRef]
8. Simor, A.E.; Phillips, E.; McGeer, A.; Konvalinka, A.; Loeb, M.; Devlin, H.R.; Kiss, A. Randomized controlled trial of chlorhexidine gluconate for washing, intranasal mupirocin, and rifampin and doxycycline versus no treatment for the eradication of methicillin-resistant *Staphylococcus aureus* colonization. *Clin. Infect. Dis.* **2007**, *44*, 178–185. [CrossRef]
9. Simor, A.E.; Stuart, T.L.; Louie, L.; Watt, C.; Ofner-Agostini, M.; Gravel, D.; Mulvey, M.; Loeb, M.; McGeer, A.; Bryce, E.; et al. Mupirocin-resistant, methicillin-resistant *Staphylococcus aureus* strains in Canadian hospitals. *Antimicrob. Agents Chemother.* **2007**, *51*, 3880–3886. [CrossRef]
10. Hawkins, J.; Kodali, S.; Matsuka, Y.V.; McNeil, L.K.; Mininni, T.; Scully, I.L.; Vernachio, J.H.; Severina, E.; Girgenti, D.; Jansen, K.U.; et al. A recombinant clumping factor A-containing vaccine induces functional antibodies to *Staphylococcus aureus* that are not observed after natural exposure. *Clin. Vaccine Immunol.* **2012**, *19*, 1641–1650. [CrossRef]
11. Anderson, A.S.; Scully, I.L.; Timofeyeva, Y.; Murphy, E.; McNeil, L.K.; Mininni, T.; Nuñez, L.; Carriere, M.; Singer, C.; Dilts, D.A.; et al. *Staphylococcus aureus* manganese transport protein C is a highly conserved cell surface protein that elicits protective immunity against S. *aureus and Staphylococcus epidermidis. J. Infect. Dis.* **2012**, *205*, 1688–1696. [CrossRef] [PubMed]
12. Brady, R.A.; Leid, J.G.; Camper, A.K.; Costerton, J.W.; Shirtliff, M.E. Identification of *Staphylococcus aureus* proteins recognized by the antibody-mediated immune response to a biofilm infection. *Infect. Immun.* **2006**, *74*, 3415–3426. [CrossRef] [PubMed]
13. Handke, L.D.; Hawkins, J.C.; Miller, A.A.; Jansen, K.U.; Anderson, A.S. Regulation of *Staphylococcus aureus* MntC Expression and Its Role in Response to Oxidative Stress. *PLoS ONE* **2013**, *8*, e77874. [CrossRef] [PubMed]
14. Nanra, J.S.; Buitrago, S.M.; Crawford, S.; Ng, J.; Fink, P.S.; Hawkins, J.; Scully, I.L.; McNeil, L.K.; Aste-Amézaga, J.M.; Cooper, D.; et al. Capsular polysaccharides are an important immune evasion mechanism for *Staphylococcus aureus. Hum. Vaccines Immunother.* **2012**, *9*, 480–487. [CrossRef]
15. Thakker, M.; Park, J.S.; Carey, V.; Lee, J.C. *Staphylococcus aureus* serotype 5 capsular polysaccharide is antiphagocytic and enhances bacterial virulence in a murine bacteremia model. *Infect. Immun.* **1998**, *66*, 5183–5189. [CrossRef]
16. Fattom, A.; Schneerson, R.; Szu, S.C.; Vann, W.F.; Shiloach, J.; Karakawa, W.W.; Robbins, J.B. Synthesis and immunologic properties in mice of vaccines composed of *Staphylococcus aureus* type 5 and type 8 capsular polysaccharides conjugated to Pseudomonas aeruginosa exotoxin A. *Infect Immun.* **1990**, *58*, 2367–2374. [CrossRef]
17. Kuklin, N.A.; Clark, D.J.; Secore, S.; Cook, J.; Cope, L.D.; McNeely, T.; Noble, L.; Brown, M.J.; Zorman, J.K.; Wang, X.M.; et al. A novel *Staphylococcus aureus* vaccine: Iron surface determinant B induces rapid antibody responses in rhesus macaques and specific increased survival in a murine S. aureus sepsis model. *Infect. Immun.* **2006**, *74*, 2215–2223. [CrossRef]
18. Joshi, A.; Pancari, G.; Cope, L.; Bowman, E.P.; Cua, D.; Proctor, R.A.; McNeely, T. Immunization with *Staphylococcus aureus* iron regulated surface determinant B (IsdB) confers protection via Th17/IL17 pathway in a murine sepsis model. *Hum. Vaccines Immunother.* **2012**, *8*, 336–346. [CrossRef]
19. Pozzi, C.; Olaniyi, R.; Liljeroos, L.; Galgani, I.; Rappuoli, R.; Bagnoli, F. Vaccines for *Staphylococcus aureus* and Target Populations. *Curr. Top. Microbiol. Immunol.* **2017**, *409*, 491–528. [CrossRef]
20. Jansen, K.U.; Girgenti, D.Q.; Scully, I.L.; Anderson, A.S. Vaccine review: "*Staphyloccocus aureus* vaccines: Problems and prospects". *Vaccine* **2013**, *31*, 2723–2730. [CrossRef]
21. Scully, I.L.; Timofeyeva, Y.; Keeney, D.; Matsuka, Y.V.; Severina, E.; McNeil, L.K.; Nanra, J.; Hu, G.; Liberator, P.A.; Jansen, K.U.; et al. Demonstration of the preclinical correlate of protection for *Staphylococcus aureus* clumping factor A in a murine model of infection. *Vaccine* **2015**, *33*, 5452–5457. [CrossRef] [PubMed]
22. Wang, L.; Lee, J.C. Murine Models of Bacteremia and Surgical Wound Infection for the Evaluation of *Staphylococcus aureus* Vaccine Candidates. *Methods Mol. Biol.* **2016**, *1403*, 409–418. [CrossRef] [PubMed]
23. Schennings, T.; Heimdahl, A.; Coster, K.; Flock, J.I. Immunization with fibronectin binding protein from *Staphylococcus aureus* protects against experimental endocarditis in rats. *Microb. Pathog.* **1993**, *15*, 227–236. [CrossRef] [PubMed]
24. Chambers, H.F.; Korzeniowski, O.M.; Sande, M.A. *Staphylococcus aureus* endocarditis: Clinical manifestations in addicts and nonaddicts. *Medicine* **1983**, *62*, 170–177. [CrossRef] [PubMed]
25. Gribenko, A.V.; Mosyak, L.; Ghosh, S.; Parris, K.; Svenson, K.; Moran, J.; Chu, L.; Li, S.; Liu, T.; Woods, V.L.; et al. Three-dimensional structure and biophysical characterization of *Staphylococcus aureus* cell surface antigen-manganese transporter MntC. *J. Mol. Biol.* **2013**, *425*, 3429–3445. [CrossRef]
26. Gribenko, A.V.; Parris, K.; Mosyak, L.; Lidia, M.; Handke, L.; Hawkins, J.C.; Severina, E.; Matsuka, Y.V.; Anderson, A.S. High Resolution Mapping of Bactericidal Monoclonal Antibody Binding Epitopes on *Staphylococcus aureus* Antigen MntC. *PLoS Pathog.* **2016**, *12*, e1005908. [CrossRef]
27. Frenck, R.W., Jr.; Creech, C.B.; Sheldon, E.; Seiden, D.J.; Kankam, M.K.; Baber, J.; Zito, E.; Hubler, R.; Eiden, J.J.; Severs, J.M.; et al. Safety, tolerability, and immunogenicity of a 4-antigen *Staphylococcus aureus* vaccine (SA4Ag): Results from a first-in-human randomised, placebo-controlled phase 1/2 study. *Vaccine* **2017**, *35*, 375–384. [CrossRef]
28. Li, X.; Wang, X.; Thompson, C.D.; Park, S.; Park, W.B.; Lee, J.C. Preclinical efficacy of clumping factor A in prevention of *Staphylococcus aureus* infection. *mBio* **2016**, *7*, e02232-15. [CrossRef]
29. Zhu, D.; Tuo, W. QS-21: A Potent Vaccine Adjuvant. *Nat. Prod. Chem. Res.* **2016**, *3*, e113. [CrossRef]

30. Bigaeva, E.; Doorn, E.; Liu, H.; Hak, E. Meta-Analysis on Randomized Controlled Trials of Vaccines with QS-21 or ISCOMATRIX Adjuvant: Safety and Tolerability. *PLoS ONE* **2016**, *11*, e0154757. [CrossRef]
31. Nissen, M.D.; Marshall, H.S.; Richmond, P.; Shakib, S.; Jiang, Q.; Cooper, D.; Rill, D.; Baber, J.; Eiden, J.J.; Gruber, W.; et al. A randomized phase I study of the safety and immunogenicity of three ascending dose levels of a 3-antigen *Staphylococcus aureus* vaccine (SA3Ag) in healthy adults. *Vaccine* **2015**, *33*, 1846–1854. [CrossRef] [PubMed]
32. Fattom, A.; Matalon, A.; Buerkert, J.; Taylor, K.; Damaso, S.; Boutriau, D. Efficacy profile of a bivalent *Staphylococcus aureus* glycoconjugated vaccine in adults on hemodialysis: Phase III randomized study. *Hum. Vaccines Immunother.* **2015**, *11*, 632–641. [CrossRef] [PubMed]
33. Fowler, V.G.; Allen, K.B.; Moreira, E.D.; Moustafa, M.; Isgro, F.; Boucher, H.W.; Corey, G.R.; Carmeli, Y.; Betts, R.; Hartzel, J.S.; et al. Effect of an investigational vaccine for preventing *Staphylococcus aureus* infections after cardiothoracic surgery: A randomized trial. *JAMA* **2013**, *309*, 1368–1378. [CrossRef] [PubMed]
34. Haley, K.P.; Skaar, E.P. A battle for iron: Host sequestration and *Staphylococcus aureus* acquisition. *Microbes Infect.* **2012**, *14*, 217–227. [CrossRef]
35. Stranger-Jones, Y.K.; Bae, T.; Schneewind, O. Vaccine assembly from surface proteins of *Staphylococcus aureus*. *Proc. Natl. Acad. Sci. USA* **2006**, *103*, 16942–16947. [CrossRef]
36. Bagnoli, F.; Fontana, M.R.; Soldaini, E.; Mishra, R.P.N.; Fiaschi, L.; Cartocci, E.; Nardi-Dei, V.; Ruggiero, P.; Nosari, S.; De Falco, M.G.; et al. Vaccine composition formulated with a novel TLR7-dependent adjuvant induces high and broad protection against *Staphylococcus aureus*. *Proc. Natl. Acad. Sci. USA* **2015**, *112*, 3680–3685. [CrossRef]
37. Torre, A.; Bacconi, M.; Sammicheli, C.; Galletti, B.; Laera, D.; Fontana, M.R.; Grandi, G.; De Gregorio, E.; Bagnoli, F.; Nuti, S.; et al. Four-Component *Staphylococcus aureus* Vaccine 4C-Staph Enhances Fcgamma Receptor Expression in Neutrophils and Monocytes and Mitigates S. aureus Infection in Neutropenic Mice. *Infect. Immun.* **2015**, *83*, 3157–3163. [CrossRef]
38. Levy, J.; Licini, L.; Haelterman, E.; Moris, P.; Lestrate, P.; Damaso, S.; Van Belle, P.; Boutriau, D. Safety and immunogenicity of an investigational 4-component *Staphylococcus aureus* vaccine with or without AS03 adjuvant: Results of a randomized phase I trial. *Hum. Vaccines Immunother.* **2015**, *11*, 620–631. [CrossRef]
39. ClinicalTrials.gov. Available online: https://clinicaltrials.gov/ct2/show/NCT02388165?term=NCT02388165&draw=2&rank=1 (accessed on 25 April 2020).
40. Pfizer Inc. Available online: https://investors.pfizer.com/investor-news/press-release-details/2018/Independent-Data-Monitoring-Committee-Recommends-Discontinuation-of-the-Phase-2b-STRIVE-Clinical-Trial-of-Staphylococcus-aureus-Vaccine-Following-Planned-Interim-Analysis/default.aspx (accessed on 25 April 2020).

microorganisms

Article

Discovery of *Staphylococcus aureus* Adhesion Inhibitors by Automated Imaging and Their Characterization in a Mouse Model of Persistent Nasal Colonization

Liliane Maria Fernandes de Oliveira [1,†,‡], Marina Steindorff [2,†], Murthy N. Darisipudi [1], Daniel M. Mrochen [1], Patricia Trübe [1], Barbara M. Bröker [1], Mark Brönstrup [2], Werner Tegge [2,*] and Silva Holtfreter [1,*]

1   Institute of Immunology and Transfusion Medicine, Department of Immunology, University Medicine Greifswald, 17475 Greifswald, Germany; liliane.fernandes@med.uni-greifswald.de (L.M.F.d.O.); venkata.darisipudi@med.uni-greifswald.de (M.N.D.); daniel.mrochen@inp-greifswald.de (D.M.M.); patricia.truebe@gmail.com (P.T.); broeker@uni-greifswald.de (B.M.B.)
2   Helmholtz Centre for Infection Research, Department of Chemical Biology, 38124 Braunschweig, Germany; Marina.Steindorff@helmholtz-hzi.de (M.S.); Mark.Broenstrup@helmholtz-hzi.de (M.B.)
*   Correspondence: Werner.Tegge@helmholtz-hzi.de (W.T.); silva.holtfreter@med.uni-greifswald.de (S.H.)
†   These authors contributed equally to this work.
‡   Current address: ZIK Plasmatis, Leibniz Institute for Plasma Science and Technology e.V. (INP), 17475 Greifswald, Germany.

Citation: Fernandes de Oliveira, L.M.; Steindorff, M.; Darisipudi, M.N.; Mrochen, D.M.; Trübe, P.; Bröker, B.M.; Brönstrup, M.; Tegge, W.; Holtfreter, S. Discovery of *Staphylococcus aureus* Adhesion Inhibitors by Automated Imaging and Their Characterization in a Mouse Model of Persistent Nasal Colonization. *Microorganisms* **2021**, *9*, 631. https://doi.org/10.3390/microorganisms9030631

Academic Editor: Rajan P. Adhikari

Received: 1 February 2021
Accepted: 14 March 2021
Published: 18 March 2021

Publisher's Note: MDPI stays neutral with regard to jurisdictional claims in published maps and institutional affiliations.

Abstract: Due to increasing mupirocin resistance, alternatives for *Staphylococcus aureus* nasal decolonization are urgently needed. Adhesion inhibitors are promising new preventive agents that may be less prone to induce resistance, as they do not interfere with the viability of *S. aureus* and therefore exert less selection pressure. We identified promising adhesion inhibitors by screening a library of 4208 compounds for their capacity to inhibit *S. aureus* adhesion to A-549 epithelial cells in vitro in a novel automated, imaging-based assay. The assay quantified DAPI-stained nuclei of the host cell; attached bacteria were stained with an anti-teichoic acid antibody. The most promising candidate, aurintricarboxylic acid (ATA), was evaluated in a novel persistent *S. aureus* nasal colonization model using a mouse-adapted *S. aureus* strain. Colonized mice were treated intranasally over 7 days with ATA using a wide dose range (0.5–10%). Mupirocin completely eliminated the bacteria from the nose within three days of treatment. In contrast, even high concentrations of ATA failed to eradicate the bacteria. To conclude, our imaging-based assay and the persistent colonization model provide excellent tools to identify and validate new drug candidates against *S. aureus* nasal colonization. However, our first tested candidate ATA failed to induce *S. aureus* decolonization.

Keywords: *Staphylococcus aureus*; colonization; mouse; JSNZ; aurintricarboxylic acid; ATA; adhesion inhibitor; mupirocin; nose

1. Introduction

Nasal colonization with *Staphylococcus aureus* is a major risk factor for invasive staphylococcal infections [1,2]. In particular, infections caused by methicillin-resistant *S. aureus* (MRSA) have limited treatment options and are associated with higher morbidity and mortality [3]. To prevent endogenous infection as well as transmission within the hospital, newly admitted patients are routinely screened for MRSA colonization and decolonized using the antibiotic mupirocin [4]. However, increasing bacterial resistance to mupirocin with a prevalence exceeding 13% for MRSA [5] and restrictions for its use have created a need for alternatives [6,7]. In the past, several alternative interventions for the clearance of *S. aureus* nasal carriage have been explored, including antibiotics, such as neomycin [8], polysporin [9], and bacitracin [10], bacteriocins such as lysostaphin [11], as well as fatty acid derivatives (lauric acid monoesters) [12], and cationic synthetic polymers [13]. However, most of these candidates have failed to completely eradicate *S. aureus* nasal colonization. Hence, new approaches to combat *S. aureus* colonization are urgently required.

The first step towards an effective colonization of *S. aureus* in the nares and on other sites of the human body is the attachment of the bacteria to human epithelial cells. This process is facilitated by multifunctional and redundant adhesins on the staphylococcal surface that bind to host cell molecules. An up-to-now underexplored possibility to prevent or eliminate *S. aureus* colonization is to specifically interfere with this bacterial adhesion. Adhesion inhibitors that do not interfere with the viability of the bacteria are promising new preventive agents, because they exert less selective pressure and therefore do not foster the development of resistances [14,15]. Initial studies have been carried out in this context, but so far, they did not lead to clinically useful compounds [16].

The identification of adhesion inhibitors from large compound libraries requires high throughput screening approaches. In former investigations on the adhesion of *S. aureus* to epithelial cells, adherent bacteria were quantified by visual inspection with a microscope and counting [17] or by employing radioactively-labeled bacteria [18]. Alternatively, the adherent bacteria were quantified by determining colony-forming unit (CFU) values after their detachment from eukaryotic cells [19]. These approaches are impractical for the screening of hundreds and thousands of compounds, and they do not detect morphologic effects of the test compounds on the eukaryotic binding partner. Recently, a procedure was reported that is amenable to higher throughput. It utilizes an ELISA to measure binding of *S. aureus* to the major eukaryotic interaction partners fibronectin, keratin, and fibrinogen [20]. However, influences of the compounds on the phenotype of the eukaryotic cells cannot be detected in this screening procedure either.

The in vivo validation of candidate adhesion inhibitors requires a robust and sustained *S. aureus* colonization model. We have previously established a persistent murine nasal colonization model using the mouse-adapted *S. aureus* strain JSNZ [21,22]. Due to its long-term adaptation to the murine host, this strain is capable of inducing persistent colonization of the murine nose and gastrointestinal tract without the need for prior antibiotic treatment [23]. This model enables researchers for the first time to study host–pathogen interaction during persistent colonization in the mouse and also to evaluate decolonization drugs.

Here we report on the development and application of a high throughput microtiter plate-based phenotypic in vitro assay that quantifies the adhesion of *S. aureus* to human epithelial cells. The procedure utilizes fluorescence labeling of eukaryotic cell nuclei and bacteria after their adhesion, followed by detection with an automated microscope with image analysis, in combination with a pipetting robot for the distribution of substance libraries and for liquid handling. We performed a medium throughput screen of more than 4000 compounds, and subsequently characterized aurintricarboxylic acid (ATA) that was identified as the most potent adhesion inhibitor in this assay. The ability of ATA to eradicate nasal *S. aureus* colonization was assessed in our persistent *S. aureus* colonization model. While the standard-of-care antibiotic mupirocin completely eliminated *S. aureus* colonization within three days of treatment, ATA did not show an anti-bacterial effect in vivo. Nevertheless, our novel microscopy-based screening approach and the mouse-adapted strain JSNZ are powerful tools to identify and validate new drug candidates against *S. aureus* nasal colonization.

2. Materials and Methods

2.1. Epithelial Cells

For the screening procedure, the human epithelial lung cell line A-549 (ACC 107) was obtained from the German Collection of Microorganisms and Cell Cultures (DSMZ, Braunschweig, Germany). For the verification of the activity of hit substances, Human Nasal Epithelial Primary Cells (HNEPC) from Provitro GmbH (Berlin, Germany) were used. Both cell lines were cultivated under conditions recommended by their respective depositors. Cell culture reagents came from Invitrogen (Carlsbad, CA, USA) and Provitro GmbH (Berlin, Germany).

2.2. S. aureus Strains

For the screening procedures, the *S. aureus* strain N315 (ST5-CC5-MRSA) was used [24], kindly provided by Prof. Dr. K. Becker, University of Greifswald, Germany. For the development of the assay conditions *S. aureus* SA113 [25] (ST8-CC8-Methicillin-sensitive *S. aureus* (MSSA)) and its adhesion-deficient deletion mutant Delta tagO were kindly provided by Prof. Dr. A. Peschel, University of Tübingen, Germany [26]. For hit validation, the wild type *S. aureus* strains 50,307,270 (rifampicin resistant), 50,128,509 (MRSA), and 50,046,981 (MSSA) were kindly provided by the Institute for Microbiology, Immunology and Hospital Hygiene from the city hospital of Braunschweig, Germany. For the in vivo experiments, the mouse-adapted *S. aureus* strain JSNZ (ST88-CC88-MSSA) was employed [21].

2.3. Compounds

Six substance collections were investigated (Table 1). Aurintricarboxylic acid (ATA) was initially part of the LOPAC collection. For further testing, it was obtained from Sigma-Aldrich/Merck (Darmstadt, Germany) as the free acid.

Table 1. Substance collections for adhesion inhibitors screenings.

Type	Entities	Stock conc.
Secondary metabolites from myxobacteria [1]	117	2 mM in DMSO
LOPAC collection of pharmacologically active compounds [2]	1408	10 mM in DMSO
VAR collection [3]	1600	5 mM in DMSO
Peptide library of the structure XXX12XXX-DKP made of D-amino acids [4]	361	2.2 mM (in 2-propanol/water 1:1)
Peptide library of the structure XXX12XXX-DKP made of L-amino acids [5]	361	4 mM (in 2-propanol/water 1:1)
Cyclic peptides of the structure [AA12AAC] made of D-amino acids [6]	361	3.5 mM in DMSO
Total	4208	

[1] Academic collection of the Helmholtz Centre for Infection Research (HZI), sourced from in-house myxobacterial research [27]. [2] Sigma-Aldrich/Merck (Darmstadt, Germany); LOPAC = abbreviation for 'Library of Pharmacologically Active Compounds'. [3] Academic collection of the HZI, sourced from multiple medicinal chemistry groups; VAR stands for 'various sources'. [4] X = mixture of all proteinogenic amino acids, 1 and 2 = defined amino acids (all proteinogenic amino acids in D-configuration), Cys excluded; DKP = diketopiperazine. [5] X = mixture of all proteinogenic amino acids, 1 and 2 = defined amino acids (all proteinogenic amino acids in L-configuration), Cys excluded; DKP = diketopiperazine; [4,5] prepared in the peptide synthesis facility of the HZI. [6] 1 and 2 = all proteinogenic amino acids in D-configuration, Cys excluded [28].

2.4. Mice

Female C57BL/6NRj mice with Specific and Opportunistic Pathogen Free status (SOPF, *S. aureus*-free, 9 weeks old) were purchased from Janvier Labs (Saint-Berthevin, France). Females were selected because they are less prone to develop genital abscesses upon JSNZ colonization than males [21]. Moreover, males tend to be more aggressive and to present fight wounds that might get infected with *S. aureus*. After delivery, the animals were acclimatized for 7 days before starting the experiments. The animals were kept in individually ventilated cages (IVC, 4 animals/cage) under SOPF conditions with litter material. Food and water (acidified with HCl) were provided ad libitum.

2.5. In Vitro Adhesion Assay

A-549 cells were resuspended in RPMI 1640 containing 10% FCS and seeded at a density of 1×10^4 cells per well in 100 µL medium into 96 well, black, optical bottom microtiter plates (sterile, cell culture-treated; order no. 165,305, Nalgene Nunc, Rochester, NY, USA). After an incubation period of 4 to 5 days under cell culture conditions at 37 °C, when the cells had formed a uniform confluent layer at the bottom of the wells, the incubation medium was removed and replaced by 75 µL infection medium (RPMI 1640 containing 1% FCS and 20 mM Hepes, pH 7.4). Test compounds were added with a pipetting robot equipped with a pin tool at a final concentration of 20 µM (pipetting robot:

Evolution P3, PerkinElmer, Waltham, MA, USA; pin tool: FP3CB, 96 floating tube pins, 0.787 mm diameter, length: 33 mm, transfer volume 80 nL, V&P Scientific, Inc., San Diego, CA, USA). In each microtiter plate, four wells each of the following controls were included: Cell culture medium containing DMSO in concentrations corresponding to the substance testing and as positive control, 50 µg/mL polyinosinic acid (Sigma-Aldrich, Darmstadt, Germany), which has been shown before to reduce the adhesion of *S. aureus* to epithelial cells by approximately 50% at this concentration [29].

25 µL of *S. aureus* N315 suspension that had been grown overnight in Brain Heart Infusion (BHI) medium (Sigma-Aldrich, Darmstadt, Germany), washed twice with phosphate-buffered saline (PBS), and re-suspended in infection medium to an optical density (OD) $OD_{600} = 1.0$ (1×10^8 CFU/mL), was added to each well, resulting in an assay concentration of $OD_{600} = 0.25$ (2.5×10^6 CFU/well). After an incubation period of 1 h at room temperature (RT), non-adherent bacteria were removed by carefully washing the cell layer three times with PBS. The A-549 cells with adherent bacteria were fixed with 4% paraformaldehyde in PBS for 20 min at RT and afterwards washed twice with PBS. For the detection of adherent bacteria 50 µL/well of a primary rabbit antibody against *S. aureus* lipoteichoic acid (Acris Antibodies GmbH, Herford, Germany) diluted 1:5000 in PBS with 1% bovine serum albumin (BSA) was added. After 35 min at RT, 50 µL of 4′,6 diamidino-2-phenylindol-dihydrochloride (DAPI-solution, Sigma-Aldrich, Darmstadt, Germany) for the detection of A-549 cell nuclei was added at a final concentration of 1 µg/mL. For some preliminary experiments in the course of evaluation, the cytoplasm was stained with CellTracker™ Red CMTPX (Molecular Probes®, Invitrogen, Carlsbad, CA, USA) by incubating with a final concentration of 5 µM together with DAPI. After 10 min at RT and three washing steps with PBS, 50 µL/well of the secondary mouse antibody anti-rabbit Alexa Fluor® 488 (Invitrogen) was added at a dilution of 1:1000 in PBS/1% BSA. After an incubation time of 45 min and washing with PBS, bacteria and A-549 cells were analyzed with the automated microscope ImageXpress Micro (IXM, Molecular Devices, Sunnyvale, CA, USA) and the dedicated software MetaXpress. The initial screening of the substance collections was carried out in duplicate in two independent experiments. In the reevaluation of the initial hits, different concentrations of active compounds were used (5, 10, 20, 50, 100 µM, for ATA also lower concentrations).

For the determination of the effect of ATA on precolonized cells, the adhesion assay was modified: After the initial 1 h incubation of the A-549 cells with bacteria without addition of compounds, the wells were washed carefully three times with PBS. Fresh infection medium and ATA were added to the cells at concentrations of 0.95 and 2.2 µg/mL. After further incubations for 1, 2, 3, 4, and 5 h, the cells were washed three times with PBS, followed by the microscope-based quantification procedure. For each time point controls without ATA were carried out, providing the reference values.

2.6. Quantification of Epithelial Cells and Bacteria

Each plate was imaged with the automated microscope. For each well, nine images from different sites with a size of 0.4×0.4 mm were acquired. A $20\times$ objective and the fluorescence filters "DAPI" (377 and 447 nm for excitation and emission, respectively) for the detection of A-549 nuclei and "FITC" (475 and 536 nm for excitation and emission, respectively) for the quantification of Alexa Fluor® 488 labeled bacteria were used. For some preliminary experiments in the course of evaluation, stained cytoplasm was visualized with the fluorescence filter "Texas Red" (560 and 624 nm for excitation and emission, respectively). Each image was analyzed with the MetaXpress software modul "Transfluor" and the mode "Vesicle area per cell". With this mode, the bacterial area per DAPI stained cell nucleus was quantified. The average was calculated from the nine different images per well. Performance of the assay was evaluated in 92 wells without added compounds; 4 wells contained 50 µg/mL polyinosinic acid as positive control. In addition, the signal from the adherent bacteria (Alexa Fluor® 488 fluorescence; excitation/emission at 485 nm/535 nm) was determined with a fluorescence microtiter plate reader (Fusion Universal Microplate Analyser, PerkinElmer, Waltham, MA, USA).

2.7. Cytotoxicity Assay

The cytotoxicity of our hit compounds on A-549 cells was quantified using a 3-(4,5-dimethylthiazol-2-yl)-2,5-diphenyltetrazolium bromide (MTT) assay following the procedure by Mosmann [30], modified by Sasse [31]. Briefly, sub-confluent A-549 cells were washed with PBS without Ca^{2+} and Mg^{2+}, trypsinized, and resuspended in DMEM containing 10% FCS. 60 µL of serial dilutions of the test compounds were added to 120 µL aliquots of a cell suspension (5000 cells) in 96 well microtiter plates in duplicate. Blank and solvent controls were incubated under identical conditions. After 24 h, 20 µL MTT in PBS was added to a final concentration of 0.5 mg/mL. After 2 h, the precipitate of formazan crystals was centrifuged and the supernatant was discarded. The precipitate was washed with 100 µL PBS and dissolved in 100 µL 2-propanol containing 0.4% hydrochloric acid. The microplates were gently shaken for 20 min to ensure a complete dissolution of the formazan, and finally the absorption was measured at 595 nm using an ELISA plate reader. The percentage of viable cells was calculated and the mean was determined with respect to the controls with medium only.

2.8. Investigation of the Antimicrobial Activity

Aliquots of 120 µL of an overnight culture of *S. aureus* N315 in BHI medium were washed, adjusted to OD_{600} = 0.015, corresponding to approximately 5×10^5 CFU/mL, and added to 60 µL of a serial dilution of the test compounds in BHI. After an incubation time of 18 h at 37 °C without shaking under moist conditions, the OD_{600} was measured with a microtiter plate reader (Fusion Universal Microplate Analyser, PerkinElmer, Waltham, MA, USA). The lowest concentration that completely suppressed growth defined the MIC values.

2.9. Preparation of the S. aureus Inoculum for In Vivo Experiments

S. aureus JSNZ strain was grown over night in BHI medium at 37 °C and 200 rpm. Thereafter, the culture was diluted to OD_{595} = 0.05 in BHI medium and cultivated at 37 °C and 200 rpm until the mid-logarithmic phase (OD_{595} = 2.0–2.5). Cells were harvested by centrifugation for 10 min at $8000\times g$, resuspended in BHI containing 14% sterile glycerine, and frozen at −80 °C. Before intranasal inoculation, bacteria stocks were thawed on ice, washed once with 40 mL PBS, and reconstituted in PBS to the desired concentration (1×10^{10} CFU/mL) based on the optical density of the suspension (OD_{595} = 1; equals 2.6×10^8 CFU/mL). The actual bacterial dose was determined by plating serial dilutions of the inoculum on LB agar plates right after intranasal colonization. Plates were incubated over night at 37 °C and CFU were counted the following day.

2.10. Preparation of Substances for Intranasal Application

For initial experiments, ATA was dissolved in three different carrier substances for intranasal application: Poloxamer 407, Softisan 649/Vaseline 9:1, and PBS. Poloxamer 407 powder (Sigma-Aldrich, Darmstadt, Germany) was dissolved in sterile distilled water to the desired concentration and homogenized over night by gently shaking at 4 °C. This substance was stored at 4–8 °C and maintained on ice before intranasal inoculation. Softisan 649/Vaseline 9:1 (*w/w*) (Fagron GmbH & Co. KG, Barsbüttel, Germany), hereafter referred to as S/V, was warmed up to 65–80 °C for 10–30 min to obtain a liquefied solution suitable for pipetting. Since mupirocin 2% (Turixin®, GlaxoSmithKline GmbH & Co, München, Germany) was also formulated with S/V, this drug was also liquefied by warming to 80 °C for 15 min to enable pipetting of small volumes. ATA (Sigma-Aldrich, Darmstadt, Germany) was dissolved in sterile distilled water to 10–20%, thereafter diluted to the required concentration with Poloxamer (20% *w/v*) or with liquefied S/V.

2.11. Treatment of Colonized Mice with Drug Carriers, ATA and Mupirocin

To exclude a direct effect of the drug carrier itself on the bacterial load, we determined the impact of the three drug carriers PBS, 20% Poloxamer 407, and Softisan 649/Vaseline 9:1 (w/w) in a persistent *S. aureus* nasal colonization model. Mice were colonized intranasally with 0.7–1.0×10^8 CFU *S. aureus* JSNZ (5 µL per nostril) under mild isoflurane anesthesia. Starting on day 3 after colonization, mice were treated once daily for 7 consecutive days with 10 µl of the respective carriers or left untreated. All substances were applied into the nasal cavity using a Hamilton© syringe (100 µL Microliter Syringe, 22s-gauge, Hamilton, Bonaduz, Switzerland) with a shortened and blunted needle. The blunted needle was inserted a few millimeters into the anterior nares to ensure application of the substances within the nasal cavity. Gastrointestinal colonization was examined via stool samples that were collected from individual mice in septic cages on day 0, 3, 6, and 10 after colonization. Mice were weighed and visually inspected for any symptoms of infection on a daily basis. On day 6 and 10 after colonization, mice were sacrificed by isoflurane overdose. The nose including the nasal cavities as well as the cecum were collected for evaluation of the bacterial load.

To determine the anti-adhesive capacity of ATA, mice were colonized with JSNZ for three days as described above and afterwards treated once per day for consecutive 7 days with ATA or left untreated. Five microliters of ATA were slowly applied in each nostril (10 µL in total) using the blunted Hamilton© syringe. The compound was applied at different concentrations (0.5%, 2%, 5%, or 10%) in two different carriers. ATA-Poloxamer was cooled on ice while ATA-S/V was warmed to 65–80 °C to turn liquid before intranasal application. The reference substance mupirocin was liquefied at 65–80 °C and applied in the clinically used dose of 2.0% (5 µL/nostril) following the same procedure. Both ATA and mupirocin were not affected by heating in this temperature range as verified by LC-MS controls and mupirocin sensitivity testing, respectively (data not shown) [8]. Animals were sacrificed by isoflurane overdose after 0, 3, or 7 days of treatment, and nose and cecum were obtained as detailed above.

2.12. Determination of the Bacterial Load

Stool samples were adjusted to 0.2 g/mL with sterile PBS followed by homogenization for 20 min at 1400 rpm and 4 °C using a Thermo Mixer C shaker (Eppendorf, Hamburg, Germany). The nose and cecum were weighed, transferred to autoclaved homogenizer tubes containing zirconium oxide beads (diameter: 1.4/2.8 mm, Precellys, France), and filled up with 1 mL sterile PBS. Cecum samples were homogenized at 6000 rpm for 2×20 s with a 15 s interval. Noses were homogenized at 6500 rpm for 2×30 s with a 5 min interval. Homogenized samples were serially diluted; 10 µL of each dilution were plated out in triplicate on *S. aureus* Chromagar plates (CHROmagar, France) and enumerated the next day.

2.13. Ethics Statement

Animal experiments received ethical approval from the responsible State authorities (Landesamt für Landwirtschaft, Lebensmittelsicherheit und Fischerei Mecklenburg-Vorpommern, 7221.3-1-018/19). The experiments were performed in accordance with the German Animal Welfare Act (Deutsches Tierschutzgesetz), the EU Directive 2010/63/EU for animal experiments and the Federation of Laboratory Animal Science Associations (FELASA). All animal experiments comply with the ARRIVE guidelines.

2.14. Statistics

Data analysis was performed using the GraphPadPrism6 package (GraphPad Software, Inc., La Jolla, California, USA). Group-wise comparisons were conducted using the Mann-Whitney test or Welch's unpaired t-test, as indicated in the particular graph. Paired samples (i.e., disease activity scores) were compared using the Friedman test and Dunn's multiple comparison test for post hoc analyses.

3. Results

3.1. Development of a Screening Method for the Identification of S. aureus Adhesion Inhibitors

Our first aim was the development of an automated microscopy-based method that allows the screening of several thousand substances for compounds that inhibit the adhesion of *S. aureus* to human lung epithelial cells (A-549) [17,32–34]. Since the bacteria were found to adhere to the plastic surface of the microtiter plates (not shown), it was important to grow the A-549 cells to a dense confluent layer before performing the adhesion assay. Initial experiments showed a linear correlation between the amount of bacteria used and the bacterial adhesion to A-549 cells (Figure S1). For the subsequent screening campaigns we found an intermediate amount (OD_{600} = 0.25, corresponding to 2.5×10^7 CFU/mL) to be optimal for the performance of the assay. An incubation time of one hour at RT was found most suitable, since longer incubation times or higher temperatures led to a bias due to bacterial proliferation and/or compounds that have an influence on bacterial growth rather than on adhesion. For the reliable detection and quantification of the bacteria in the automated microscope, fluorescence labeling was employed, and several methods were evaluated, including pre-incubation of the bacteria with fluorescein isothiocyanate (FITC) and Syto 9. The most suitable approach was paraformaldehyde fixation at the end of the incubation period, followed by the staining of adherent bacteria with a commercial primary antibody against *S. aureus* lipoteichoic acid and an Alexa Fluor 488-conjugated secondary antibody, together with DAPI staining of the eukaryotic cell nuclei.

During assay development, we investigated the adhesion of *S. aureus* N315 to A-549 cells in relation to the growth phase. *S. aureus* was cultivated for 2 h, 4 h, 6 h, 8 h, 19 h, and 24 h in BHI, harvested, washed, and resuspended in infection medium. The adhesion of overnight grown bacteria was comparable to the adhesion of mid-log phase bacteria (Figure S2). For technical reasons we preferred to use overnight cultures in our approach.

The adherence was measured with an automated microscope by a quantitative detection of the 'vesicle area per cell', the "vesicles" representing the Alexa Fluor-labeled bacteria and the "cells" the DAPI-stained nuclei of the A-549 lung epithelia. In parallel, adherent bacteria were quantified by scanning the plates with a fluorescence microtiter plate reader. The microscopic approach led to more reliable and reproducible values than the data obtained with the microtiter plate reader. The reliability of both assays was assessed by performing the adhesion test in 92 wells without addition of compounds, including controls (Figure S3). For the microscopy-based test, the Z′ factor was determined to be 0.54, which was considered to be of sufficient quality for the screening procedure [35], whereas the Z′ factor for the microtiter plate-based test was only 0.12 and considered insufficient.

The functionality of the assay in terms of the detection of influences on adhesion was validated with the *S. aureus* wild type strain SA113 and its isogenic mutant SA113 Delta tagO, whose efficiency of adhesion was reduced by approximately 50% as shown before [34]. Different ratios of bacteria to A-549 cells showed the expected differences between the strains (Figure S1).

The cell nuclei and the adherent bacteria were reliably identified by automated microscopy and image analysis. As illustrated in Figure 1, there was a good match between the microscopic pictures of the bacteria and the A-549 cells after adhesion, fixation, and fluorescence staining (upper pictures) and the results from the automated image analysis and the labeling of the recognized structures (lower pictures). By staining the cytoplasm of the A-549 cells in addition to the cell nuclei and the bacteria, we verified during the assay development that the bacteria did not migrate from their initial attachment site on the A-549 cells to sites on the well surface where the eukaryotic cells have been detached during the washing procedures (Figure S4). To conclude, the established automated microscope-based approach enabled the quantitative investigation of adhesion processes in microtiter plates and additionally revealed morphological changes that are induced to the cells upon exposure to the tested substances.

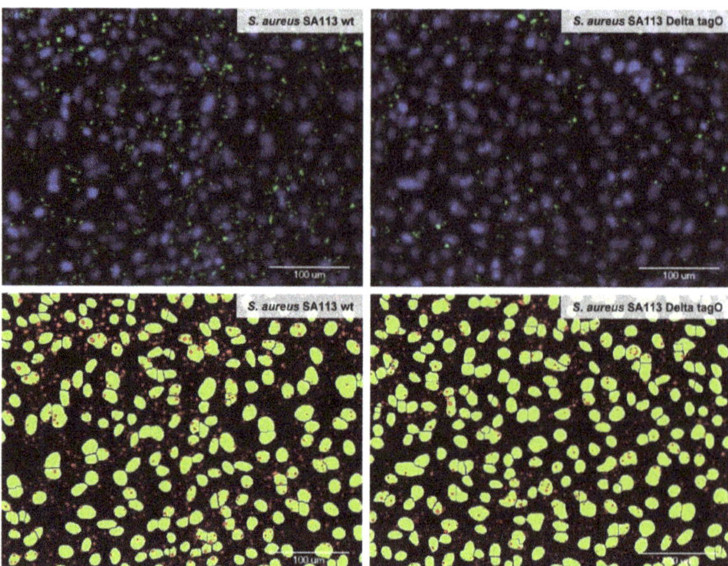

Figure 1. Identification of adherent bacteria by automated microscopy and image analysis. Adhesion of wild type *Staphylococcus aureus* SA113 (**left**) and *S. aureus* SA113 Delta tagO (**right**) to A-549 cells is depicted. Cell nuclei were stained with DAPI (**upper panel**, blue and **lower panel**, green) and bacteria were stained with a primary antibody against teichoic acid and an Alexa 488-labeled secondary antibody (**upper panel**, green and **lower panel**, red). Upper panel: Images acquired with the automated microscope after washing and staining; lower panel: Bacterial and epithelial cell areas as identified by the imaging software MetaXpress with the module "Transfluor" and the mode "Vesicle area per cell", based on the microscopic images shown in the upper panel. A bacterial density of OD_{600} = 0.1 has been used (for quantification please refer to Figure S1).

3.2. Screening of Compound Libraries for S. aureus Adhesion Inhibitors

After the optimization and validation of the assay conditions, bacterial adhesion was investigated in a screening approach with 4208 compounds (Table 1) at 20 µM concentration in duplicate. The compound collection consisted of six different parts. Our in-house collection of secondary metabolites from myxobacteria (part 1) is a unique source of highly bio-active compounds. Many of those compounds have shown antibacterial, antiherbal, cytotoxic, and other properties and are the focus of ongoing investigations by us and many others [27,29]. The LOPAC collection of pharmacologically active compounds (part 2) is a broad collection of a large number of drugs and drug-like molecules, most of which have already been studied in preclinical and clinical investigations. Hits that would be derived from the collection have the advantage that an approval as a drug can be aided by prior data that is already available, which may speed up the process considerably and reduce cost ("drug repurposing"). Our front-runner ATA was part of the LOPAC collection. The VAR collection (part 3) is also unique at our institute and provides a variety of unusual structures that are not found in most commercial substance collections and extended the "structure space" of our screening campaign considerably. The linear and cyclic peptides (part 4–6) are consisting of sublibraries with defined and randomized positions. Such libraries have been used successfully in other screening approaches by employing an iterative stepwise procedure to enhance biological activities. We considered this combination of unique and drug-like structures a solid basis for our project.

The adhesion inhibitor polyinosinic acid was used at 50 µg/mL as a positive control in each microtiter plate [36]. In the primary screen, 62 compounds were found to be active with at least 30% reduction of adhesion (Supplementary Tables S1 and S2). Substances that caused substantial loss of eukaryotic cells during the incubation were excluded from further testing. An advantage of using the automated microscope was the possibility to reinvestigate wells individually by visual inspection of the pictures that were recorded. Since the ratio of bacteria to cell nuclei was determined, the approach was, contrary to the microtiter plate readings, insensitive to partial losses of the epithelial cells during the washing procedures. The antimicrobial effect of the initial hits was evaluated in a broth dilution assay (MIC determinations). Substances with MIC values below 100 µM were excluded from further testing. After the re-evaluation process three compounds were identified that reliably and selectively reduced bacterial adhesion without detaching the eukaryotic cells and with MIC values > 100 µM (Table S3 and Figure S5; for structures see Figure 2B and Figure S6).

3.3. ATA Inhibited S. aureus Adhesion to Epithelial Cells In Vitro

The most active substance found in the screen was polyaromatic aurintricarboxylic acid (ATA, Figure 2B), which reproducibly reduced bacterial adhesion with a half-maximal inhibitory concentration (IC50) of 0.95 µg/mL. The maximum effect size was a reduction of adhesion to ca. 20% of the original value; this was reached at a compound concentration of approximately 2.2 µg/mL (Figure 2A).

Over the course of the adhesion experiment, ATA did not cause any apparent morphological effect on the A-549 cells. In addition, bacterial growth was not impaired by ATA up to a concentration of 42 µg/mL (100 µM, data not shown). In order to assess whether ATA exerts cytotoxic effects, an MTT test with A-549 cells was carried out. After 24 h of incubation, ATA caused less than half-maximum inhibition of metabolic activity of the A-549 cells up to the highest test concentration of 370 µg/mL (data not shown). Toxicities for ATA in mice have been reported before. A dose of 100 mg/kg/day per day i.p. was found to be lethal, whereas 30 mg/kg/day was not [37].

To analyze the activity of ATA towards different S. aureus strains, three clinical isolates of S. aureus were used, strain 50,307,270 (rifampicin resistant), strain 50,128,509 (MRSA), and strain 50,046,981 (MSSA). ATA could inhibit the adhesion of all strains with a potency that was equal to or higher than that found for N315 (Figure 2C).

Next, the effect of ATA on HNEPC was investigated. The results obtained with the immortalized epithelial cell line A-549 could be confirmed: The adhesion of S. aureus N315 to the primary cells was reduced with an IC50 of approximately 1 µg/mL (Figure 2D). Again, ATA did not completely block adhesion, but reduced it to approximately 40% of the original level.

In the primary screening assay, the compounds were already present when bacteria were added to the host cells. Thus, the assay mimics a preventive setting. In order to investigate whether ATA might also work in a therapeutic setting, its effect on already adherent bacteria was investigated next. For this purpose, A-549 cells were precolonized with S. aureus N315 for one hour, followed by the removal of non-adherent and planktonic bacteria by washing. Further incubations from one to five hours with 0.95 µg/mL and 2.2 µg/mL ATA (corresponding to the IC50 and maximal effective concentrations, respectively) were carried out. In comparison to the controls without an active compound, ATA was able to reduce the number of bacteria from precolonized A-549 cells in a dose- and time-dependent manner (Figure 2E). This demonstrates that ATA can disrupt an already existing bacterial adhesion, which suggests that ATA might be used in both preventive and therapeutic settings.

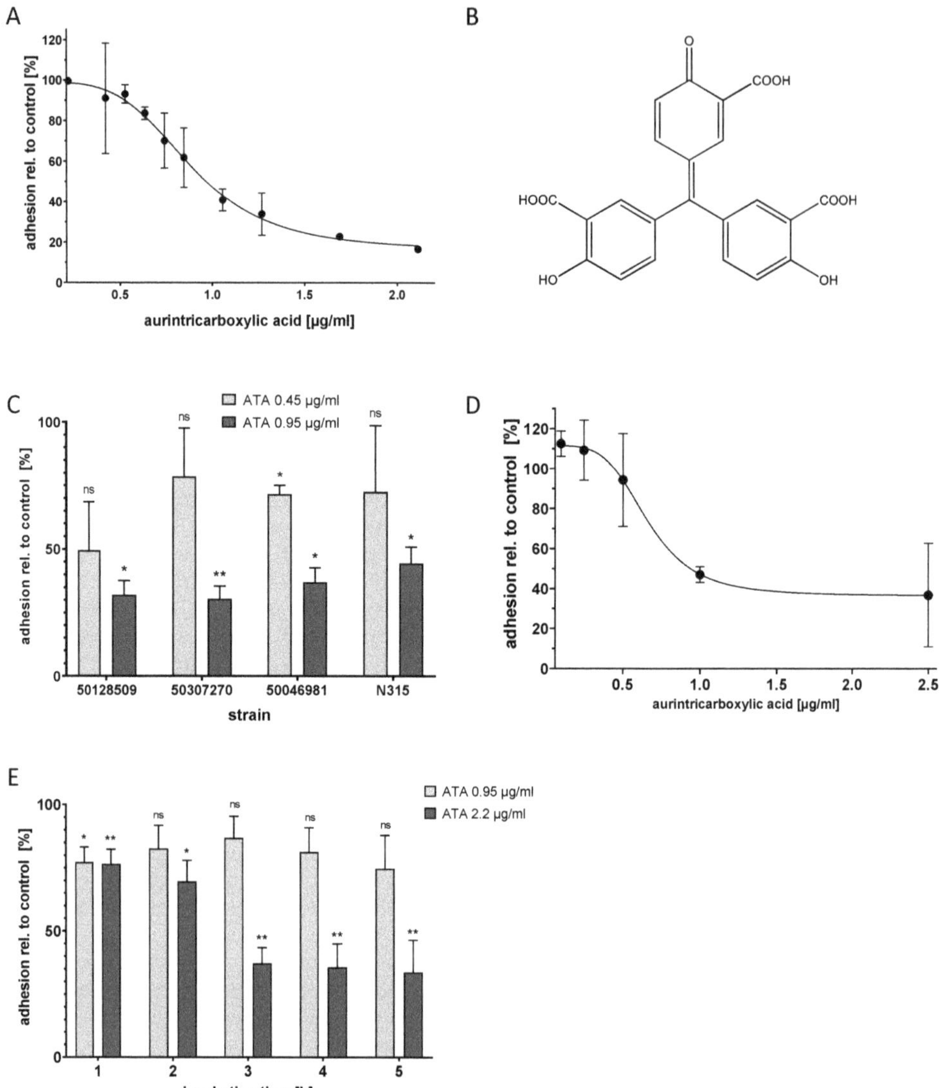

Figure 2. Aurintricarboxylic acid (ATA) inhibited *S. aureus* adhesion to A-549 cells and to human primary epithelial cells in a dose-dependent manner and reduced precolonization. (**A**) Adhesion of *S. aureus* N315 to A-549 cells was determined in the presence of different concentrations of ATA using the microscopic adhesion assay; Mean and standard deviation for three replicates are depicted. (**B**) Structure of ATA in its monomeric form. (**C**) ATA reduced the adhesion of *S. aureus* clinical isolates to A-549 cells. Mean and standard deviation for two replicates are depicted. * $p < 0.4$, ** $p < 0.003$, ns = not significant, as compared to control according to Welch's unpaired *t*-test. (**D**) Adhesion of *S. aureus* to primary epithelial cells from human nares (HNEPC) in presence of ATA. Mean and standard deviation for three replicates are depicted. (**E**) Application of ATA to A-549 cells that were pre-colonized with *S. aureus* N315 resulted in the time-dependent detachment of the bacteria; incubation times are indicated. The values for each time point are related to controls without ATA. Mean and standard deviation for three replicates are depicted. * $p < 0.014$, ** $p < 0.01$, ns = not significant, as compared to control according to Welch's unpaired *t*-test.

3.4. Drug Carriers Did Not Interfere with S. aureus Colonization

A suitable drug carrier has to distribute the active compound throughout the nasal cavity to facilitate *S. aureus* elimination [38], without affecting the bacterial load itself. In a pilot experiment, we compared the intranasal distribution of three drug carriers, Poloxamer 407, S/V, and PBS. We mixed them with Evans blue to visually inspect their distribution in the nasal cavity. Poloxamer 407 and S/V were chosen because of their very different viscosity and physical properties, which could influence their distribution in the nasal compartments. Whereas S/V is highly viscous at room temperature and needs to be warmed up to become liquid and pipettable, Poloxamer 407 is liquid at low temperature and turns into a gel at 37 °C. All substances were applied into the nasal cavity by inserting a Hamilton© syringe with a shortened and blunted needle a few millimeters into the anterior nares. All three substances spread equally well throughout the entire nasal cavity 30 min after application. They were found in the ventral region close to the nares, along all nasal turbinates, and also in the most dorsal regions, such as the nasopharynx (data not shown).

Next we studied the influence of the carrier substances on *S. aureus* colonization in vivo using a new mouse model of *S. aureus* nasal colonization/decolonization established by our group [21,22]. Mice were intranasally inoculated with the mouse-adapted *S. aureus* JSNZ strain to induce persistent colonization. Mice were colonized with JSNZ for three days and subsequently treated daily with Poloxamer, S/V or PBS without the active compound for 3 to 7 days. As expected, *S. aureus* JSNZ persistently colonized the nasopharynx and gastrointestinal tract of mice throughout the experiment (Figure 3). The tested drug carriers did not impact on *S. aureus* colonization, as reflected by constant bacterial loads in the nose, cecum, and feces. This advantageous result is in strong contrast to our preliminary findings in the cotton rat model (see below). A complete elimination of *S. aureus* in the nose and cecum was only observed in a single mouse treated for 3 days with S/V. These experiments demonstrated that all tested carrier substances are suitable for the delivery of candidate drugs in our *S. aureus* nasal colonization/decolonization model.

3.5. ATA Failed to Induce S. aureus Decolonization, While Mupirocin Was Highly Effective

Finally, we investigated whether the hit compound ATA reduces *S. aureus* nasal burden. We used Poloxamer and S/V as drug carriers. After three days of bacterial colonization, mice were treated for 7 consecutive days with ATA ointment in two different doses (0.5% and 2%). Its effect on the *S. aureus* load was compared with the human therapeutic agent mupirocin. As expected, the antibiotic mupirocin eradicated *S. aureus* from the nasal region even in a short treatment regime of 3 days (Figure 4A). Although applied only to the nasal cavity, mupirocin also eliminated *S. aureus* from the gastrointestinal tract of most mice within 7 days of treatment (Figure 4B,C). In contrast, ATA did not reduce the *S. aureus* burden in the nose, cecum or feces. This was true at concentrations of 0.5% (2.5 mg/kg body weight) or 2% (10 mg/kg body weight) and in combination with Poloxamer or S/V. Even very high ATA concentrations of up to 10% just reaffirmed the inefficacy of the compound in reducing *S. aureus* colonization. On the contrary, ATA treatment even increased the median nasal burden of *S. aureus* JSNZ (Figure 4). Moreover, mice treated with 10% ATA showed a slight reduction of weight when compared with untreated mice, resulting in a significantly higher disease activity index of this group ($p = 0.0473$ for 10% ATA-S/V; Figure S7). Altogether, the in vivo experiments demonstrate the robustness and suitability of our murine *S. aureus* nasal colonization/decolonization model for testing candidate drugs for *S. aureus* decolonization. ATA, however, failed to induce *S. aureus* decolonization.

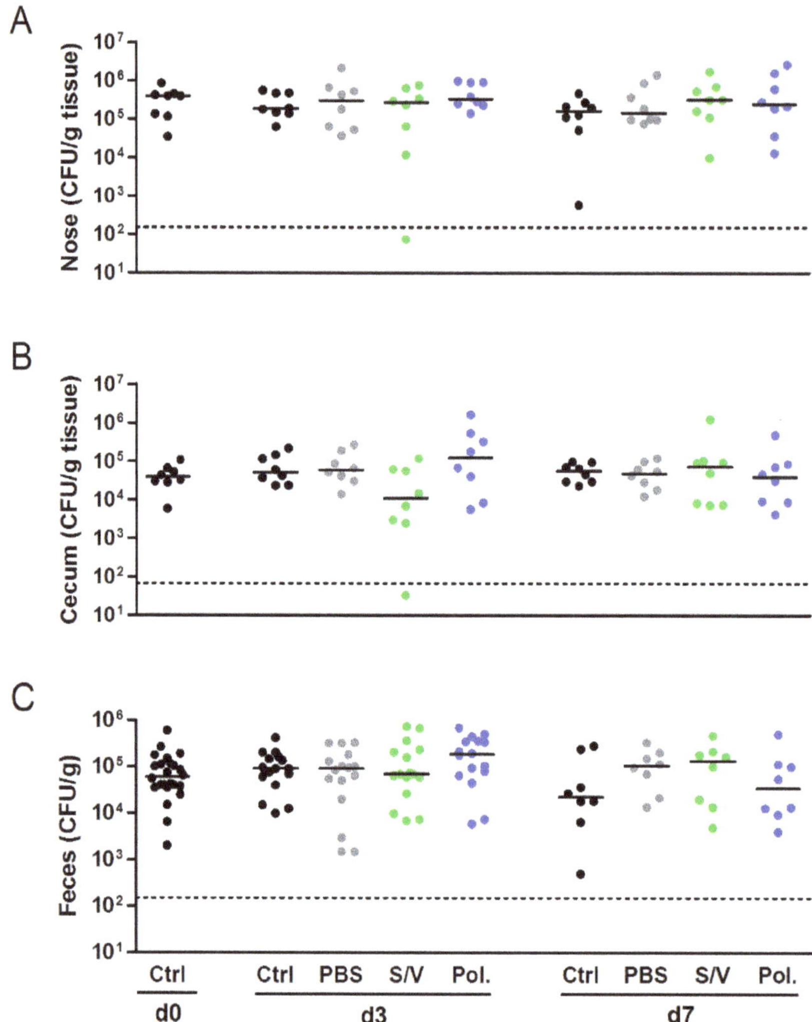

Figure 3. Drug carriers did not interfere with the bacterial burden in a persistent *S. aureus* colonization model. Female C57BL/6N mice were colonized intranasally with $0.7–1.0 \times 10^8$ CFU *S. aureus* JSNZ. Starting on day 3 after colonization, mice were treated once daily for 7 consecutive days with 10 μL of the carrier substances PBS, Softisan 649/Vaseline, Poloxamer 407 or left untreated. The *S. aureus* burden was determined in the homogenized nose (**A**), cecum (**B**), and feces (**C**) 0, 3, and 7 days after starting the treatment. The detection limit is indicated by a dashed line; medians are indicated. Data were pooled from two independent experiments with 4 mice/group. Abbreviations: Ctrl—control; PBS—phosphate-buffered saline; Pol.—Poloxamer 407; S/V—Softisan 649/Vaseline 9:1 (*w*/*w*).

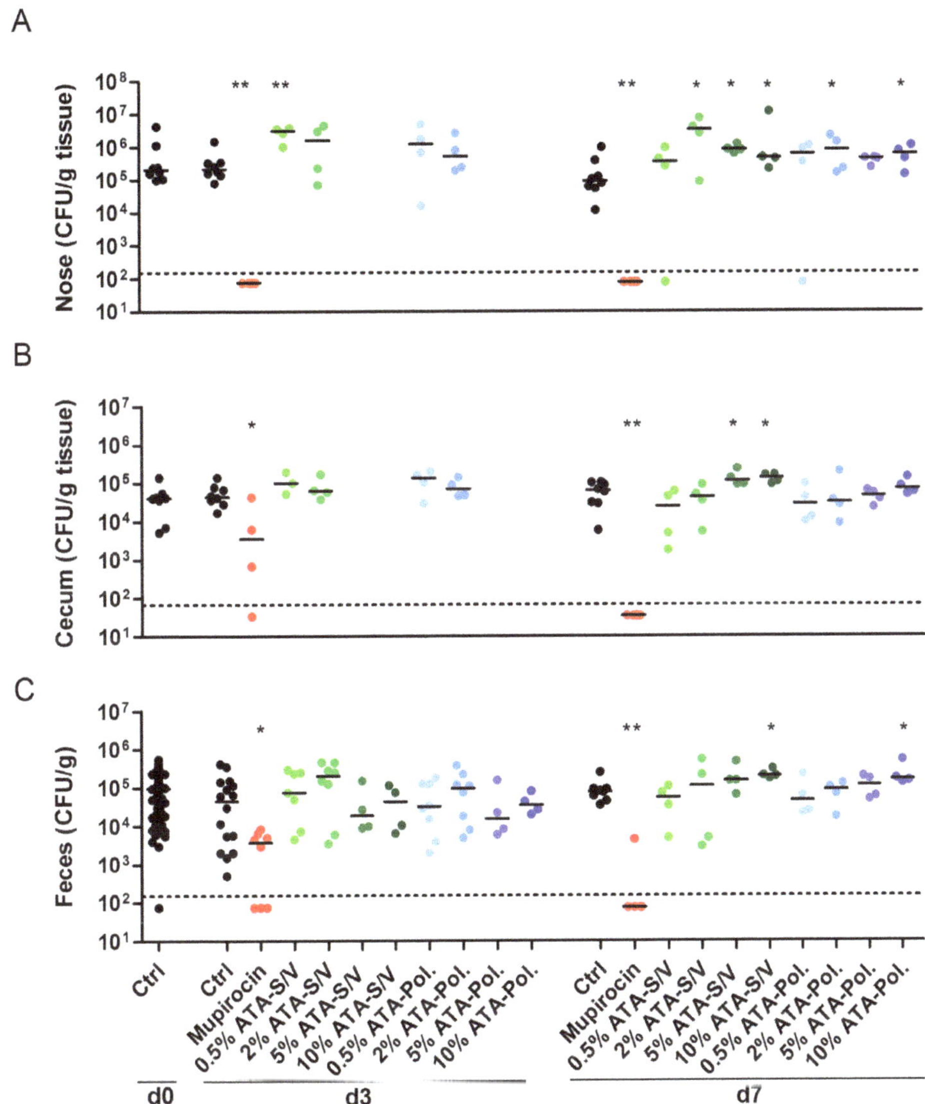

Figure 4. Mupirocin eradicated *S. aureus* colonization, while ATA was not effective. Female C57BL/6N mice were colonized intranasally with $0.7–1.0 \times 10^8$ CFU *S. aureus* JSNZ. After three days of colonization, mice were treated on a daily basis for 7 days with 10 µL 2% Mupirocin or ATA in different concentrations (0.5%, 2.5%, 5% or 10%) using Softisan 649/Vaseline or Poloxamer 407 as drug carrier. Groups receiving 5% or 10% ATA were only analyzed at day 7. A control group remained untreated. The *S. aureus* bacterial load was determined in the homogenized nose (**A**), cecum (**B**), and feces (**C**) 0, 3, and 7 days after starting the treatment. The detection limit is indicated by a dashed line; medians are indicated. Data were pooled from two independent experiments with 4 mice/group. Statistics: Groups were compared with Ctrl using Mann-Whitney test, * $p < 0.05$, ** $p < 0.01$. Abbreviations: Ctrl—control; Pol.—Poloxamer 407; S/V—Softisan 649/Vaseline 9:1 (*w/w*).

4. Discussion

Due to increasing numbers of mupirocin-resistant *S. aureus*, new decolonization approaches are urgently required. In this study, we investigated pathoblockers, which means compounds that reduce the pathogenicity of the bacteria without interfering with their viability. Hence, adhesion inhibitors should exert less selection pressure than antibiotics and reduce the development of resistances. In accordance with this goal, we have sorted out compounds that showed antimicrobial effects (MIC values) below 100 µM in our further evaluation of the initial hits.

Our first task was the development of a method that allows the screening of several thousand substances for compounds that inhibit the adhesion of *S. aureus* to epithelial cells. For this purpose, an automated microscope was employed that allows the quantitative investigation of adhesion processes in microtiter plates and that can in addition be used to reveal morphological changes that are induced to the cells by the influence of the substances. Studies of the adhesion of bacteria to respiratory epithelial cells are frequently carried out with the human lung epithelial cell line A-549 [17,24–32], which was also employed in the present investigation.

In the human nose *S. aureus* is in a constant process of exponential growth due to ongoing excretion and shedding [39]. Such a situation suggests good chances for a therapeutic application of adhesion inhibitors. We report a medium throughput screening of more than 4.000 compounds for *S. aureus* adhesion inhibitors and the subsequent characterization of ATA, which was identified as the most potent compound in this assay. Our method works with unlabeled bacteria and eukaryotic cells during the adhesion process, which offers the possibility to use unmodified wild type bacterial strains and to capture any influences of the test compounds on the morphology of the eukaryotic cells in the initial screening campaign. This is an advantage over previously published procedures, as it allows the rapid exclusion of compounds that exert their effect mainly through toxic effects on the eukaryotic cells. It is likely that the approach for identifying adhesion inhibitors presented in this study can be applied to other settings, where the adhesion of pathogens to epithelial cells or eukaryotic cells in general is of interest. Bacterial adhesion to epithelial cells is the important first step in many clinically significant conditions, like the infection of the urinary tract with *Escherichia coli* [40], the colonization of the lungs with *Pseudomonas aeruginosa* [41] or the adherence of Streptococci to the pharynx [42]. By employing specific and adequate staining procedures, the approach described in this study can be adopted to the particular setting.

The number of bacteria that may have internalized into the A-549 cells has not been determined in our current protocol. Previous studies reported a low internalization rate with only 15% of A-549 cells being infected and low bacterial numbers per cell after 1.5 h of incubation [43]. Therefore, we do not expect a relevant influence of internalization events on our results.

For ATA several biological functions have been described [44–50], but the finding that ATA potently prevents the adhesion of *S. aureus* to epithelial cells was new. It should be noted that ATA, which is shown in its monomeric form in Figure 2B, has a strong tendency to oligomerize, which results in polyanionic structures [51]. We speculate that the activity of ATA to reduce adhesion of *S. aureus* to epithelial cells might be linked to its acidic functions, because several other polyanionic substances like polyinosinic acid and sulfated polysaccharides are known to inhibit the cellular adhesion of *S. aureus* [35,52–54]. Whereas the polypharmacological properties and the limited chemical stability render a systemic use of ATA challenging, a topical administration for the decolonization of the nose appeared attractive based on the cellular data.

Apart from the anti-adhesive effect reported here, ATA can act as a *S. aureus* pathoblocker by potent inhibition of Stp1 [55]. Stp1 is a Ser/Thr-phosphatase involved in the global regulation of staphylococcal virulence factors, such as alpha-hemolysin and leukocidins, immune-evasion molecules, such as SCIN and CHIPS, as well as capsular polysaccharide synthesis enzymes [56–59]. Hence, Stp1 inhibition switches *S. aureus* from an invasive, virulent state to a more silent, immune-evasive state. Administration of ATA significantly reduced the severity of staphylococcal infection in a murine abscess formation model [55].

ATA is also well-known as a small molecule inhibitor of nucleases and nucleic acid-binding enzymes [60]. Diverse studies have reported that ATA inhibits the replication of a variety of viruses in vitro by interference with viral polymerases and RNA-binding proteins [61–63]. Moreover, ATA compromised bacterial biofilm formation by limiting protein-nucleic acid interaction [64]. Whether this is also relevant to *S. aureus* biofilms, which are also composed of eDNA, remains to be investigated.

Despite being the top candidate in the in vitro screening approach, ATA was ineffective in promoting *S. aureus* decolonization in our mouse *S. aureus* colonization model, but even increased the bacterial burden when applied intranasally in a higher dose (10% ATA, equals 50 mg/kg body weight/day). We suggest that the anti-adhesive effect of ATA observed in our in vitro assay could have been masked in vivo by its anti-inflammatory activity [65–68]. It is well known that the innate and adaptive immune responses are critical for the clearance or persistence of *S. aureus* nasal colonization [69–71]. ATA, however, can inhibit innate and adaptive immune responses in many ways. The compound interferes with JAK-STAT signaling, thereby inhibiting IFN-γ-induced iNOS expression [68]. Moreover, ATA is able to block chemotaxis of dendritic cells and T cells [67], to inhibit complement activation in vitro [65], and to convert naïve CD4+ T cells into FoxP3+ regulatory T cells [66]. Collectively, these mechanisms could reduce the immune response against the colonizing staphylococci and thereby enhance bacterial persistence in the nasal cavity. In consequence, such an anti-inflammatory effect could override the anti-adhesive properties of ATA in the nasal colonization model. ATA's anti-inflammatory effect could also explain other cases where a protective effect shown in vitro could not be reproduced in the living organism. In line with our data, intraperitoneal injection of ATA enhanced disease severity in murine models of vaccinia virus and orbivirus infection [44,72].

The in vivo evaluation of novel decolonization drugs requires a robust and persistent *S. aureus* colonization model with high discriminatory power between treated and untreated animals. In the past, researchers usually colonized laboratory mice with human-adapted clinical *S. aureus* isolates. Due to poor adaptation to the murine environment, human *S. aureus* isolates usually do not consistently colonize mice and are frequently eliminated from the murine nasal cavity within days [12,21,73,74]. Nevertheless, several drug candidates were investigated using this model [8,12]. However, the variable colonization rate and the rapid natural decolonization strongly reduced the signal-to-noise-ratio in these studies [12].

In search for a more robust and reliable animal model, Kokai-Kun et al. established a nasal colonization model in cotton rats [11]. In cotton rats there is consistent and persistent high-level (~5000 CFUs/nose) nasal colonization by *S. aureus* and consequently a high discriminatory power. This model has been used over the past two decades to evaluate antimicrobials, e.g., lysostaphin [11], cationic methacrylate polymers [13], epidermicin [75], and antimicrobial peptides [76]. Cotton rats are highly excitable rodents and hence much more difficult to handle than mice [77]. We carried out initial studies with ATA in the cotton rat model using human-adapted *S. aureus* strains. However, several drug carriers without drug led to a drastic decrease of the nasal bacterial load within just a few days, which made it impossible to discern any additional drug-related effect (data not shown). Hence, it is essential that the drug carrier distributes the active compound throughout the nasal cavity to facilitate *S. aureus* elimination [38], but does not affect the bacterial load itself. In our study, all three tested substances (Poloxamer 407, S/V, and PBS) spread equally well throughout the entire nasal cavity 30 min after application and did not impact on the *S. aureus* density in the nose.

We have recently reported that laboratory and wild mice are natural hosts of *S. aureus*, and that their colonizing strains show features of adaptation to their murine host. The prototype of these mouse-adapted strains, JSNZ, can colonize the murine nose and gastrointestinal tract for several weeks [21,59]. Colonization was induced in all inoculated animals with average nasal colonization loads even higher than in the cotton rat model (2×10^5 CFU/g nose tissue, corresponding to 4×10^4 CFU/nose). Our consistent and reproducible *S. aureus* colonization model now enables researchers to reliably evaluate novel decolonization drugs in the mouse.

The validity of our animal model is underlined by the results of mupirocin treatment, which served as benchmark for *S. aureus* decolonization. After three days of treatment nasal decolonization was 100%, decolonization of the gastrointestinal tract took a few days longer. In studies with mice that were colonized with human *S. aureus* strains or that used the cotton rat model, mupirocin frequently failed to completely decolonize the nose even at the clinically effective concentration of 2% mupirocin [8,12,13,76]. Apart from the animal model, this might also be due to differences in the experimental settings such as different application modes and schemes. Moreover, the employed carrier substances could have an impact on the spatial distribution of mupirocin within the nasal cavity and its release over time [11].

5. Conclusions

Adhesion inhibitors are a promising, but yet underexplored option to prevent or eliminate *S. aureus* colonization. The microscopy-based screening presented here is a powerful method to identify and test new *S. aureus* adhesion inhibitors. Our adhesion assay works with unlabeled bacteria and eukaryotic cells, which offers the possibility to use unmodified wild type bacterial strains and to capture any influences of the test compounds on the morphology of the eukaryotic cells in the initial screening campaign. Hence, it allows the rapid exclusion of compounds that exert their effect mainly through toxic effects on the eukaryotic cells. Importantly, this approach can be applied to other settings where the adhesion of pathogens to epithelial cells or eukaryotic cells in general is of interest.

The mouse-adapted strain JSNZ induces persistent nasal colonization and therefore provides an excellent tool for testing drug candidates against *S. aureus* nasal colonization in mice. In contrast to previous models using human-adapted *S. aureus* strains in mice, this model has a high discriminatory power between treated and untreated animals. While cotton rats can also be persistently colonized with *S. aureus*, their handling is much more difficult and less molecular tools are available to study the host response in detail. Even though our first candidate ATA failed to decolonize *S. aureus* from the murine nose, we are optimistic that our screening approach and persistent colonization model will provide a powerful tool for identifying effective antibacterial drugs in the future.

Supplementary Materials: The following are available online at https://www.mdpi.com/2076-2607/9/3/631/s1, Figure S1: Adherence of the *S. aureus* SA113 Delta tagO mutant vs. the wild type strain, Figure S2: Adhesion of *S. aureus* N315 to A-549 cells after precultivation of the bacteria in BHI for different time periods Figure S3: Assessment of the performance of the adhesion test, Figure S4: Visualization of the adhesion of *S. aureus* N315 to A-549 cells, Figure S5: Reevaluation of the adhesion-reducing effect of the three most active compounds that were identified in the screening campaigns. Figure S6: Structures of the active compounds HZI10676D08 (methyl 7-hydroxy-9-methyl-6-oxo-6H-oxepino [2,3-b]chromene-5-carboxylate) and HZI10687B10 (pseudohypericin). Figure S7: Evaluation of the health status of the *S. aureus*-colonized mice after treatment with ATA or mupirocin.

Author Contributions: Conceptualization, M.B., W.T., B.M.B., and S.H.; formal analysis, L.M.F.d.O., and M.S., methodology, L.M.F.d.O., M.S., W.T., and S.H.; investigation, L.M.F.d.O., M.S., W.T., M.N.D., D.M.M., P.T., and S.H.; data curation, L.M.F.d.O., and M.S.; writing—original draft preparation, L.M.F.d.O., M.S., W.T., and S.H.; writing—review and editing, all co-authors; supervision, W.T., M.B., B.M.B., and S.H.; funding acquisition, M.B., W.T., and S.H. All authors have read and agreed to the published version of the manuscript.

Funding: This research was funded by the Bundesministerium für Bildung und Forschung (BMBF), grant number 01Kl0710 (SkinStaph), the Deutsche Forschungsgemeinschaft (DFG), grant number FKZ GRK1870/1, and the Bundesministerium für Bildung und Forschung (BMBF), InfectControl 2020, project InVAC, FKZ 03ZZ0806B to D.M.M.

Acknowledgments: We thank A. Peschel, University of Tübingen, Germany, K. Becker, University of Greifswald, Germany, and W. Bautsch, Institute for Microbiology, Immunology and Hospital Hygiene, Braunschweig, Germany, for kindly providing *S. aureus* strains. We thank G. Höfle for providing test compounds. Furthermore, we thank Randi Diestel for excellent support concerning the operation of the IXM, Tabea Ellebracht for technical assistance in performing the screening experiments, Raimo Franke for help with statistical evaluation, Kilian Wietschel and Grazyna Domanska for support with animal dissection, Susan Mouchantat for her support regarding the intranasal application of ointments, and Susanne Neumeister, Erika Friebe, Fawaz Al'Sholui, and Sabine Prettin for technical assistance. We thank Carsten Peukert for support regarding the generation of the graphical abstract.

Conflicts of Interest: The authors declare no conflict of interest. The funders had no role in the design of the study; in the collection, analyses, or interpretation of data; in the writing of the manuscript, or in the decision to publish the results.

References

1. Safdar, N.; Bradley, E.A. The Risk of Infection after Nasal Colonization with *Staphylococcus aureus*. *Am. J. Med.* **2008**, *121*, 310–315. [CrossRef]
2. Sakr, A.; Brégeon, F.; Mège, J.L.; Rolain, J.M.; Blin, O. *Staphylococcus aureus* nasal colonization: An update on mechanisms, epidemiology, risk factors, and subsequent infections. *Front. Microbiol.* **2018**, *9*, 2419. [CrossRef] [PubMed]
3. Van Hal, S.J.; Jensen, S.O.; Vaska, V.L.; Espedido, B.A.; Paterson, D.L.; Gosbell, I.B. Predictors of mortality in *Staphylococcus aureus* bacteremia. *Clin. Microbiol. Rev.* **2012**, *25*, 362–386. [CrossRef] [PubMed]
4. Bode, L.G.M.; Kluytmans, J.A.J.W.; Wertheim, H.F.L.; Bogaers, D.; Vandenbroucke-Grauls, C.M.J.E.; Roosendaal, R.; Troelstra, A.; Box, A.T.A.; Voss, A.; Van Der Tweel, I.; et al. Preventing surgical-site infections in nasal carriers of *Staphylococcus aureus*. *N. Engl. J. Med.* **2010**, *362*, 9–17. [CrossRef]
5. Dadashi, M.; Hajikhani, B.; Darban-Sarokhalil, D.; Van Belkum, A.; Goudarzi, M. Mupirocin resistance in *Staphylococcus aureus*: A systematic review and meta-analysis. *J. Glob. Antimicrob. Resist.* **2020**, *20*, 238–247. [CrossRef]
6. Poovelikunnel, T.; Gethin, G.; Humphreys, H. Mupirocin resistance: Clinical implications and potential alternatives for the eradication of MRSA. *J. Antimicrob. Chemother.* **2015**, *70*, 2681–2692. [CrossRef] [PubMed]
7. Madden, G.R.; Sifri, C.D. Antimicrobial Resistance to Agents Used for *Staphylococcus aureus* Decolonization: Is There a Reason for Concern? *Curr. Infect. Dis. Rep.* **2018**, *20*, 1–11. [CrossRef] [PubMed]
8. Blanchard, C.; Brooks, L.; Beckley, A.; Colquhoun, J.; Dewhurst, S.; Dunman, P.M. Neomycin sulfate improves the antimicrobial activity of mupirocin-based antibacterial ointments. *Antimicrob. Agents Chemother.* **2016**, *60*, 862–872. [CrossRef]
9. Fung, S.; O'Grady, S.; Kennedy, C.; Dedier, H.; Campbell, I.; Conly, J. The utility of polysporin ointment in the eradication of methicillin-resistant *Staphylococcus aureus* colonization: A pilot study. *Infect. Control Hosp. Epidemiol.* **2000**, *21*, 653–655. [CrossRef]
10. Soto, N.E.; Vaghjimal, A.; Stahl-Avicolli, A.; Protic, J.R.; Lutwick, L.I.; Chapnick, E.K. Bacitracin versus mupirocin for *Staphylococcus aureus* nasal colonization. *Infect. Control Hosp. Epidemiol.* **1999**, *20*, 351–353. [CrossRef]
11. Kokai-Kun, J.F.; Walsh, S.M.; Chanturiya, T.; Mond, J.J. Lysostaphin cream eradicates *Staphylococcus aureus* nasal colonization in a cotton rat model. *Antimicrob. Agents Chemother.* **2003**, *47*, 1589–1597. [CrossRef] [PubMed]
12. Rouse, M.S.; Rotger, M.; Piper, K.E.; Steckelberg, J.M.; Scholz, M.; Andrews, J.; Patel, R. In vitro and in vivo evaluations of the activities of lauric acid monoester formulations against *Staphylococcus aureus*. *Antimicrob. Agents Chemother.* **2005**, *49*, 3187–3191. [CrossRef] [PubMed]
13. Thoma, L.M.; Boles, B.R.; Kuroda, K. Cationic Methacrylate Polymers as Topical Antimicrobial Agents against *Staphylococcus aureus* Nasal Colonization. *Biomacromolecules* **2014**, *15*, 2933–2943. [CrossRef] [PubMed]
14. Allen, R.C.; Popat, R.; Diggle, S.P.; Brown, S.P. Targeting virulence: Can we make evolution-proof drugs? *Nat. Rev. Microbiol.* **2014**, *12*, 300–308. [CrossRef] [PubMed]
15. Dickey, S.W.; Cheung, G.Y.C.; Otto, M. Different drugs for bad bugs: Antivirulence strategies in the age of antibiotic resistance. *Nat. Rev. Drug Discov.* **2017**, *16*, 457–471. [CrossRef]
16. Leonard, A.C.; Petrie, L.E.; Cox, G. Bacterial Anti-adhesives: Inhibition of *Staphylococcus aureus* Nasal Colonization. *ACS Infect. Dis.* **2019**, *5*, 1668–1681. [CrossRef] [PubMed]
17. Hawley, C.A.; Watson, C.A.; Orth, K.; Krachler, A.M. A MAM7 peptide-based inhibitor of *Staphylococcus aureus* adhesion does not interfere with in vitro host cell function. *PLoS ONE* **2013**, *8*, e81216. [CrossRef] [PubMed]
18. Le Blay, G.; Fliss, I.; Lacroix, C. Comparative detection of bacterial adhesion to Caco-2 cells with ELISA, radioactivity and plate count methods. *J. Microbiol. Methods* **2004**, *59*, 211–221. [CrossRef] [PubMed]
19. Liang, X.; Yu, C.; Sun, J.; Liu, H.; Landwehr, C.; Holmes, D.; Ji, Y. Inactivation of a two-component signal transduction system, SaeRS, eliminates adherence and attenuates virulence of *Staphylococcus aureus*. *Infect. Immun.* **2006**, *74*, 4655–4665. [CrossRef] [PubMed]
20. Petrie, L.E.; Leonard, A.C.; Murphy, J.; Cox, G. Development and validation of a high-throughput whole cell assay to investigate *Staphylococcus aureus* adhesion to host ligands. *JBC* **2020**, in press, Manuscript RA120.015360. [CrossRef] [PubMed]

21. Holtfreter, S.; Radcliff, F.J.; Grumann, D.; Read, H.; Johnson, S.; Monecke, S.; Ritchie, S.; Clow, F.; Goerke, C.; Bröker, B.M.; et al. Characterization of a Mouse-Adapted *Staphylococcus aureus* Strain. *PLoS ONE* **2013**, *8*, e71142. [CrossRef] [PubMed]
22. Sun, Y.; Emolo, C.; Holtfreter, S.; Wiles, S.; Kreiswirth, B.; Missiakas, D.; Schneewind, O. Staphylococcal Protein A Contributes to Persistent Colonization of Mice with *Staphylococcus aureus*. *J. Bacteriol.* **2018**, *200*, e00735-17. [CrossRef] [PubMed]
23. Trübe, P.; Hertlein, T.; Mrochen, D.M.; Schulz, D.; Jorde, I.; Krause, B.; Zeun, J.; Fischer, S.; Wolf, S.A.; Walther, B.; et al. Bringing together what belongs together: Optimizing murine infection models by using mouse-adapted *Staphylococcus aureus* strains. *Int. J. Med. Microbiol.* **2018**, *309*, 26–38. [CrossRef] [PubMed]
24. Kuroda, M.; Ohta, T.; Uchiyama, I.; Baba, T.; Yuzawa, H.; Kobayashi, I.; Cui, L.; Oguchi, A.; Aoki, K.; Nagai, Y.; et al. Whole genome sequencing of meticillin-resistant *Staphylococcus aureus*. *Lancet Infect. Dis.* **2001**, *357*, 1225–1240. [CrossRef]
25. Iordanescu, S. Host controlled restriction mutants of *Staphylococcus aureus*. *Arch. Roum. de Pathol. Exp. et de Microbiol.* **1975**, *34*, 55–58.
26. Weidenmaier, C.; Peschel, A.; Xiong, Y.Q.; Kristian, S.A.; Dietz, K.; Yeaman, M.R.; Bayer, A.S. Lack of wall teichoic acids in *Staphylococcus aureus* leads to reduced interactions with endothelial cells and to attenuated virulence in a rabbit model of endocarditis. *J. Infect. Dis.* **2005**, *191*, 1771–1777. [CrossRef] [PubMed]
27. Weissman, K.J.; Müller, R. Myxobacterial secondary metabolites: Bioactivities and modes-of-action. *Nat. Prod. Rep.* **2010**, *27*, 1276–1295. [CrossRef]
28. Tegge, W.; Bautsch, W.; Frank, R. Synthesis of cyclic peptides and peptide libraries on a new disulfide linker. *J. Pept. Sci.* **2007**, *13*, 693–699. [CrossRef]
29. Herrmann, J.; Abou Fayad, A.; Müller, R. Natural products from myxobacteria: Novel metabolites and bioactivities. *Nat. Prod. Rep.* **2017**, *34*, 135–160. [CrossRef]
30. Mosmann, T. Rapid colorimetric assay for cellular growth and survival: Application to proliferation and cytotoxicity assays. *J. Immunol. Methods* **1983**, *65*, 55–63. [CrossRef]
31. Sasse, F.; Steinmetz, H.; Schupp, T.; Petersen, F.; Memmert, K.; Hofmann, H.; Heusser, C.; Brinkmann, V.; Von Matt, P.; Höfle, G.; et al. Argyrins, immunosuppressive cyclic peptides from myxobacteria—I. Production, isolation, physico-chemical and biological properties. *J. Antibiot.* **2002**, *55*, 543–551. [CrossRef] [PubMed]
32. Ahmed, G.F.; Elkhatib, W.F.; Noreddin, A.M. Inhibition of *Pseudomonas aeruginosa* PAO1 adhesion to and invasion of A549 lung epithelial cells by natural extracts. *J. Infect. Public Health* **2014**, *7*, 436–444. [CrossRef] [PubMed]
33. Arita, Y.; Joseph, A.; Koo, H.C.; Li, Y.; Palaia, T.A.; Davis, J.M.; Kazzaz, J.A. Superoxide dismutase moderates basal and induced bacterial adherence and interleukin-8 expression in airway epithelial cells. *Am. J. Physiol. Lung Cell. Mol. Physiol.* **2004**, *287*, L1199–L1206. [CrossRef] [PubMed]
34. Weidenmaier, C.; Kokai-Kun, J.F.; Kristian, S.A.; Chanturiya, T.; Kalbacher, H.; Gross, M.; Nicholson, G.; Neumeister, B.; Mond, J.J.; Peschel, A. Role of teichoic acids in *Staphylococcus aureus* nasal colonization, a major risk factor in nosocomial infections. *Nat. Med.* **2004**, *10*, 243–245. [CrossRef] [PubMed]
35. Harding, H.P.; Zhang, Y.; Ron, D. Protein translation and folding are coupled by an endoplasmic-reticulum-resident kinase. *Nature* **1999**, *397*, 271–274. [CrossRef] [PubMed]
36. Weidenmaier, C.; Kokai-Kun, J.F.; Kulauzovic, E.; Kohler, T.; Thumm, G.; Stoll, H.; Gotz, F.; Peschel, A. Differential roles of sortase-anchored surface proteins and wall teichoic acid in *Staphylococcus aureus* nasal colonization. *Int. J. Med. Microbiol.* **2008**, *298*, 505–513. [CrossRef] [PubMed]
37. Smee, D.F.; Hurst, B.L.; Wong, M.H. Lack of efficacy of aurintricarboxylic acid and ethacrynic acid against vaccinia virus respiratory infections in mice. *Antivir. Chem. Chemother.* **2010**, *20*, 201–205. [CrossRef] [PubMed]
38. Weidenmaier, C.; Goerke, C.; Wolz, C. *Staphylococcus aureus* determinants for nasal colonization. *Trends Microbiol.* **2012**, *20*, 243–250. [CrossRef] [PubMed]
39. Krismer, B.; Peschel, A. Does *Staphylococcus aureus* nasal colonization involve biofilm formation? *Future Microbiol.* **2011**, *6*, 489–493. [CrossRef]
40. Nielubowicz, G.R.; Mobley, H.L.T. Host-pathogen interactions in urinary tract infection. *Nat. Rev. Urol.* **2010**, *7*, 430–441. [CrossRef]
41. Doig, P.; Todd, T.; Sastry, P.A.; Lee, K.K.; Hodges, R.S.; Paranchych, W.; Irvin, R.T. Role of pili in adhesion of Pseudomonas aeruginosa to human respiratory epithelial cells. *Infect. Immun.* **1988**, *56*, 1641–1646. [CrossRef] [PubMed]
42. Manetti, A.G.O.; Zingaretti, C.; Falugi, F.; Capo, S.; Bombaci, M.; Bagnoli, F.; Gambellini, G.; Bensi, G.; Mora, M.; Edwards, A.M.; et al. *Streptococcus pyogenes* pili promote pharyngeal cell adhesion and biofilm formation. *Mol. Microbiol.* **2007**, *64*, 968–983. [CrossRef] [PubMed]
43. Surmann, K.; Simon, M.; Hildebrandt, P.; Pförtner, H.; Michalik, S.; Stentzel, S.; Steil, L.; Dhople, V.M.; Bernhardt, J.; Schlüter, R.; et al. A proteomic perspective of the interplay of *Staphylococcus aureus* and human alveolar epithelial cells during infection. *J. Proteom.* **2015**, *128*, 203–217. [CrossRef]
44. Alonso, C.; Utrilla-Trigo, S.; Calvo-Pinilla, E.; Jimenez-Cabello, L.; Ortego, J.; Nogales, A. Inhibition of Orbivirus Replication by Aurintricarboxylic Acid. *Int. J. Mol. Sci.* **2020**, *21*, 7294. [CrossRef]
45. Boukhalfa, A.; Miceli, C.; Avalos, Y.; Morel, E.; Dupont, N. Interplay between primary cilia, ubiquitin-proteasome system and autophagy. *Biochimie* **2019**, *166*, 286–292. [CrossRef] [PubMed]

46. Obrecht, A.S.; Urban, N.; Schaefer, M.; Röse, A.; Kless, A.; Meents, J.E.; Lampert, A.; Abdelrahman, A.; Müller, C.E.; Schmalzing, G.; et al. Identification of aurintricarboxylic acid as a potent allosteric antagonist of P2X1 and P2X3 receptors. *Neuropharmacology* **2019**, *158*, 107749. [CrossRef]
47. Park, S.; Guo, Y.; Negre, J.; Preto, J.; Smithers, C.C.; Azad, A.K.; Overduin, M.; Murray, A.G.; Eitzen, G. Fgd5 is a Rac1-specific Rho GEF that is selectively inhibited by aurintricarboxylic acid. *Small GTPases* **2021**, *12*, 147–160. [CrossRef]
48. Qiu, Q.; Zheng, Z.; Chang, L.; Zhao, Y.S.; Tan, C.; Dandekar, A.; Zhang, Z.; Lin, Z.; Gui, M.; Li, X.; et al. Toll-like receptor-mediated IRE1alpha activation as a therapeutic target for inflammatory arthritis. *EMBO J.* **2013**, *32*, 2477–2490. [CrossRef] [PubMed]
49. Roos, A.; Dhruv, H.D.; Mathews, I.T.; Inge, L.J.; Tuncali, S.; Hartman, L.K.; Chow, D.; Millard, N.; Yin, H.H.; Kloss, J.; et al. Identification of aurintricarboxylic acid as a selective inhibitor of the TWEAK-Fn14 signaling pathway in glioblastoma cells. *Oncotarget* **2017**, *8*, 12234–12246. [CrossRef] [PubMed]
50. Shadrick, W.R.; Mukherjee, S.; Hanson, A.M.; Sweeney, N.L.; Frick, D.N. Aurintricarboxylic acid modulates the affinity of hepatitis C virus NS3 helicase for both nucleic acid and ATP. *Biochemistry* **2013**, *52*, 6151–6159. [CrossRef] [PubMed]
51. Wang, P.; Kozlowski, J.; Cushman, M. Isolation and structure elucidation of low molecular weight components of aurintricarboxylic acid (ATA). *J. Org. Chem.* **1992**, *57*, 3861–3866. [CrossRef]
52. Baba, M.; Nakajima, M.; Schols, D.; Pauwels, R.; Balzarini, J.; De Clercq, E. Pentosan polysulfate, a sulfated oligosaccharide, is a potent and selective anti-HIV agent in vitro. *Antiv. Res.* **1988**, *9*, 335–343. [CrossRef]
53. Lee, J.-H.; Shim, J.S.; Lee, J.S.; Kim, M.-K.; Chung, M.-S.; Kim, K.H. Pectin-like acidic polysaccharide from Panax ginseng with selective antiadhesive activity against pathogenic bacteria. *Carbohydr. Res.* **2006**, *341*, 1154–1163. [CrossRef] [PubMed]
54. Pieters, R.J. Intervention with bacterial adhesion by multivalent carbohydrates. *Med. Res. Rev.* **2007**, *27*, 796–816. [CrossRef] [PubMed]
55. Zheng, W.; Cai, X.; Xie, M.; Liang, Y.; Wang, T.; Li, Z. Structure-Based Identification of a Potent Inhibitor Targeting Stp1-Mediated Virulence Regulation in *Staphylococcus aureus*. *Cell. Chem. Biol.* **2016**, *23*, 1002–1013. [CrossRef] [PubMed]
56. Burnside, K.; Lembo, A.; De Los Reyes, M.; Iliuk, A.; Binhtran, N.T.; Connelly, J.E.; Lin, W.J.; Schmidt, B.Z.; Richardson, A.R.; Fang, F.C.; et al. Regulation of hemolysin expression and virulence of *Staphylococcus aureus* by a serine/threonine kinase and phosphatase. *PLoS ONE* **2010**, *5*, e11071. [CrossRef] [PubMed]
57. Cameron, D.R.; Ward, D.V.; Kostoulias, X.; Howden, B.P.; Moellering, R.C., Jr.; Eliopoulos, G.M.; Peleg, A.Y. Serine/threonine phosphatase Stp1 contributes to reduced susceptibility to vancomycin and virulence in *Staphylococcus aureus*. *J. Infect. Dis.* **2012**, *205*, 1677–1687. [CrossRef] [PubMed]
58. Ohlsen, K.; Donat, S. The impact of serine/threonine phosphorylation in *Staphylococcus aureus*. *Int. J. Med. Microbiol.* **2010**, *300*, 137–141. [CrossRef] [PubMed]
59. Sun, F.; Ding, Y.; Ji, Q.; Liang, Z.; Deng, X.; Wong, C.C.; Yi, C.; Zhang, L.; Xie, S.; Alvarez, S.; et al. Protein cysteine phosphorylation of SarA/MgrA family transcriptional regulators mediates bacterial virulence and antibiotic resistance. *Proc. Natl. Acad. Sci. USA* **2012**, *109*, 15461–15466. [CrossRef]
60. Benchokroun, Y.; Couprie, J.; Larsen, A.K. Aurintricarboxylic acid, a putative inhibitor of apoptosis, is a potent inhibitor of DNA topoisomerase II in vitro and in Chinese hamster fibrosarcoma cells. *Biochem. Pharmacol.* **1995**, *49*, 305–313. [CrossRef]
61. He, R.; Adonov, A.; Traykova-Adonova, M.; Cao, J.; Cutts, T.; Grudesky, E.; Deschambaul, Y.; Berry, J.; Drebot, M.; Li, X. Potent and selective inhibition of SARS coronavirus replication by aurintricarboxylic acid. *Biochem. Biophys. Res. Commun.* **2004**, *320*, 1199–1203. [CrossRef]
62. Chen, Y.; Bopda-Waffo, A.; Basu, A.; Krishnan, R.; Silberstein, E.; Taylor, D.R.; Talele, T.T.; Arora, P.; Kaushik-Basu, N. Characterization of aurintricarboxylic acid as a potent hepatitis C virus replicase inhibitor. *Antivir. Chem. Chemother.* **2009**, *20*, 19–36. [CrossRef]
63. Hung, H.C.; Chen, T.C.; Fang, M.Y.; Yen, K.J.; Shih, S.R.; Hsu, J.T.; Tseng, C.P. Inhibition of enterovirus 71 replication and the viral 3D polymerase by aurintricarboxylic acid. *J. Antimicrob. Chemother.* **2010**, *65*, 676–683. [CrossRef]
64. Deng, B.; Ghatak, S.; Sarkar, S.; Singh, K.; Das Ghatak, P.; Mathew-Steiner, S.S.; Roy, S.; Khanna, S.; Wozniak, D.J.; McComb, D.W.; et al. Novel Bacterial Diversity and Fragmented eDNA Identified in Hyperbiofilm-Forming *Pseudomonas aeruginosa* Rugose Small Colony Variant. *iScience* **2020**, *23*, 100827. [CrossRef]
65. Lipo, E.; Chasman, S.M.; Kumar-Singh, R. Aurintricarboxylic Acid Inhibits Complement Activation, Membrane Attack Complex, and Choroidal Neovascularization in a Model of Macular Degeneration. *Investig. Ophthalmol. Vis. Sci.* **2013**, *54*, 7107–7114. [CrossRef] [PubMed]
66. Lim, D.G.; Park, Y.H.; Kim, S.E.; Kim, Y.H.; Park, C.S.; Kim, S.C.; Park, C.G.; Han, D.J. Aurintricarboxylic acid promotes the conversion of naive CD4+CD25- T cells into Foxp3-expressing regulatory T cells. *Int. Immunol.* **2011**, *23*, 583–592. [CrossRef] [PubMed]
67. Zhang, F.; Wei, W.; Chai, H.; Xie, X. Aurintricarboxylic acid ameliorates experimental autoimmune encephalomyelitis by blocking chemokine-mediated pathogenic cell migration and infiltration. *J. Immunol.* **2013**, *190*, 1017–1025. [CrossRef] [PubMed]
68. Chen, C.W.; Chao, Y.; Chang, Y.H.; Hsu, M.J.; Lin, W.W. Inhibition of cytokine-induced JAK-STAT signalling pathways by an endonuclease inhibitor aurintricarboxylic acid. *Br. J. Pharmacol.* **2002**, *137*, 1011–1020. [CrossRef] [PubMed]
69. Brown, A.F.; Leech, J.M.; Rogers, T.R.; McLoughlin, R.M. *Staphylococcus aureus* Colonization: Modulation of Host Immune Response and Impact on Human Vaccine Design. *Front. Immunol.* **2014**, *4*, 507. [CrossRef]

70. Mulcahy, M.E.; McLoughlin, R.M. Host-Bacterial Crosstalk Determines *Staphylococcus aureus* Nasal Colonization. *Trends Microbiol.* **2016**, *24*, 872–886. [CrossRef]
71. Broker, B.M.; Holtfreter, S.; Bekeredjian-Ding, I. Immune control of *Staphylococcus aureus*—Regulation and counter-regulation of the adaptive immune response. *Int. J. Med. Microbiol.* **2014**, *304*, 204–214. [CrossRef] [PubMed]
72. Myskiw, C.; Deschambault, Y.; Jefferies, K.; He, R.; Cao, J. Aurintricarboxylic acid inhibits the early stage of vaccinia virus replication by targeting both cellular and viral factors. *J. Virol.* **2007**, *81*, 3027–3032. [CrossRef] [PubMed]
73. Kiser, K.B.; Cantey-Kiser, J.M.; Lee, J.C. Development and characterization of a *Staphylococcus aureus* nasal colonization model in mice. *Infect. Immun.* **1999**, *67*, 5001–5006. [CrossRef]
74. Mrochen, D.M.; De Oliveira, L.M.F.; Raafat, D.; Holtfreter, S. *Staphylococcus aureus* Host Tropism and Its Implications for Murine Infection Models. *Int. J. Mol. Sci.* **2020**, *21*, 7061. [CrossRef] [PubMed]
75. Halliwell, S.; Warn, P.; Sattar, A.; Derrick, J.P.; Upton, M. A single dose of epidermicin NI01 is sufficient to eradicate MRSA from the nares of cotton rats. *J. Antimicrob. Chemother.* **2017**, *72*, 778–781. [CrossRef]
76. Desbois, A.P.; Sattar, A.; Graham, S.; Warn, P.A.; Coote, P.J. MRSA decolonization of cotton rat nares by a combination treatment comprising lysostaphin and the antimicrobial peptide ranalexin. *J. Antimicrob. Chemother.* **2013**, *68*, 2569–2575. [CrossRef] [PubMed]
77. Niewiesk, S.; Prince, G. Diversifying animal models: The use of hispid cotton rats (*Sigmodon hispidus*) in infectious diseases. *Lab. Anim.* **2002**, *36*, 357–372. [CrossRef]

Article

Impact of the Histidine-Containing Phosphocarrier Protein HPr on Carbon Metabolism and Virulence in *Staphylococcus aureus*

Linda Pätzold [1], Anne-Christine Brausch [1], Evelyn-Laura Bielefeld [1], Lisa Zimmer [1], Greg A. Somerville [2], Markus Bischoff [1,*] and Rosmarie Gaupp [1]

[1] Institute of Medical Microbiology and Hygiene, Saarland University, D-66421 Homburg, Germany; Linda.Paetzold@uks.eu (L.P.); anne-christine.brausch@web.de (A.-C.B.); evelyn_kirch@yahoo.de (E.-L.B.); zimmer.lisa@yahoo.de (L.Z.); rosmariegaupp@gmail.com (R.G.)

[2] School of Veterinary Medicine and Biomedical Sciences, University of Nebraska, Lincoln, NE 68588, USA; gsomerville3@unl.edu

* Correspondence: markus.bischoff@uks.eu; Tel.: +49-6841-162-39-63

Abstract: Carbon catabolite repression (CCR) is a common mechanism pathogenic bacteria use to link central metabolism with virulence factor synthesis. In gram-positive bacteria, catabolite control protein A (CcpA) and the histidine-containing phosphocarrier protein HPr (encoded by *ptsH*) are the predominant mediators of CCR. In addition to modulating CcpA activity, HPr is essential for glucose import via the phosphotransferase system. While the regulatory functions of CcpA in *Staphylococcus aureus* are largely known, little is known about the function of HPr in CCR and infectivity. To address this knowledge gap, *ptsH* mutants were created in *S. aureus* that either lack the open reading frame or harbor a *ptsH* variant carrying a thymidine to guanosine mutation at position 136, and the effects of these mutations on growth and metabolism were assessed. Inactivation of *ptsH* altered bacterial physiology and decreased the ability of *S. aureus* to form a biofilm and cause infections in mice. These data demonstrate that HPr affects central metabolism and virulence in *S. aureus* independent of its influence on CcpA regulation.

Keywords: *Staphylococcus aureus*; physiology; metabolism; carbon catabolite repression; CcpA; HPr

Citation: Pätzold, L.; Brausch, A.-C.; Bielefeld, E.-L.; Zimmer, L.; Somerville, G.A.; Bischoff, M.; Gaupp, R. Impact of the Histidine-Containing Phosphocarrier Protein HPr on Carbon Metabolism and Virulence in *Staphylococcus aureus*. *Microorganisms* **2021**, *9*, 466. https://doi.org/10.3390/microorganisms9030466

Academic Editor: Rajan P. Adhikari

Received: 7 February 2021
Accepted: 19 February 2021
Published: 24 February 2021

Publisher's Note: MDPI stays neutral with regard to jurisdictional claims in published maps and institutional affiliations.

1. Introduction

Carbon catabolite repression (CCR) is a common regulatory mechanism of bacteria to coordinate central metabolism with available carbon source(s) [1]. By modulating transcription of genes encoding proteins involved in the import and catabolism of carbon metabolites, bacterial CCR facilitates the efficient use of available carbon sources [1]. In pathogenic bacteria, regulators of CCR often affect transcription of virulence factors that are important for the exploitation of host-derived nutrient sources [2].

Staphylococcus aureus is a gram-positive opportunistic pathogen and a frequent cause of nosocomial infections in which central metabolism and infectivity are linked by numerous regulatory factors, including the catabolite control proteins A (CcpA) and E (CcpE), CodY, Rex, RpiRc, and SrrAB [3]. CcpA, a member of the GalR-LacI repressor family [4], is thought to be the major factor regulating CCR in *S. aureus* by binding catabolite-responsive element (*cre*) sequences of target genes [5]. Depending on the *cre* sequence location in the promotor region, the binding of CcpA results in either activation or repression of transcription [6]. Studies using *Bacillus megaterium* and *Streptococcus pyogenes* demonstrated that the binding affinity of CcpA for *cre* sites is low, but can be increased drastically by complex formation with the histidine-containing phosphocarrier protein (HPr), encoded by *ptsH* [7,8]. Electrophoretic mobility shift assays suggest this is also true in *S. aureus* [6], although CcpA can also bind to *cre* sites in the absence of HPr [9]. Activity of HPr is dependent on at least two phosphorylation sites, namely amino acids histidine 15 (His-15) and serine 46 (Ser-46) [1]. For complex formation with CcpA, HPr must be phosphorylated

Microorganisms **2021**, *9*, 466. https://doi.org/10.3390/microorganisms9030466　　　　https://www.mdpi.com/journal/microorganisms

on Ser-46 [7]. This ATP-requiring process is catalyzed by the HPr-kinase/phosphorylase (HPrK/P), which is regulated in a dose-dependent manner by the glycolytic intermediate fructose-1,6-bisphosphate (FBP) [10]. For this reason, the amount of Ser-46 phosphorylated HPr (P-Ser-HPr) is closely connected with glycolytic activity of the cell and the uptake of sugars. Sugar uptake in bacteria is predominantly mediated by the phosphotransferase system (PTS), consisting of three main components: HPr, enzyme I (EI), and enzyme II (EII) [11]. In a first step, HPr is phosphorylated at His-15 (P-His-HPr) by E1, using the glycolytic intermediate phosphoenolpyruvate as the phosphate donor. The phosphate group is transferred to the substrate by EII, which translocates and phosphorylates the sugar into the cell at the same time. Activated glucose, namely glucose 6-phosphate, then enters glycolysis [12]; hence, HPr connects glycolytic activity with CCR via its dual role in sugar uptake through the PTS and as an activator of CcpA [13,14].

Numerous genes have been identified to be regulated on the transcriptional level by CcpA in *S. aureus* [15–17]. In addition to genes/operons involved in carbon catabolism, the synthesis of factors associated with biofilm formation and virulence of *S. aureus* are also influenced by CcpA [6,15–21]. Specifically, CcpA promotes transcription of the *ica*-operon and *cidA* [19], encoding proteins needed for polysaccharide intercellular adhesion (PIA) synthesis and extracellular DNA release, respectively [22,23]. These observations are consistent with the fact that deletion of *ccpA* abrogates biofilm formation under glucose-rich conditions. [19]. Furthermore, inactivation of *ccpA* in *S. aureus* reduces the formation of liver and skin abscesses in mouse models of infection [6,24,25]. Taken together, these observations demonstrate the linkage between CcpA, glucose catabolism, and virulence in *S. aureus*; however, the function of HPr remains largely unknown. Here, we characterize the function of HPr of *S. aureus* in the context of carbon metabolism, growth kinetics, biofilm formation, and in vivo infectivity in different murine infection models.

2. Materials and Methods

2.1. Bacterial Strains and Plasmids

The bacterial strains and plasmids used in this study are listed in Table 1. All mutant strains generated for this study were confirmed by sequencing of the affected region, and by assessing gene transcription by quantitative real-time reverse transcriptase PCR (qRT-PCR).

Table 1. Strains and plasmids used in this study.

Strain	Description [1]	Reference or Source
S. aureus		
Newman	Mouse pathogenic laboratory strain (ATCC 25904)	[30]
RN4220	NCTC8325-4 derivative, acceptor of foreign DNA	[31]
SA113	PIA-dependent biofilm producer (ATCC 35556), *agr rsbU*	[32]
Nm *ccpA*	MST14; Newman Δ*ccpA*::*tet*(L); Tc^R	[15]
Nm *ptsH*	Newman Δ*ptsH*::lox72	This study
Nm *ptsH-aph*	Newman Δ*ptsH*::lox66-*aphaIII*-lox71; Kan^R	This study
Nm *ptsH::ptsH*	Newman Δ*ptsH*::pBT *ptsH*; Tc^R	This study
Nm *ptsH**	Newman Δ*ptsH*::lox72 pBT*ptsH**; Tc^R	This study
Nm *ccpA_ptsH*	Newman Δ*ccpA*::*tet*(L) Δ*ptsH*::lox72; Tc^R	This study
RN4220 *ptsH*	RN4220 Δ*ptsH*::lox72	This study
SA113 *ccpA*	KS66; SA113 Δ*ccpA*::*tet*(L); Tc^R	[19]
SA113 *ptsH*	SA113 Δ*ptsH*::lox72	This study
SA113 *ptsH::ptsH*	SA113 Δ*ptsH*::pBT *ptsH*; Tc^R	This study
SA113 *ptsH**	SA113 Δ*ptsH*::lox72 pBT*ptsH**; Tc^R	This study
SA113 *ccpA_ptsH*	SA113 Δ*ccpA*::*tet*(L) Δ*ptsH*::lox72; Tc^R	This study
E. coli		
DH5α	Cloning strain	Invitrogen
DC10B	Δ*dcm* in the DH10B background; Dam methylation only	[28]

Table 1. Cont.

Strain	Description [1]	Reference or Source
Plasmids		
pBT	*S. aureus* suicide plasmid; *tet*(L)	[27]
pBT *lox-aph*	pBT derivative harboring lox66-*aphAIII*-lox71; *tet*(L), *aphIII*	[26]
pRAB1	Temperature sensitive *E. coli-S. aureus* shuttle plasmid, expression of *cre* in staphylococci; *cat, bla*	[29]
pBT '*ptsI*	pBT derivative harboring a C-terminal *ptsI* fragment; *tet*(L)	This study
pBT *ptsH1*	pBT derivative harboring a T136G *ptsH* variant; *tet*(L)	This study
pBT *ptsH* KO	pBT derivative harboring the genomic regions flanking *ptsH* and lox66-*aphAIII*-lox71 of pBT *lox-aph*; *aphIII, tet*(L)	This study

[1] Kan^R, kanamicin-resistant; PIA, polyintercellular adhesin; Tc^R, tetracycline-resistant.

2.2. Bacterial Growth Conditions

S. aureus strains were grown in tryptic soy broth (TSB) containing 0.25% (*w/v*) glucose (BD, Heidelberg, Germany) or on TSB plates containing 1.5% agar (TSA). Antibiotics were only used for strain construction and phenotypic selection at the following concentrations: tetracycline, 2.5 µg/mL; erythromycin, 2.5 µg/mL; kanamycin, 15 µg/mL; and chloramphenicol, 10 µg/mL. Bacteria from overnight cultures were diluted in pre-warmed TSB to an optical density at 600 nm (OD$_{600}$) of 0.05. All bacterial cultures were incubated at 37 °C and 225 rpm with a flask-to-medium ratio of 10:1. Samples for determination of the OD$_{600}$, pH, and metabolites were taken every hour. The growth rate (μ) of *S. aureus* strains was calculated by the formula (ln OD_2–ln OD_1)/(t_2–t_1), with OD_1 and OD_2 being the OD calculated from the exponential growth phase at time t_1 and t_2, respectively. The generation time of each strain was determined using the formula ln $2/\mu$.

2.3. Mutant Construction

For the *S. aureus ptsH* deletion mutants, 1.4- and 1.1-kb fragments (nucleotides 1053472-1054838 and 1055049-1056169 of GenBank accession no. AP009351.1, respectively), containing the flanking regions of the *ptsH* open reading frame (ORF), were amplified by PCR from chromosomal DNA of *S. aureus* strain Newman using primer pairs MBH-94/MBH-112 and MBH-113/MBH-114, respectively (Supplementary Table S1). The PCR products were digested with KpnI/EcoRI and BamHI/XbaI, respectively, and cloned together with the EcoRI/BamHI-digested lox66-*aphAIII*-lox71 resistance cassette obtained from pBT *lox-aph* [26] into KpnI/XbaI-digested suicide vector pBT [27] to generate plasmid pBT *ptsH* KO. Plasmid pBT *ptsH* KO was propagated in *E. coli* strain DC10B [28] and subsequently electroporated directly into *S. aureus* strain Newman to obtain strain Newman Δ*ptsH*-*aph*, in which nucleotides 8 to 225 of the 267-bp spanning *ptsH* ORF were replaced by the lox66-*aphAIII*-lox71 cassette by allelic replacement. The deletion of *ptsH* in Newman Δ*ptsH*-*aph* was confirmed by PCR, and the strain was then used as a donor for transducing the lox66-*aphAIII*-lox71 tagged *ptsH* deletion into *S. aureus* strains SA113 and RN4220. Resistance marker-free Δ*ptsH*::lox72 derivatives were constructed by treatment with a Cre recombinase expressed from the temperature-sensitive vector pRAB1 [29], which was subsequently removed from the *aphIII*-cured derivatives by culturing the strains at 42 °C.

For the cis-complementation of the Δ*ptsH*::lox72 mutants, a 1-kb fragment (nucleotides 1055996-1056973 of GenBank accession no. AP009351.1) of the C-terminal region of the *ptsI* ORF and the annotated terminator region of the *ptsHI* operon was amplified by primers MBH-427/MBH-428 (Supplementary Table S1), digested with EcoRI/KpnI and cloned into EcoRI/KpnI-digested suicide vector pBT [27] to generate plasmid pBT '*ptsI*. The plasmid was electroporated into *S. aureus* strain RN4220, and a tetracycline-resistant RN4220 derivative that integrated pBT '*ptsI* in its chromosome at the *ptsI* locus was used as donor to phage-transduce the *tet*(L)-tagged *ptsI* allele into Nm Δ*ptsH* and SA113 Δ*ptsH*, respectively, thereby replacing the *ptsH::lox72* deletion with the *ptsI::*pBT '*ptsI* genomic region containing a functional *ptsHI* operon.

For the construction of *S. aureus ptsH* variants harboring a T to G exchange of nucleotide 136 of the *ptsH* ORF (termed *ptsH**), 0.6-kb and 1.1-kb fragments, containing either the promoter region of *ptsH* and the N-terminal part of the *ptsH* ORF (nucleotides 1054446-1054998 of GenBank accession no. AP009351.1) or the C-terminal part of the *ptsH* ORF and an N-terminal fragment of the *ptsI* ORF (nucleotides 1054976-1056121 of GenBank accession no. AP009351.1) were amplified by PCR from chromosomal DNA of *S. aureus* strain Newman using primer pairs MBH-484/MBH-485 and MBH-86/MBH-20, respectively (Supplementary Table S1). Primer MBH-484 contains a non-complementary base that introduces a point mutation in the PCR fragment leading to the T136G exchange of the *ptsH* ORF. Both PCR products were digested with StyI and subsequently ligated with T4 DNA-ligase. The ~1.7-kb ligation product was gel-purified, digested with KpnI/PstI, and cloned into KpnI/PstI-digested pBT to generate plasmid pBT *ptsH1* (Table 1). Presence of the T136G exchange in *ptsH** harbored by plasmid pBT *ptsH1* was confirmed by sequencing, the plasmid propagated in *E. coli* strain DH5α, electroporated into RN4220 Δ*ptsH*, and selected for tetracycline-resistance. A tetracycline-resistant RN4220 derivative that integrated pBT *ptsH1* at the *ptsHI* locus was used as donor to transduce the *tet*(L)-tagged *ptsH** allele into the Δ*ptsH* mutants.

S. aureus double mutants lacking *ptsH* and *ccpA* were created by transducing the *tet*(L)-tagged *ccpA* deletion of MST14 into Δ*ptsH* derivatives.

2.4. RNA Isolation and Purification, cDNA Synthesis and qRT-PCR

S. aureus strains were cultivated in TSB as described above. Bacterial pellets were collected after 2 h and 8 h of incubation by centrifugation at 5000 rpm at 4 °C for 5 min, and immediately suspended in 100 µL ice-cold TE-buffer (10 mM Tris-HCl, 1 mM EDTA, pH 8). Bacteria were disrupted, total RNA isolated, transcribed into cDNA, and qRT-PCRs carried out as described previously [33] using the primers listed in Supplementary Table S1. Transcriptional levels of target genes were normalized against the mRNA concentration of housekeeping gene *gyrB* according to the $2^{-\Delta CT}$ method.

2.5. Measurement of pH, Glucose, Acetate, and Ammonium in Culture Supernatants

Aliquots (1.5 mL) of bacterial cultures were centrifuged for 2 min at $10,000 \times g$, and supernatants were removed, pH measured, and stored at -20 °C until further use. Glucose, acetate, and ammonia concentrations were determined with kits purchased from R-Biopharm (Pfungstadt, Germany) and used according to the manufacturer's directions. The metabolite concentrations were measured from at least three independent experiments.

2.6. Biofilm Assays

Biofilm formation under static conditions was assessed as described [19]. Briefly, overnight cultures were diluted to an OD_{600} of 0.05 in fresh TSB medium supplemented with glucose to a final concentration of 0.75 % (*w/v*), and 200 µL of the cell suspension was used per well to inoculate sterile, flat-bottom 96-well polystyrene microtiter plates (BD). After incubation for 24 h at 37 °C without shaking, the plate wells were washed twice with phosphate-buffered saline (pH 7.2) and dried in an inverted position. Adherent cells were safranin-stained (30 sec with 0.1% safranin; Merck, Darmstadt, Germany) and the absorbance of stained biofilms was measured at 490 nm after resolving the stain with 100 µL 30 % (*v/v*) acetic acid, using a microtiter plate reader (Victor^2 1420 Multilabel Counter; Perkin Elmer, Rodgau, Germany).

Biofilm formation under flow conditions was performed as described [34], with minor modifications: Bacteria from overnight cultures were diluted to an OD_{600} of 0.05 in fresh TSB medium supplemented with glucose to a total concentration of 0.75% (*w/v*) and cultivated for 2 h at 37 °C with shaking at 150 rpm. Flow cells (Stovall Life Science) were filled with pre-warmed TSB medium supplemented with glucose to a total concentration of

0.75% (w/v), attached to a peristaltic pump (Ismatec REGLO Digital; Postnova, Landsberg am Lech, Germany) and inoculated with 0.5 mL of the bacterial cultures. Thirty minutes after inoculation, the flowrate was set to 0.5 mL/min and chamber. Biofilm formation was visually documented at different times.

For the assessment of biofilm formation on medical devices under dynamic conditions, peripheral venous catheter (PVC, Venflon Pro Safety 18 G; BD) fragments of 1 cm length were placed into reaction tubes filled with 1 mL of TSB and inoculated with 5×10^5 CFU of TSB-washed bacterial cells obtained from exponential growth phase (inoculation of TSB from overnight cultures to an OD_{600} of 0.05 and incubation for 2.5 h at 37 °C and 225 rpm). The PVC fragments were incubated under non-nutrient limited conditions for five days at 37 °C and 150 rpm, and the media were replaced with fresh media every 24 h. PVC fragments were placed five days post inoculation into fresh reaction tubes filled with 1 mL of TSB, biofilms were detached from the catheter surface and resolved by sonication (50 watt for 5 min) followed by 1 min of vortexing. CFU rates and biomasses of resolved biofilms and culture supernatants at day five post inoculation were determined by plate counting and OD_{600} measurements, respectively.

2.7. Primary Attachment Assay on Polystyrene

The primary attachment of bacterial cells to polystyrene surfaces was performed as described [35], with minor modifications. Briefly, bacteria from the exponential growth phase (inoculation of TSB from overnight cultures to an OD_{600} of 0.05 and incubation for 2.5 h at 37 °C and 225 rpm) were diluted in TSB to 3000 CFU/mL. 100 µL of the bacterial inoculum was poured onto polystyrene petri dishes (Sarstedt, Nümbrecht, Germany) and incubated under static conditions at 37 °C for 30 min. After incubation, petri dishes were rinsed gently three times with 5 mL of sterile PBS (pH 7.5), and subsequently covered with 15 mL of TSB containing 0.8% agar maintained at 48 °C. Plates were incubated at 37 °C for 24 h. Bacterial attachment to polystyrene was defined as the number of CFU remaining on the petri dish bottom after washing compared to the number of CFU remaining on the petri dish bottom without washing.

2.8. Animal Models

All animal experiments were performed with approval of the local State Review Board of Saarland, Germany (project identification codes 60/2015 [approved 21.12.2015], and 34/2017 [approved 09.11.2017]), and conducted following the national and European guidelines for the ethical and human treatment of animals. PBS-washed bacterial cells obtained from exponential growth phase cultures were used as inoculum.

For the murine abscess model, infection of animals was carried out as described [33], with minor modifications; specifically, 8- to 12- week-old female C57BL/6N mice (Charles River, Sulzfeld, Germany) were anesthetized by isoflurane inhalation (3.5%; Baxter, Unterschleißheim, Germany) and 100 µL bacterial suspension containing 5×10^7 CFU were administered intravenously by retro bulbar injection. Immediately after infection, animals were treated with a single dose of carprofen (5 mg/kg; Zoetis, Berlin, Germany). Behavior and weight of mice was monitored daily, and four days post-infection, mice were sacrificed, and livers and kidneys were removed. The bacterial loads in liver and kidney tissues were determined by homogenization of weight-adjusted organs in PBS (pH 7.4), followed by serial dilutions on sheep blood agar plates and plate counting after 24 h incubation at 37 °C.

For the *S. aureus* based murine foreign body infection model, implantation of catheter fragments and infection of animals was carried out as described [36], with minor modifications: 8- to 12-week-old female C57BL/6N mice (Charles River) were anesthetized by intraperitoneal injection of 0.05 mg/kg body weight fentanyl (Hexal, Holzkirchen, Germany), 5 mg/kg midazolam (Hameln Pharma Plus, Hameln, Germany) and 0.5 mg/kg medetomidine (Orion Pharma, Hamburg, Germany). After treatment with a dose of carprofen (5 mg/kg, Zoetis), the animals were shaved with an animal trimmer (BBraun, Melsungen, Germany) and depilated with asid-med hair removal cream (Asid Bonz, Herrenberg, Germany) on both flanks. The depilated skin was disinfected with ethanol (70%) and 1 cm catheter fragments (PVC, 14G, Sarstedt) were implanted subcutaneously and inoculated with 1×10^4 CFU of the respective *S. aureus* strains. Wounds were closed with staples (Fine Science Tools, Heidelberg, Germany) and anesthesia was antagonized with 1.2 mg/kg body weight naloxone (Inresa, Freiburg im Breisgau, Germany), 0.5 mg/kg flumazenil (Inresa) and 2.5 mg/kg atipamezole (Orion Pharma). Behavior and weight of the animals was monitored daily. Ten days post infection, animals were sacrificed, edema sizes were measured and photo documented, and catheter fragments with surrounding tissue were harvested for microbial analyses. Excised tissues were homogenized in 1 mL TSB with a hand disperser (POLYTRON PT 1200 E; Kinematica, Eschbach, Germany), and biofilms were detached from the PVC fragments and resolved by sonication (50 watt for 5 min) followed by vortexing (1 min). CFU rates in tissue and of biofilm formed on the catheter were determined by plating serial dilutions on sheep blood agar plates and plate counting after 24 h of incubation at 37 °C.

2.9. Statistical Analyses

The statistical significance of changes between groups was assessed by one-way ANOVA followed by Holm-Sidak's post-hoc tests for experiments containing ≥ 5 biological replicates using the GraphPad software package Prism 6.01 (San Diego, CA 92108, USA). *p* values < 0.05 were considered statistically significant.

3. Results and Discussion

3.1. Growth, pH Characteristics, and Metabolite Profiles Differ between ptsH and ccpA Mutants

To determine if inactivation of *ptsH* in *S. aureus* leads to changes in growth and carbon catabolism, mutants were constructed in the *S. aureus* laboratory strain Newman (Table 1) and growth and physiology were assessed (Figure 1). In detail, a mutant lacking *ptsH* (Δ*ptsH*) was constructed and cis-complemented (*ptsH::ptsH*), and a Δ*ccpA ptsH* double mutant was created (*ccpA_ptsH*). In addition, a *ptsH* mutant harboring a point mutation in the *ptsH* gene (T136G) leading to the substitution of serine to alanine at position 46 of HPr (HPr-S46A) was constructed (*ptsH**). The phosphorylation at this amino acid represents a known prerequisite for HPr to activate CcpA in other gram-positive bacteria (14), while its activity in the phosphotransferase uptake system (PTS) should be unaffected. The parental strain Newman and the cis-complemented *ptsH* derivative displayed similar growth characteristics and comparable generation times, respectively. In contrast, all *ptsH* mutants (*ptsH*, *ccpA_ptsH*, and *ptsH**) had reduced growth rates in the exponential (1–3 h) and the transition phase (4–6 h) relative to the wild type (Figure 1 and Table 2). Interestingly, the growth rate of the isogenic *ccpA* deletion mutant was only slightly diminished relative to that of strain Newman (Figure 1 and Table 2), and differed significantly from the wild type only during the transition phase. After 12 h of cultivation, growth yields were comparable for all strains (Figure 1b), suggesting that neither the lack of CcpA nor HPr has a clear long-term effect on biomass production of *S. aureus* cultured in rich medium. This is in line with earlier findings regarding CcpA [15].

Table 2. Generation time of *S. aureus* strains cultivated in TSB under aerobic conditions.

Strain	Generation Time (min) [1]	*p* Value [2]
Newman	28.6 ± 1.9	
Nm *ptsH*	34.8 ± 1.6	<0.01
Nm *ptsH::ptsH*	27.8 ± 1.3	0.34
Nm *ptsH**	33.3 ± 1.4	<0.01
Nm *ccpA*	30.0 ± 1.3	0.19
Nm *ccpA_ptsH*	34.3 ± 1.0	<0.01

[1] Data are presented as mean ± SD (n = 6). [2] *p* values were determined by one-way ANOVA and Holm-Sidak's multiple comparison test.

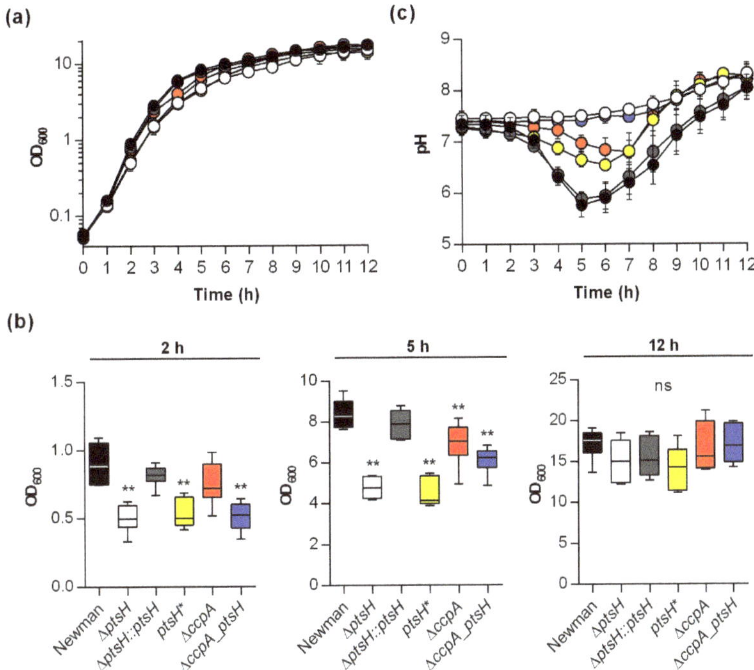

Figure 1. Impact of *ptsH* and/or *ccpA* on growth and pH profiles of *S. aureus* TSB cultures. Bacteria were inoculated to an OD_{600} of 0.05 in TSB and cultured aerobically at 37 °C and 225 rpm. OD_{600} (**a,b**) and pH measurements (**c**) of the culture media were determined hourly. Symbols represent: strains Newman (black symbols), Nm *ptsH* (white symbols), Nm *ptsH::ptsH* (grey symbols), Nm *ptsH** (yellow symbols), Nm *ccpA* (red symbols), and Nm *ccpA_ptsH* (blue symbols). The results are the mean ± SD of at least five independent experiments. (**b**) OD_{600} readings of the cell cultures at 2, 5, and 12 h of growth, respectively. The data are presented as box and whisker plot showing the interquartile range (25–75%, box), the median (horizontal line), and the standard deviation (bars) of 5–6 independent experiments. **, $p < 0.01$ (one-way ANOVA and Holm-Sidak's multiple comparison test. Only differences between Newman and mutants are shown).

To assess the bacterial acid production during growth, the pH of culture supernatants was measured over time (Figure 1c). The pH of culture supernatants from wild type and cis-complemented *ptsH* mutant cultures were similar. In contrast, pH values of culture supernatants from the *ptsH* and *ccpA_ptsH* deletion mutants indicated that little acid was produced during growth. The pH profiles of the *ptsH** and *ccpA* mutant cultures were between these two extremes but indicated that acidic end-products were produced and consumed during growth. Taken together, these data indicate that inactivation of *ptsH*, or interference with HPr phosphorylation, delays growth and medium acidification to a greater extent than does deletion of *ccpA*. In addition, the small differences in physiological parameters (i.e., growth and pH kinetics) between the *ptsH* and the *ccpA_ptsH* mutants and between the *ccpA* and *ptsH** mutants indicate additional, CcpA-independent functions of HPr.

To get an idea about the metabolic processes that are active in Newman wild type and mutant cells cultured in TSB, the concentrations of glucose, acetate, and ammonia were determined in culture supernatants over time (Figure 2). Strain Newman and the cis-complemented *ptsH* mutant (*ptsH::ptsH*) depleted all available glucose in the medium within the first 5 h of cultivation (Figure 2a). In contrast, glucose depletion in Δ*ptsH* and Δ*ccpA_ptsH* mutant cultures was severely delayed, and low concentrations of glucose were still detectable in culture supernatants even after 10 h of growth. In Δ*ccpA* and *ptsH** mutant cultures, glucose levels decreased slower than in wild type cultures, and no glucose was detectable after 7 h of growth (Figure 2a).

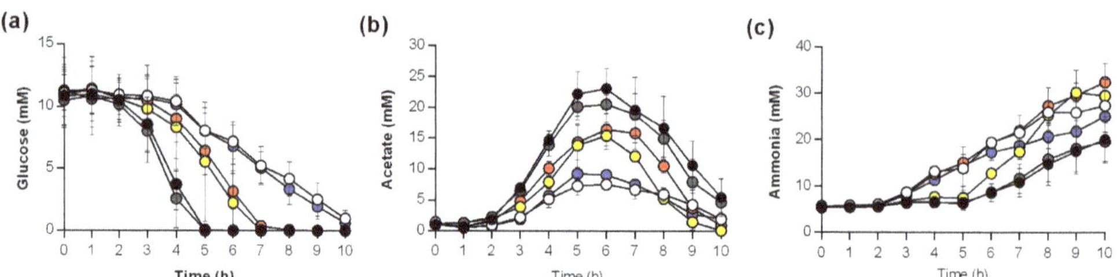

Figure 2. Impact of *ptsH* and/or *ccpA* on glucose consumption, acetate and ammonia production of *S. aureus* Newman during in vitro growth. *S. aureus* strains Newman (black symbols), Nm *ptsH* (white symbols), Nm *ptsH::ptsH* (grey symbols), Nm *ptsH** (yellow symbols), Nm *ccpA* (red symbols), and Nm *ccpA_ptsH* (blue symbols) were cultivated in TSB, and glucose (**a**), acetate (**b**), and ammonia (**c**) concentrations in culture supernatants were determined hourly. Results are presented as the average and standard deviation of at least three independent experiments.

When *S. aureus* is cultured aerobically in a glucose-containing medium, cells produce and secrete acetate as long as glucose is available [37]. Consistent with this fact, increasing acetate concentrations in the culture supernatants were observed during the first 5–6 h of growth for all strains (Figure 2b). However, while all strains accumulated acetate in the medium, the maximum concentrations differed; specifically, the wild type and the cis-complemented *ptsH* derivative accumulated up to 22 mM of acetate. In contrast, supernatants from Δ*ptsH* and Δ*ccpA_ptsH* mutant strain cultures had approximately one-third of the concentration of that from the wild type strain. Similar to that seen in the pH profiles (Figure 1c), the acetate profiles of *ptsH** and Δ*ccpA* mutant cultures centered in between those two extremes (Figure 2b). At 7–8 h post inoculation, acetate levels decreased in the supernatants of all cultures, irrespective of the fact that glucose was present in Δ*ptsH* and Δ*ccpA_ptsH* mutant cultures (Figure 2a).

S. aureus also utilizes amino acids as carbon sources for growth, a process that requires deamination of the amino acids, resulting in the secretion of ammonia into the culture supernatant [20]. The uptake and catabolism of amino acids in *S. aureus* is subject to CCR [20]. While glucose was present in the medium, ammonia levels remained low in the wild type and *ptsH::ptsH* culture supernatants, followed by a steady increase in the ammonia concentrations (Figure 2c). In contrast, cultures of strains Nm *ptsH* and Nm *ccpA_ptsH* began to accumulate ammonia beginning at 3 h of cultivation. Interestingly, the ammonia concentration in the supernatant of the Δ*ccpA* mutant closely resembled that of the *ptsH* deletion mutants, while the *ptsH** mutant resembled the late induction of the wild type strain (Figure 2c).

Taken together, these data show that the inactivation of *ptsH* or *ccpA* results in distinct differences in glucose consumption, acetate accumulation and reutilization, and ammonia secretion in *S. aureus*. Furthermore, the exchange of an amino acid critical for the interaction of HPr with CcpA in the *ptsH** mutant resulted in metabolite profiles (i.e., glucose and acetate) comparable to the Δ*ccpA* mutant, while some alterations in the growth profile, generation time, and ammonia secretion were observed. Importantly, after 12 h of growth, the biomass of *S. aureus* Newman was independent of *ccpA* and *ptsH*, suggesting that *S. aureus* has other means to utilize carbon sources in the growth medium. Specifically, *ptsH* mutants were able to utilize glucose from the growth medium—although much slower than the wild type—demonstrating that *S. aureus* can transport glucose independent of the group translocation PTS [38]. A likely compensatory transporter would be one of the many ATP binding cassette transporters identified in *S. aureus* [39].

3.2. Inactivation of ptsH and/or ccpA Alters Transcription of TCA Cycle and Virulence Factor Genes

CcpA is known to affect transcription of a large number of central carbon metabolism and virulence genes [15–17]. For this reason, the effect of *ptsH* deletion on transcription of genes regulated by CcpA such as *citB* (encoding the TCA cycle key enzyme aconitase), *pckA* (encoding the gluconeogenesis key enzyme phosphoenolpyruvate carboxykinase), and *hla* (encoding α-hemolysin) was assessed. Specifically, mRNA levels were determined in cells from the exponential (i.e., 2 h) and post-exponential growth phases (i.e., 8 h) by qRT-PCR (Figure 3).

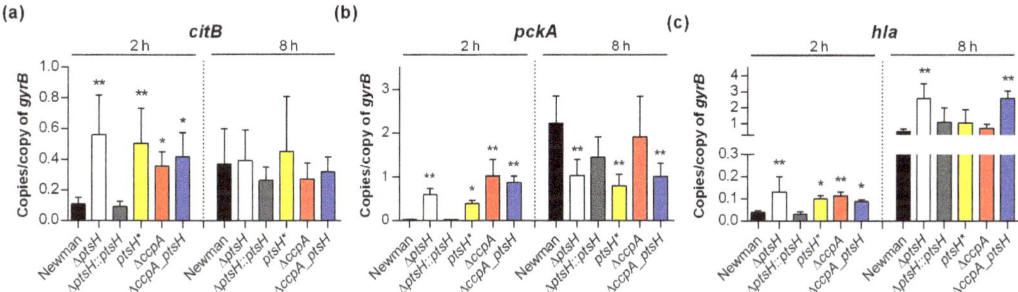

Figure 3. Effect of *ptsH* and/or *ccpA* mutations on the transcription of *S. aureus*. Newman wild type and mutant cells were cultured aerobically in TSB, as outlined in Materials and Methods. Cells were harvested at the time points indicated, total RNAs isolated, and qRT-PCRs performed for *citB* (**a**), *pckA* (**b**), and *hla* (**c**). Transcripts were quantified in reference to the transcription of gyrase B. Data are presented as mean + SD of five biological replicates. *, *p* < 0.05; **, *p* < 0.01 (one-way ANOVA and Holm-Sidak's multiple comparison test; only differences between Newman and mutants are shown).

Consistent with our previous observation that *S. aureus* transcription of *citB* and *pckA* is repressed by CcpA when cultivated with glucose [16], deletion of *ccpA* significantly increased the level of *citB* and *pckA* mRNA in exponential growth phase cells relative to wild type cells (Figure 3a,b). As expected, in the post-exponential growth phase, comparable *citB* and *pckA* transcript levels were observed in the Δ*ccpA* mutant and the wild type. Similar to the Δ*ccpA* mutant, exponential growth phase cells of the *ptsH* mutants (Δ*ptsH*, *ptsH**, and Δ*ccpA_ptsH*) had comparable *citB* and *pckA* mRNA levels, while the cis-complemented *ptsH* derivative (*ptsH::ptsH*) had *citB* and *pckA* transcript levels comparable to the wild type strain. In contrast to exponential growth phase cultures, all three *ptsH* mutants produced significantly lower levels of *pckA* mRNA than the wild type at 8 h, suggesting that *pckA* transcription is affected by HPr at later growth stages in a way that is independent of CcpA. This differed from the results for *citB* in which all *ptsH* mutants (Δ*ptsH*, *ptsH**, and Δ*ccpA_ptsH*) had comparable transcript levels to that of the Δ*ccpA* mutant and the wild type after 8 h of growth. The fact that the *ptsH** mutant produced *pckA* transcript levels similar to the *ptsH* mutant, but not similar to the *ccpA* mutant suggests that HPr phosphorylated at serine 46 acts in part independent of CcpA. This observation is consistent with that found in other gram-positive bacteria, where the serine 46-phosphorylated HPr exerted effects on CCR via CcpA and inducer exclusion [40]. However, it cannot be excluded that differences in *pckA* transcription between the *ptsH** and Δ*ccpA* mutants are due to differences in protein stability. The reason why protein stability cannot be excluded is because phosphorylation of the *B. subtilis* HPr homolog at Ser-46 stabilized the protein [41], while a serine to alanine exchange of Ser-46 in the *E. coli* HPr homolog was found to decrease the stability of the protein [42].

CcpA represses transcription of *hla* during the exponential growth phase when bacteria are cultured in presence of glucose [15,16]. Similarly, levels of *hla* mRNA from the exponentially growing Δ*ccpA* mutant and all *ptsH* mutants (Figure 3c) were de-repressed, while the cis-complemented *ptsH* deletion mutant produced *hla* transcript levels that were comparable to the wild type. During the post-exponential growth phase, only the *hla* transcript levels of the Δ*ptsH* and the Δ*ccpA_ptsH* double mutant were significantly increased (Figure 3c), suggesting that HPr affects expression of α-hemolysin in a CcpA-dependent and -independent manner. Taken together, these data suggest that exponential growth phase *S. aureus* was cultured in the presence of glucose, HPr affects the transcription primarily via activation of CcpA, while in the post-exponential growth phase cells of *S. aureus*, HPr is likely to affect gene transcription by CcpA-independent mechanism(s).

3.3. Impact of ptsH Deletion on Biofilm Formation of S. aureus SA113

CcpA is important for polysaccharide intercellular adhesin (PIA)-dependent biofilm formation by staphylococci under glucose-rich in vitro conditions [19,24,43]. The importance of HPr on sugar import and gene regulation suggests that HPr might influence biofilm formation of *S. aureus*. Strain Newman is a weak biofilm producer in glucose-rich medium under in vitro conditions [34], hence we transduced the *ptsH* mutations into *S. aureus* strain SA113, which forms a strong biofilm under these conditions [19,22]. The ability of SA113 mutant strains were analyzed using a semi-quantitative static biofilm assay (Figure 4a) and in biofilm flow cells (Figure 4b).

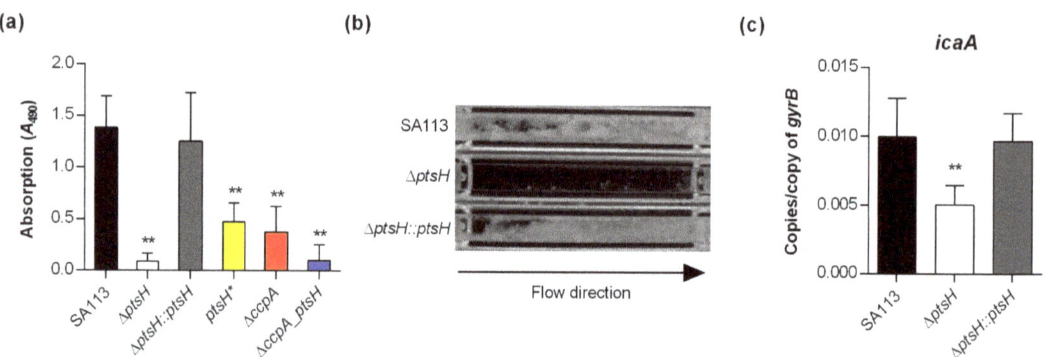

Figure 4. Mutations of *ptsH* affect biofilm formation of *S. aureus* under static and flow conditions. (**a**) Biofilm growth of *S. aureus* strains in a static 96-well microplate assay. The data show the mean + SD of five biological replicates. (**b**) Flow cell chambers were inoculated with *S. aureus* strains as indicated, allowed to attach to the surfaces for 30 min, and incubated under constant flow for 24 h. The results shown are representative of two independent experiments. (**c**) Effect of the *ptsH* mutation on the transcription of *icaA* in *S. aureus* strain SA113. Cells of SA113, the Δ*ptsH* mutant, and the cis-complemented *ptsH::ptsH* derivative were cultured aerobically in TSB. After 2 h of growth, cells were harvested, total RNAs isolated, and qRT-PCRs performed for *icaA*. Transcripts were quantified in reference to gyrase B mRNA. Data are presented as mean + SD of five biological replicates. **, $p < 0.01$ (one-way ANOVA and Holm-Sidak's multiple comparison test; only differences between SA113 and mutants are shown).

Under static conditions, the Δ*ptsH* and Δ*ccpA_ptsH* mutants of SA113 displayed drastic decreases in their biofilm formation capacities on polystyrene surfaces, whereas the cis-complemented derivative (*ptsH::ptsH*) formed biofilms that were comparable to the ones seen with the wild type (Figure 4a). Deletion of *ccpA* or the S46A mutation of HPr in SA113 (*ptsH**) also significantly reduced biofilm formation, however, not to the extent seen with the Δ*ptsH* mutant, supporting our hypothesis that a functional PTS is important for *S. aureus* to form a biofilm in this type of assay. In the flow chamber assay, the Δ*ptsH* mutant failed to produce a clear biofilm within the microchannel after 24 h of constant flow, while both, the wild type and the cis-complemented *ptsH* derivative, almost completely filled the microchannel with biomass (Figure 4b), suggesting that HPr is also important for biofilm formation under shear flow. To exclude that the latter phenotype was caused by a decreased capacity of the Δ*ptsH* mutant to attach to the microchannel surface, the primary attachment capacities of the strains were determined. Here, no clear differences in attachment towards polystyrene surfaces were obtained for the strain triplet, suggesting that the observed lack of biofilm formation of the SA113 Δ*ptsH* mutant is likely due to a deficiency in biofilm maturation. To determine whether this effect might be due to a decreased capacity of the mutant to produce PIA, we assayed the transcription of *icaA*, which is part of the *icaADBC* polycistronic mRNA that encodes proteins needed for PIA synthesis [22]. Consistent with the reduced ability of the SA113 *ptsH* mutant to form a biofilm under static and flow conditions, we observed significantly decreased levels of *icaA* transcripts in the *ptsH* deletion mutant relative to the wild type and the cis-complemented mutant (Figure 4c). Together, these data suggest that HPr, in part, promotes biofilm formation of *S. aureus* by enhancing the expression of the PIA synthesis machinery.

In a third biofilm assay intended to resemble the in vivo situation more closely, we studied the ability of SA113 and its derivatives to form biofilms on peripheral venous catheter (PVC) fragments under non-nutrient limited conditions (Figure 5).

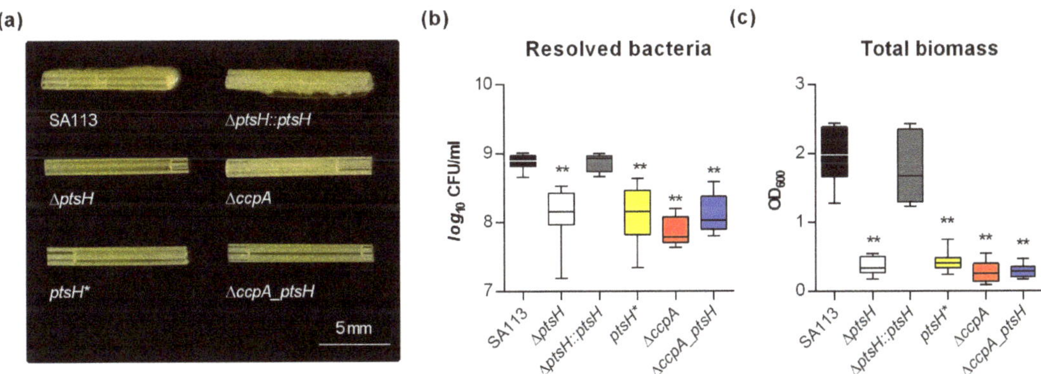

Figure 5. Inactivation of *ptsH* and/or *ccpA* reduces the biofilm formation capacity of *S. aureus* on medical devices. (**a**) Images of *S. aureus*-loaded catheter fragments at day 5 post inoculation (6.3-fold magnification). The results are representative of three independent experiments. (**b,c**) Colony forming units (CFU) and total biomass of detached biofilms were determined by plate counting (**b**) and measuring the OD_{600} of the TSB solutions (**c**). The data are presented as box and whisker plot showing the interquartile range (25–75%, box), the median (horizontal line), and the standard deviation (bars) of nine independent experiments. **, $p < 0.01$ (one-way ANOVA and Holm-Sidak's multiple comparison test; only differences between SA113 and mutants are shown).

Using this assay, a strong biofilm was macroscopically detectable on catheter fragments inoculated with the wild type or the cis-complemented *ptsH* derivative at 5 days post inoculation (Figure 5a). In contrast, on catheter fragments inoculated with either the Δ*ptsH* mutant, the Δ*ccpA* mutant, the *ptsH** mutant, or the Δ*ccpA_ptsH* double mutant, almost no biofilm was visible. These observations were further supported by CFU and OD_{600} determinations of TSB solutions harboring the detached biofilms (Figure 5b,c). A significant reduction in viable bacteria (~1 log) was observed on all fragments inoculated with mutants when compared to the wild type inoculated fragments (Figure 5b). Similar to the CFU data, the OD_{600} values were approximately 10-fold lower in the detached biofilms formed by the mutants (Figure 5c), suggesting that *ptsH* and *ccpA* deletions elicit rather comparable effects on the biofilm forming capacity of *S. aureus* on PVC surfaces under non-nutrient limited conditions.

3.4. HPr Contributes to Infectivity and Biofilm Formation of S. aureus SA113 in a Murine Foreign Body Infection Model

CcpA is not required for biofilm formation of *S. aureus* and *S. epidermidis* on implanted catheter fragments in normoglycemic mice [24,43], but this did not address the function of HPr in vivo. In order to address this question, the ability of strains SA113, the Δ*ptsH* mutant, and the cis-complemented *ptsH* derivative to form biofilms on implanted catheter fragments was assessed in the murine foreign body infection model [36] with normoglycemic mice (Figure 6).

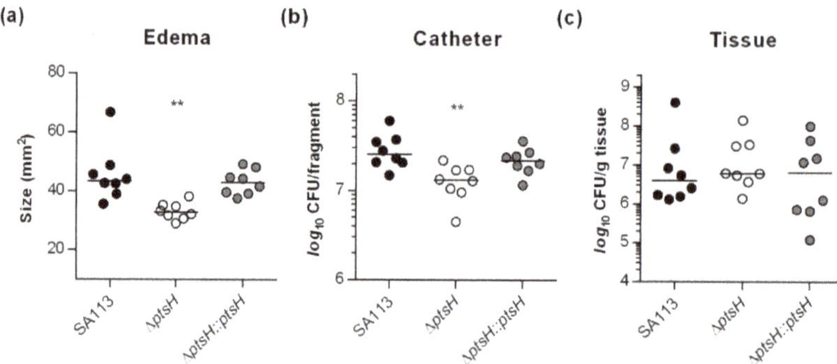

Figure 6. Inactivation of *ptsH* reduces the infectivity of *S. aureus* SA113 in a murine foreign body infection model. Catheter fragments were implanted subcutaneously into the back of normoglycemic mice and inoculated with cells of *S. aureus* strains SA113 (black symbols), its *ptsH* deletion mutant (white symbols), and the cis-complemented *ptsH* mutant (grey symbols), respectively (*n* = eight animals per group). Ten days post infection, animals were euthanized, edema sizes around the implanted catheters were measured (**a**), and the catheters and surrounding tissues were explanted. Bacterial loads from catheter detached biofilms (**b**) and in surrounding tissue homogenates (**c**) were determined by CFU counting. The data represent the values of every individual animal (symbols) and the median (horizontal line). **, *p* < 0.01 (one-way ANOVA and Holm-Sidak's multiple comparison test).

Mice challenged with the Δ*ptsH* mutant displayed a clear reduction in edema sizes around the implanted catheter fragments (Figure 6a) and a small but significant reduction (~2-fold) in detached bacteria (Figure 6b), when compared to animals infected with wild type bacteria or the cis-complemented *ptsH* derivative. In contrast, no significant differences in bacterial loads of tissues surrounding the catheter fragments were obtained (Figure 6c). These findings suggest that HPr, unlike CcpA, has a small but important function on the biofilm formation capacity of PIA producing *S. aureus* in normoglycemic mice. This difference is probably due to a reduced sugar uptake capacity of the *ptsH* mutant, which might interfere with the enhanced carbon and energy demand of *S. aureus* during biofilm maturation.

3.5. HPr and CcpA Are Both Required for Full Infectivity of S. aureus in a Murine Liver Abscess Model

The formation of liver abscesses is one of the clinical manifestations caused by *S. aureus* in which CcpA exerts a strong effect on disease progression in normoglycemic mice [6]. To determine how P-Ser-HPr affects infectivity of *S. aureus* in a murine liver abscess model, the bacterial loads in livers four days post infection were assessed (Figure 7).

Consistent with previous observations [6,24], we observed a nearly 3 log reduction in bacterial loads in liver tissue of C57BL/6 mice challenged with the *ccpA* mutant bacteria (median 2.2 × 10⁵ CFU/g tissue) relative to mice infected with the wild type strain (median 7.6 × 10⁷ CFU/g tissue). Importantly, a greater reduction in CFU/g liver was observed (~4 log; median 1.4 × 10⁴ CFU/g tissue), when mice were challenged with the Δ*ptsH* mutant. Infection of mice with the cis-complemented *ptsH* derivative resulted in a bacterial burden in the liver (median 4.2×10^7 CFU/g tissue) comparable to that seen in wild type infected mice, demonstrating that the decreased CFU rates determined in liver tissues of $\Delta ptsH$ infected mice were due to the lack of HPr. Notably, mice challenged with the *ptsH** mutant carrying the S46A exchange in HPr also caused an almost ~4 log reduction in bacterial loads in liver tissue (median 2.1×10^4 CFU/g tissue), suggesting that both, the deletion of *ptsH* and a mutation of serine 46 of HPr, alter the virulence of *S. aureus* in this murine infection model in a CcpA-independent manner. The lowest CFU rates in liver tissues were observed when mice were challenged with the $\Delta ccpA\_ptsH$ double mutant (median 4.8×10^3 CFU/g tissue), suggesting that CcpA might also exert some effects on virulence of *S. aureus* in this infection model independently of HPr.

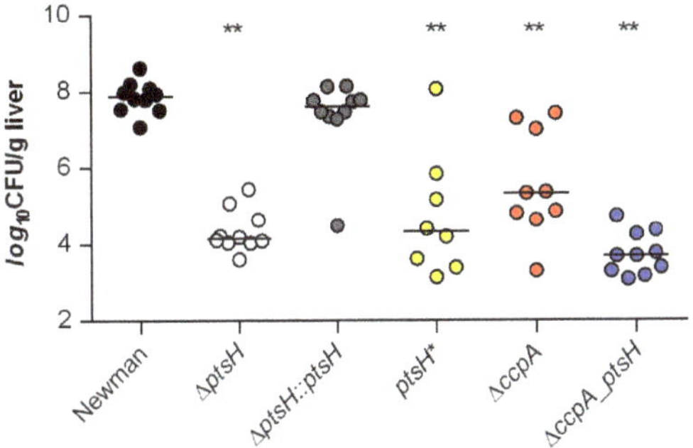

Figure 7. Inactivation of *ptsH* and/or *ccpA* results in decreased bacterial burden of *S. aureus* Newman in a murine abscess model. C57BL/6N mice were infected retro bulbar with 5×10^7 CFU of *S. aureus* strains Newman (black symbols), its $\Delta ptsH$ (white symbols) and $\Delta ccpA$ (red symbols) mutants, the *ptsH** mutant (yellow symbols), the $\Delta ccpA\_ptsH$ double mutant (blue symbols), and the cis-complemented *ptsH* derivative (grey symbols), respectively. Bacterial loads in liver tissue homogenates were determined four days post infection. The data display the median (horizontal lines) and individual values of every animal (dots; n = 8–10 animals per group). **, $p < 0.01$ (one-way ANOVA and Holm-Sidak's multiple comparison test; only differences between Newman and mutants are shown).

4. Conclusions

Central carbon metabolism and virulence factor synthesis are tightly linked in *S. aureus* and controlled by several transcription factors [3]. Notably, CcpA is the only transcription factor known to enhance infectivity of *S. aureus* [6,24,25], while other regulators such as CcpE, CodY, and RpiRc are thought to attenuate rather than to promote infectivity of this bacterium in mice [33,44–47]. We show here that HPr contributes positively to infectivity of *S. aureus* in mice, presumably by affecting central carbon metabolism and virulence factor synthesis in a CcpA-dependent and -independent manner. These effects are likely mediated through changes in sugar transport and carbon metabolization that alter biofilm formation [24]. It is also possible that HPr in *S. aureus* acts like the HPr homolog of *E. coli* to modulate quorum sensing by interacting with autoinducer-2 (AI-2) modifying factors [48]. Given the importance of HPr on biofilm formation and virulence in *S. aureus*, this phosphocarrier protein could be a promising drug target for the development of novel anti-staphylococcal compounds.

Supplementary Materials: The following are available online at https://www.mdpi.com/2076-260 7/9/3/466/s1, Table S1: Primers used in this study.

Author Contributions: Conceptualization, M.B., R.G. and G.A.S.; methodology, L.P., E.-L.B. and R.G.; investigation, L.P., A.-C.B., E.-L.B. and L.Z.; writing—original draft preparation, L.P. and M.B.; writing—review and editing, M.B., R.G. and G.A.S.; visualization, M.B. and R.G.; supervision, R.G.;

project administration, M.B. and R.G.; funding acquisition, M.B. All authors have read and agreed to the published version of the manuscript.

Funding: This research was funded by the German Research Foundation (DFG), grant numbers BI 1350/1-2 and SFB1027. We acknowledge the support by the DFG and Saarland University within the funding program Open Access Publishing.

Data Availability Statement: The datasets generated and analyzed during the current study are available from the corresponding author on reasonable request.

Acknowledgments: The authors are grateful to Karin Hilgert and Benjamin Kastell for excellent technical assistance.

Conflicts of Interest: The authors declare no conflict of interest.

References

1. Deutscher, J. The mechanisms of carbon catabolite repression in bacteria. *Curr. Opin. Microbiol.* **2008**, *11*, 87–93. [CrossRef]
2. Görke, B.; Stülke, J. Carbon catabolite repression in bacteria: Many ways to make the most out of nutrients. *Nat. Rev. Microbiol.* **2008**, *6*, 613–624. [CrossRef]
3. Richardson, A.R. Virulence and Metabolism. *Microbiol. Spectr.* **2019**, *7*. [CrossRef]
4. Henkin, T.M.; Grundy, F.J.; Nicholson, W.L.; Chambliss, G.H. Catabolite repression of α amylase gene expression in *Bacillus subtilis* involves a trans-acting gene product homologous to the *Escherichia coli lacI* and *galR* repressors. *Mol. Microbiol.* **1991**, *5*, 575–584. [CrossRef] [PubMed]
5. Weickert, M.J.; Chambliss, G.H. Site-directed mutagenesis of a catabolite repression operator sequence in *Bacillus subtilis*. *Proc. Natl. Acad. Sci. USA* **1990**, *87*, 6238–6242. [CrossRef] [PubMed]
6. Li, C.; Sun, F.; Cho, H.; Yelavarthi, V.; Sohn, C.; He, C.; Schneewind, O.; Bae, T. CcpA mediates proline auxotrophy and is required for *Staphylococcus aureus* pathogenesis. *J. Bacteriol.* **2010**, *192*, 3883–3892. [CrossRef]
7. Deutscher, J.; Saier, M.H. ATP-dependent protein kinase-catalyzed phosphorylation of a seryl residue in HPr, a phosphate carrier protein of the phosphotransferase system in *Streptococcus pyogenes*. *Proc. Natl. Acad. Sci. USA* **1983**, *80*, 6790. [CrossRef] [PubMed]
8. Schumacher, M.A.; Allen, G.S.; Diel, M.; Seidel, G.; Hillen, W.; Brennan, R.G. Structural Basis for Allosteric Control of the Transcription Regulator CcpA by the Phosphoprotein HPr-Ser46-P. *Cell* **2004**, *118*, 731–741. [CrossRef]
9. Leiba, J.; Hartmann, T.; Cluzel, M.E.; Cohen-Gonsaud, M.; Delolme, F.; Bischoff, M.; Molle, V. A novel mode of regulation of the *Staphylococcus aureus* catabolite control protein A (CcpA) mediated by Stk1 protein phosphorylation. *J. Biol. Chem.* **2012**, *287*, 43607–43619. [CrossRef]
10. Ramstrom, H.; Sanglier, S.; Leize-Wagner, E.; Philippe, C.; Van Dorsselaer, A.; Haiech, J. Properties and regulation of the bifunctional enzyme HPr kinase/phosphatase in *Bacillus subtilis*. *J. Biol. Chem.* **2003**, *278*, 1174–1185. [CrossRef]
11. Hengstenberg, W.; Penberthy, W.K.; Hill, K.L.; Morse, M.L. Phosphotransferase system of *Staphylococcus aureus*: Its requirement for the accumulation and metabolism of galactosides. *J. Bacteriol.* **1969**, *99*, 383–388. [CrossRef]
12. Postma, P.W.; Lengeler, J.W.; Jacobson, G.R. Phosphoenolpyruvate:carbohydrate phosphotransferase systems of bacteria. *Microbiol. Rev.* **1993**, *57*, 543. [CrossRef] [PubMed]
13. Deutscher, J.; Küster, E.; Bergstedt, U.; Charrier, V.; Hillen, W. Protein kinase-dependent HPr/CcpA interaction links glycolytic activity to carbon catabolite repression in Gram-positive bacteria. *Mol. Microbiol.* **1995**, *15*, 1049–1053. [CrossRef] [PubMed]
14. Deutscher, J.; Francke, C.; Postma, P.W. How Phosphotransferase System-Related Protein Phosphorylation Regulates Carbohydrate Metabolism in Bacteria. *Microbiol. Mol. Biol. Rev.* **2006**, *70*, 939. [CrossRef]
15. Seidl, K.; Stucki, M.; Ruegg, M.; Goerke, C.; Wolz, C.; Harris, L.; Berger-Bächi, B.; Bischoff, M. *Staphylococcus aureus* CcpA affects virulence determinant production and antibiotic resistance. *Antimicrob. Agents Chemother.* **2006**, *50*, 1183–1194. [CrossRef] [PubMed]
16. Seidl, K.; Müller, S.; François, P.; Kriebitzsch, C.; Schrenzel, J.; Engelmann, S.; Bischoff, M.; Berger-Bächi, B. Effect of a glucose impulse on the CcpA regulon in *Staphylococcus aureus*. *BMC Microbiol.* **2009**, *9*, 95. [CrossRef]
17. Reed, J.M.; Olson, S.; Brees, D.F.; Griffin, C.E.; Grove, R.A.; Davis, P.J.; Kachman, S.D.; Adamec, J.; Somerville, G.A. Coordinated regulation of transcription by CcpA and the *Staphylococcus aureus* two-component system HptRS. *PLoS ONE* **2018**, *13*, e0207161. [CrossRef] [PubMed]
18. Seidl, K.; Bischoff, M.; Berger-Bächi, B. CcpA mediates the catabolite repression of *tst* in *Staphylococcus aureus*. *Infect. Immun.* **2008**, *76*, 5093–5099. [CrossRef] [PubMed]
19. Seidl, K.; Goerke, C.; Wolz, C.; Mack, D.; Berger-Bächi, B.; Bischoff, M. *Staphylococcus aureus* CcpA affects biofilm formation. *Infect. Immun.* **2008**, *76*, 2044–2050. [CrossRef]
20. Halsey, C.R.; Lei, S.; Wax, J.K.; Lehman, M.K.; Nuxoll, A.S.; Steinke, L.; Sadykov, M.; Powers, R.; Fey, P.D. Amino Acid Catabolism in *Staphylococcus aureus* and the Function of Carbon Catabolite Repression. *mBio* **2017**, *8*. [CrossRef]
21. Nuxoll, A.S.; Halouska, S.M.; Sadykov, M.R.; Hanke, M.L.; Bayles, K.W.; Kielian, T.; Powers, R.; Fey, P.D. CcpA regulates arginine biosynthesis in *Staphylococcus aureus* through repression of proline catabolism. *PLoS Pathog.* **2012**, *8*, e1003033. [CrossRef] [PubMed]

22. Cramton, S.E.; Gerke, C.; Schnell, N.F.; Nichols, W.W.; Götz, F. The intercellular adhesion (*ica*) locus is present in *Staphylococcus aureus* and is required for biofilm formation. *Infect. Immun.* **1999**, *67*, 5427–5433. [CrossRef] [PubMed]
23. Mann, E.E.; Rice, K.C.; Boles, B.R.; Endres, J.L.; Ranjit, D.; Chandramohan, L.; Tsang, L.H.; Smeltzer, M.S.; Horswill, A.R.; Bayles, K.W. Modulation of eDNA release and degradation affects *Staphylococcus aureus* biofilm maturation. *PLoS ONE* **2009**, *4*, e5822. [CrossRef] [PubMed]
24. Bischoff, M.; Wonnenberg, B.; Nippe, N.; Nyffenegger-Jann, N.J.; Voss, M.; Beisswenger, C.; Sunderkötter, C.; Molle, V.; Dinh, Q.T.; Lammert, F.; et al. CcpA Affects Infectivity of *Staphylococcus aureus* in a Hyperglycemic Environment. *Front. Cell. Infect. Microbiol.* **2017**, *7*, 172. [CrossRef]
25. Liao, X.; Yang, F.; Wang, R.; He, X.; Li, H.; Kao, R.Y.; Xia, W.; Sun, H. Identification of Catabolite Control Protein A from *Staphylococcus aureus* as a Target of Silver Ions. *Chem. Sci.* **2017**, *8*, 8061–8066. [CrossRef]
26. Hartmann, T.; Zhang, B.; Baronian, G.; Schulthess, B.; Homerova, D.; Grubmüller, S.; Kutzner, E.; Gaupp, R.; Bertram, R.; Powers, R.; et al. Catabolite control protein E (CcpE) is a LysR-type transcriptional regulator of tricarboxylic acid cycle activity in *Staphylococcus aureus*. *J. Biol. Chem.* **2013**, *288*, 36116–36128. [CrossRef]
27. Giachino, P.; Engelmann, S.; Bischoff, M. Sigma(B) activity depends on RsbU in *Staphylococcus aureus*. *J. Bacteriol.* **2001**, *183*, 1843–1852. [CrossRef]
28. Monk, I.R.; Shah, I.M.; Xu, M.; Tan, M.W.; Foster, T.J. Transforming the untransformable: Application of direct transformation to manipulate genetically *Staphylococcus aureus* and *Staphylococcus epidermidis*. *mBio* **2012**, *3*. [CrossRef]
29. Leibig, M.; Krismer, B.; Kolb, M.; Friede, A.; Gotz, F.; Bertram, R. Marker removal in staphylococci via Cre recombinase and different lox sites. *Appl. Environ. Microbiol.* **2008**, *74*, 1316–1323. [CrossRef]
30. Duthie, E.S. Variation in the antigenic composition of staphylococcal coagulase. *J. Gen. Microbiol.* **1952**, *7*, 320–326. [CrossRef]
31. Kreiswirth, B.N.; Lofdahl, S.; Betley, M.J.; O'Reilly, M.; Schlievert, P.M.; Bergdoll, M.S.; Novick, R.P. The toxic shock syndrome exotoxin structural gene is not detectably transmitted by a prophage. *Nature* **1983**, *305*, 709–712. [CrossRef]
32. Iordanescu, S.; Surdeanu, M. Two Restriction and Modification Systems in *Staphylococcus aureus* NCTC8325. *Microbiology* **1976**, *96*, 277–281. [CrossRef]
33. Gaupp, R.; Wirf, J.; Wonnenberg, B.; Biegel, T.; Eisenbeis, J.; Graham, J.; Herrmann, M.; Lee, C.Y.; Beisswenger, C.; Wolz, C.; et al. RpiRc Is a Pleiotropic Effector of Virulence Determinant Synthesis and Attenuates Pathogenicity in *Staphylococcus aureus*. *Infect. Immun.* **2016**, *84*, 2031–2041. [CrossRef]
34. Beenken, K.E.; Blevins, J.S.; Smeltzer, M.S. Mutation of *sarA* in *Staphylococcus aureus* limits biofilm formation. *Infect. Immun.* **2003**, *71*, 4206–4211. [CrossRef] [PubMed]
35. Zhu, Y.; Nandakumar, R.; Sadykov, M.R.; Madayiputhiya, N.; Luong, T.T.; Gaupp, R.; Lee, C.Y.; Somerville, G.A. RpiR homologues may link *Staphylococcus aureus* RNAIII synthesis and pentose phosphate pathway regulation. *J. Bacteriol.* **2011**, *193*, 6187–6196. [CrossRef] [PubMed]
36. Rupp, M.E.; Ulphani, J.S.; Fey, P.D.; Bartscht, K.; Mack, D. Characterization of the importance of polysaccharide intercellular adhesin/hemagglutinin of *Staphylococcus epidermidis* in the pathogenesis of biomaterial-based infection in a mouse foreign body infection model. *Infect. Immun.* **1999**, *67*, 2627–2632. [CrossRef]
37. Ferreira, M.T.; Manso, A.S.; Gaspar, P.; Pinho, M.G.; Neves, A.R. Effect of oxygen on glucose metabolism: Utilization of lactate in *Staphylococcus aureus* as revealed by in vivo NMR studies. *PLoS ONE* **2013**, *8*, e58277. [CrossRef] [PubMed]
38. Vitko, N.P.; Grosser, M.R.; Khatri, D.; Lance, T.R.; Richardson, A.R. Expanded glucose import capability affords *Staphylococcus aureus* optimized glycolytic flux during infection. *mBio* **2016**, *7*. [CrossRef] [PubMed]
39. Gill, S.R.; Fouts, D.E.; Archer, G.L.; Mongodin, E.F.; Deboy, R.T.; Ravel, J.; Paulsen, I.T.; Kolonay, J.F.; Brinkac, L.; Beanan, M.; et al. Insights on evolution of virulence and resistance from the complete genome analysis of an early methicillin-resistant *Staphylococcus aureus* strain and a biofilm-producing methicillin-resistant *Staphylococcus epidermidis* strain. *J. Bacteriol.* **2005**, *187*, 2426–2438. [CrossRef] [PubMed]
40. Deutscher, J.; Herro, R.; Bourand, A.; Mijakovic, I.; Poncet, S. P-Ser-HPr—a link between carbon metabolism and the virulence of some pathogenic bacteria. *Biochim. Biophys. Acta* **2005**, *1754*, 118–125. [CrossRef]
41. Pullen, K.; Rajagopal, P.; Branchini, B.R.; Huffine, M.E.; Reizer, J.; Saier, M.H., Jr.; Scholtz, J.M.; Klevit, R.E. Phosphorylation of serine-46 in HPr, a key regulatory protein in bacteria, results in stabilization of its solution structure. *Protein Sci. A Publ. Protein Soc.* **1995**, *4*, 2478–2486. [CrossRef]
42. Thapar, R.; Nicholson, E.M.; Rajagopal, P.; Waygood, E.B.; Scholtz, J.M.; Klevit, R.E. Influence of N-cap mutations on the structure and stability of *Escherichia coli* HPr. *Biochemistry* **1996**, *35*, 11268–11277. [CrossRef] [PubMed]
43. Sadykov, M.R.; Hartmann, T.; Mattes, T.A.; Hiatt, M.; Jann, N.J.; Zhu, Y.; Ledala, N.; Landmann, R.; Herrmann, M.; Rohde, H.; et al. CcpA coordinates central metabolism and biofilm formation *in Staphylococcus epidermidis*. *Microbiology* **2011**, *157*, 3458–3468. [CrossRef] [PubMed]
44. Hartmann, T.; Baronian, G.; Nippe, N.; Voss, M.; Schulthess, B.; Wolz, C.; Eisenbeis, J.; Schmidt-Hohagen, K.; Gaupp, R.; Sunderkötter, C.; et al. The catabolite control protein E (CcpE) affects virulence determinant production and pathogenesis of *Staphylococcus aureus*. *J. Biol. Chem.* **2014**, *289*, 29701–29711. [CrossRef]
45. Ding, Y.; Liu, X.; Chen, F.; Di, H.; Xu, B.; Zhou, L.; Deng, X.; Wu, M.; Yang, C.G.; Lan, L. Metabolic sensor governing bacterial virulence in *Staphylococcus aureus*. *Proc. Natl. Acad. Sci. USA* **2014**, *111*, E4981–E4990. [CrossRef]

46. Balasubramanian, D.; Ohneck, E.A.; Chapman, J.; Weiss, A.; Kim, M.K.; Reyes-Robles, T.; Zhong, J.; Shaw, L.N.; Lun, D.S.; Ueberheide, B.; et al. *Staphylococcus aureus* coordinates leukocidin expression and pathogenesis by sensing metabolic fluxes via RpiRc. *mBio* **2016**, *7*. [CrossRef] [PubMed]
47. Montgomery, C.P.; Boyle-Vavra, S.; Roux, A.; Ebine, K.; Sonenshein, A.L.; Daum, R.S. CodY deletion enhances in vivo virulence of community-associated methicillin-resistant *Staphylococcus aureus* clone USA300. *Infect. Immun.* **2012**, *80*, 2382–2389. [CrossRef]
48. Ha, J.H.; Hauk, P.; Cho, K.; Eo, Y.; Ma, X.; Stephens, K.; Cha, S.; Jeong, M.; Suh, J.Y.; Sintim, H.O.; et al. Evidence of link between quorum sensing and sugar metabolism in *Escherichia coli* revealed via cocrystal structures of LsrK and HPr. *Sci. Adv.* **2018**, *4*, eaar7063. [CrossRef]

 *microorganisms*

Article

Role of SrtA in Pathogenicity of *Staphylococcus lugdunensis*

Muzaffar Hussain [1] , **Christian Kohler** [2] **and Karsten Becker** [1,2,*]

1   Institute of Medical Microbiology, University Hospital Münster, 48149 Münster, Germany; muzaffa@uni-muenster.de

2   Friedrich Loeffler-Institute of Medical Microbiology, University Medicine Greifswald, 17475 Greifswald, Germany; christian.kohler@med.uni-greifswald.de

*   Correspondence: karsten.becker@med.uni-greifswald.de; Tel.: +49-3834-86-5560

Received: 6 November 2020; Accepted: 9 December 2020; Published: 11 December 2020

Abstract: Among coagulase-negative staphylococci (CoNS), *Staphylococcus lugdunensis* has a special position as causative agent of aggressive courses of infectious endocarditis (IE) more reminiscent of IEs caused by *Staphylococcus aureus* than those by CoNS. To initiate colonization and invasion, bacterial cell surface proteins are required; however, only little is known about adhesion of *S. lugdunensis* to biotic surfaces. Cell surface proteins containing the LPXTG anchor motif are covalently attached to the cell wall by sortases. Here, we report the functionality of *Staphylococcus lugdunensis* sortase A (SrtA) to link LPXTG substrates to the cell wall. To determine the role of SrtA dependent surface proteins in biofilm formation and binding eukaryotic cells, we generated SrtA-deficient mutants (Δ*srtA*). These mutants formed a smaller amount of biofilm and bound less to immobilized fibronectin, fibrinogen, and vitronectin. Furthermore, SrtA absence affected the gene expression of two different adhesins on transcription level. Surprisingly, we found no decreased adherence and invasion in human cell lines, probably caused by the upregulation of further adhesins in Δ*srtA* mutant strains. In conclusion, the functionality of *S. lugdunensis* SrtA in anchoring LPXTG substrates to the cell wall let us define it as the pathogen's housekeeping sortase.

Keywords: *Staphylococcus lugdunensis*; sortase A; surface proteins; LPXTG

---

1. Introduction

As known for *Staphylococcus aureus* and several coagulase-negative staphylococci (CoNS), *Staphylococcus lugdunensis* is also part of the human microbiota colonizing miscellaneous skin surface habitats [1–4]. Infections due to this opportunistic pathogen resemble those caused by *S. aureus* rather than "classical" CoNS infections [5]. In particular, aggressive and highly destructive courses of native and prosthetic valve infectious endocarditis (IE) have been reported [6]. Despite its clinical impact, only a few factors contributing to the pathogenicity of *S. lugdunensis* have been described, including a fibrinogen binding protein (Fbl), a von Willebrand-factor binding protein (vWbl), and a multifunctional autolysin (AtlL) [7–11].

To initiate invasive infections, staphylococcal cells irrespective of the species must adhere to cells of the host tissue or to the extracellular matrix (ECM) [12]. This complex multifactorial process is mediated by strong interactions of cell wall-anchored (CWA) proteins with host structures. In *S. aureus*, four distinct categories of CWA proteins have been identified, and the microbial surface components recognizing adhesive matrix molecules (MSCRAMMs) were elucidated as the largest group [13,14]. CWA proteins are characterized by a sorting signal containing a carboxy-terminal LPXTG anchor motif [15]. In *S. aureus* and many other Gram-positive species, the LPXTG-sorting signal is cleaved by the membrane-bound transpeptidase sortase. Sortase A (SrtA) of *S. aureus* is a 206 amino acid peptide that catalyzes a transpeptidation process [16–19] and is considered the housekeeping sortase [20]. The cleavage between the threonine and glycine residues is followed by an amide linkage between the carboxyl group of threonine and the amino group of a pentaglycine cross-bridge of the cell membrane-attached peptidoglycan [16,19]. Mutants lacking the *srtA* gene are defective in the cell wall anchoring of LPXTG proteins [16,17,21]. Thus, cells are unable to anchor cell surface proteins of *S. aureus* important for the adherence to eukaryotic cell structures. In consequence, Δ*srtA* mutants were attenuated to establish infections as shown in a murine model [22]. In addition to SrtA, a further *S. aureus* sortase isoform, sortase B (SrtB), has been described [23]. More recently, a SrtA-deficient mutant of *S. lugdunensis* was significantly less virulent than the parental strain in a rat IE model [24]. Therefore, the use of SrtA inhibitors might represent a promising anti-virulence therapy strategy to disrupt the anchoring of bacterial surface proteins, which are critical for the pathogen´s adherence to the host.

Here, we report the functionality of *S. lugdunensis* SrtA to anchor LPXTG substrates to the cell wall and defined it as the pathogen´s housekeeping sortase. Generating *srtA*-deficient mutants, we determined the role of *S. lugdunensis* SrtA-dependent surface proteins in biofilm formation and invasion of eukaryotic cells. Finally, we confirmed the influence of a functional SrtA on the gene expression of further LPXTG proteins.

2. Materials and Methods

2.1. Bacterial Strains and Culture Media

Lysogeny broth (LB) or agar were used for cultivation of *Escherichia coli* and staphylococci were cultivated in tryptic soy broth (TSB) or agar (TSA) (Difco, Detroit, Detroit, MI, USA), brain heart infusion (BHI) broth or agar (Merck, Darmstadt, Darmstadt, Germany) and Mueller Hinton (MH) broth or agar (Mast, Merseyside, Bootle, UK). Antibiotics were added to MH agar in appropriate amounts (ampicillin, 100 µg/mL, Sigma; erythromycin, 10 µg/mL Serva; and chloramphenicol, 10 µg/mL, Serva, Heidelberg, Germany) for selection of resistance in *E. coli* or *S. lugdunensis*. All bacterial strains and plasmids used in this study are presented in Table 1.

Table 1. Bacterial strains used in this study.

Strains	Relevant Genotype or Plasmid	Properties	Reference or Source
	S. lugdunensis strains		
Sl48		Clinical isolate	Germany [b]
Sl44		Clinical isolate	Germany [b]
Mut7	Sl48 *srtA::EmR*	Sl48 deficient in sortase-A	This study
Mut47	Sl44 *srtA::EmR*	Sl44 deficient in sortase-A	This study
SL241		Clinical isolate	Germany [b]
SL253		Clinical isolate	Germany [b]
	S. aureus strain		
S. aureus Cowan 1 (ATCC 12598)		Reference isolate from septic arthritis	ATCC
	E. coli strains		
DH5α	*supE44ΔlacU169 (φ80 lacZΔM15) hsdR17 recA1 end A1 gyrA96 thi-1 relA1*	Cloning host	Stratagene
TG1	*supE hsdΔ5 thiΔ(lac-proAB) F′(traD36 proAB⁺ lacIᑫ lacZΔM15)*	Cloning host	Stratagene
DH5α (pBT37)	pBT9*atlL::EmR*	Shuttle vector pBT9 containing *atlL::EmR*	This study
	Eukaryotic strains		
EA.hy 926 cells			[25]
A549 fibroblast			[26]
Human bladder carcinoma cell line 5637			[27]

[b] kindly provided by F. Szabados and S. Gatermann (Bochum, Germany).

2.2. Growth Characteristics

Bacteria were grown in BHI overnight followed by dilution of the culture in 100 mL fresh BHI in 500 mL flask to an optical density 578 nm (OD_{578}) of 0.05. Flasks were incubated at 160 rpm at 37 °C and the OD_{578} were determined hourly for a period of 10 h followed by final sampling after 24 h to establish growth curves. For growth experiments in little volumes, bacteria were grown in 5 mL BHI in a glass tube for 18 h at 37 °C with shaking at 160 rpm. Next day, OD_{578} were measured against un-inoculated BHI after vortex of growth.

2.3. Characterization of Agglutination

The Pastorex Staph Plus (Sanofi Diagnostic Pasteur, S.A., Marnes la Coquette, France), a rapid agglutination test for the simultaneous detection of the *S. aureus* fibrinogen affinity antigen (clumping factor), protein A, and capsular polysaccharides were used to differentiate wild type and mutant strains.

2.4. Biofilm Formation

Bacteria were grown in 5 mL BHI, TSB, BHI + 0.5% glucose or TSB + 0.5% glucose in glass test tubes for 8 h at 37 °C with shaking at 160 rpm. Afterwards, the cells were diluted 1:100 with the same type of fresh medium and 100 µL were added to wells in quadruplicate in a 96 well microtiterplate. Plates were incubated at 37 °C without shaking overnight. Next day bacteria were removed, plates were washed with PBS, and the biofilm bacteria mix were fixed with ice cold methanol for 10 min at −20 °C. After washing once with PBS, the adhered biofilm bacteria were stained with crystal violet for 10 min at room temperature. Excess stain was removed by washing 3x with PBS. The adhered bacteria were brought into solution by the addition of 100 µL of 35% acetic acid to each well. Finally, the plates were read at OD_{595} in the iMark Microplate Reader (Bio Rad).

2.5. DNA Manipulations and Transformations

Staphylococcal cells were lysed with recombinant lysostaphin (20 U/mL, Applied Micro, New York, NY, USA). Genomic DNA isolated by using QIAamp DNA Mini Blood Kit (Qiagen, Hilden, Germany) and plasmid DNA were prepared using the Qiagen Plasmid Mini kit (Qiagen). DNA fragments were isolated from agarose gels using the QIAEX II Gel extraction kit (Qiagen).

2.6. Construction of a srtA-Deficient Mutant

The method for the generation of a *srtA* lacking mutant was same as described before [7]. In brief, a primer set 1 (SrtA1FH, and SrtA1RE,) and primer set 2 (SrtA2FE, and SrtAl2RB,) were used to amplify PCR products of 809 and 793 bp, respectively, from genomic DNA of *S. lugdunensis* strain Sl48. The PCR products were ligated into the shuttle vector pBT9 and the ligation mixes were transformed into *E. coli* TG1 cells followed by incubation on LB plates containing ampicillin. A clone containing *srtA* gene as an insert was designated as pBT*srtA*. The restriction analysis and sequence data of clone pBT*srtA* confirmed *srtA* gene as an insert. A 1,479 bp *ermB* fragment was PCR-amplified using primers ermbF, and ermbR from the plasmid pEC4 containing the staphylococcal transposon Tn551 in a ClaI restriction site [28]. The *ermB* primers were designed from NCBI accession # AF239773. The *ermB* cassette was restricted with EcoRI and ligated into the EcoRI-restricted pBT*srtA*. The freshly prepared *E. coli* TG1 cells were transformed with the ligation mix and cultivated on ampicillin- and erythromycin-containing LB plates. Clones with the plasmid conferring resistance to both antibiotics were designated as pBT*srtAE* and were selected for further work. For the construction of the *srtA* allelic replacement mutant, protoplasts of *S. lugdunensis* strains Sl44 and Sl48 transformed with pBT*srtAE* as described by Palma et al. [29]. Chloramphenicol-sensitive and erythromycin-resistant colonies were detected by replica plating protocol onto plates containing chloramphenicol or erythromycin at 37 °C overnight. Clones that were sensitive for chloramphenicol and resistant to erythromycin were designated Mut7 (wild-type strain Sl44) or Mut47 (Wild-type strain Sl48) and were selected for further analyses. Correct insertion of the *ermB* was confirmed by using srtA1FH, srtA2RB, ermBF and ermBR primers.

2.7. The Complementation of a srtA-Deficient Mutant

A PCR product of *srtA* gene including the ribosomal binding site was PCR amplified using primer set (SrtA1FB and SrtA2RH) from genomic DNA of *S. lugdunensis* Sl48. The PCR product was restricted with HindIII and BamHI and ligated into the pCU1 plasmid. Freshly prepared *E. coli* TG1 cells were transformed with ligation mix. A plasmid containing *srtA* as an insert was designated pCU*srtA*. The isolated plasmid pCU*srtA* transformed by protoplast method into Mut7 and Mut47. Transformants were grown on TSA plates containing 10 μg of chloramphenicol and 10 μg of erythromycin per mL. Clones expressing SrtA were designated as Mut7C and Mut47C.

2.8. Cell Protein Preparations

Briefly, bacteria were grown overnight in BHI at 37 °C with shaking (160 rpm) and cells were harvested by centrifugation and washed twice with phosphate buffered saline. Then the cell surface proteins from PBS washed bacterial cells pellets were extracted by following methods: In the first method, bacterial cell surface proteins were generated by the LiCl method as described earlier [30]. In a second method we used the hydroxylamine hydrochloride method as described by Ton-That et al. [17]. In brief, bacteria were grown in two tubes containing 5 mL BHI overnight at 37 °C and 160 rpm. After centrifugation of the cells (4000 rpm 4 °C for 10 min), 1.5 mL of 50 mM Tris HCl pH 7.0 was added to the control pellet. To the other pellet we added 1.5 mL 100 mM hydroxylamine HCl in 50 mM Tris HCl pH 7.0. Both sets were stirred at 600 rpm at 37 °C in a heating block for 1 h. After centrifugation of the cells (13,000 rpm 4 °C for 10 min, the proteins in both supernatants were precipitated by addition of TCA to a final concentration of 10%. After centrifugation of the precipitated proteins (13,000 rpm 4 °C for 10 min), 200 μL of a 2× standard Laemmli SDS Page Sample buffer were added to the pellets and heated at 95 °C for 3 min. Samples were ready to load on SDS page gel. Whole cell lysates were prepared by the suspensions of cell pellets solved in 50 mM Tris/HCl pH 8.0 containing 15 mM NaCl. Lysostaphin (20 U/mL) and protease inhibitor cocktail (Sigma) were added and the mixture was incubated at 37 °C for 30 min. Then DNase was added to break DNA and, after centrifugation, liquid supernatants were used as whole cell lysate. For the generation of cell wall, cytoplasmic, and membrane proteins, *S. lugdunensis* were grown overnight in BHI at 37 °C with shaking (160 rpm) and cells were harvested by centrifugation and washed twice with phosphate buffered saline, and adjusted to an optical density at 600 nm (OD_{600}) of 20. A 1-mL portion of the bacterial suspension was pelleted and resuspended in 200 μL of digestion buffer (50 mM Tris-HCl, 20 mM $MgCl_2$, 30% [wt/vol] raffinose; pH 7.5) containing complete mini-EDTA-free protease inhibitors (Roche). Cell wall proteins were solubilized by digestion with lysostaphin (500 μg/mL) at 37 °C for 30 min. Protoplasts were harvested by centrifugation (5000× *g*, 15 min) and the supernatant was retained as the cell wall fraction. Protoplast pellets were washed once in digestion buffer, sedimented (5000× *g*, 15 min), and resuspended in ice-cold lysis buffer (50 mM Tris-HCl [pH 7.5]) containing protease inhibitors and DNase (80 μg/mL). Protoplasts were lysed on ice by vortexing. Complete lysis was confirmed by phase-contrast microscopy. The membrane fractions were obtained by centrifugation at 18,500× *g* for 1 h at 4 °C. The pellets (membrane fraction) were washed once and resuspended in ice-cold lysis buffer. Cell wall fractions and protoplast suspensions were centrifuged under the same conditions and the pellets were resuspended in 200 μL of lysis buffer. The liquid supernatant from protoplast suspension retained as cytosolic fraction.

2.9. SDS-PAGE and Ligand Overlay Analysis

The prepared cell surface proteins were separated in SDS-PAGE mini gel approaches. For Western ligand blot analysis, fibronectin (Fn) (Chemicon, Temecula, CA, USA), fibrinogen (Fg) (Calbiochem, San Diego, CA, USA), collagen type I (Cn) (Sigma; Sigma product #7774) or vitronectin (Vn) were purified by the method of Yatohgo et al. [28] and labeled with biotin (Roche, Mannheim, Germany). The cell surface proteins separated by SDS-PAGE were transferred electrophoretically (Trans-blot SD, Bio-Rad, Munich, Germany) onto nitrocellulose membranes (Schleicher and Schüll, Dassel, Germany) and were blocked with 3% BSA (bovine serum albumin fraction V, Sigma). The nitrocellulose membranes with blotted proteins were exposed to biotinylated ligands, treated with avidin alkaline phosphatase and subsequently bands were revealed using NBT (Nitrotetrazolium Blue chloride) and BCIP (5-Brom-4-chlor-3-indoxylphosphat) as recommended by the manufacturer´s protocol (Bio-Rad).

2.10. Expression of Recombinant Sortase-A

The *srtA* gene was PCR amplified from genomic DNA of *S. lugdunensis* using the primer set srtAF and srtAR. The *srtA* gene was ligated into the pQE30 expression vector and the vector was transformed into *E. coli* TG1 cells. *E. coli* TG1 cells were cultivated in LB medium plus 100 µg/mL ampicillin at 37 °C and 150 rpm and IPTG (Isopropyl-β-D-thiogalactopyranosid, Sigma) 1 mM per mL was added at OD_{578} 0.5 to induce the expression of SrtA. *E. coli* cells were lysed by lysozyme (1 mg/mL lysis buffer) and SrtA was purified in a single step under native conditions using Ni-nitrilotriacetic acid (NINTA) resin column according to the manufacturer's recommendation (Qiagen).

2.11. Polyclonal Antibodies

Polyclonal antibodies against the recombinant SrtA were raised commercially (Genosphere Biotechnologies, Paris, France) in two rabbits by applying standard procedures of 70 days with 4 immunizations. The IgG fraction from the crude antiserum was obtained on protein-A column (Pierce, Rockford, IL, USA). In IgG preparations, naturally occurring anti-staphylococcal antibodies were removed by mixing the sera with 10 volumes of LiCl cell surface extracts from strain Mut47, which does not produce the SrtA, and immune-complexes were partially removed by centrifugation at 14,000 rpm at 4 °C for 15 min. For control Western immunoblotting, cell proteins of the wild type were separated on a SDS-Page, transferred onto a nitrocellulose membrane (Schleicher and Schuell, Dassel, Germany), and probed with anti-SrtA sera. Bound rabbit immunoglobulin G was detected with alkaline phosphatase-conjugated goat anti-rabbit immunoglobulins (Dako Diagnostika, Hamburg, Germany) with NBT/BCIP color reaction (Bio-Rad).

2.12. ELISA Adherence assays

The employed ELISA adherence assays of staphylococci to ECM proteins and host cells were done as described earlier [7].

2.13. Cell Culture and Flow Cytometric Invasion Assay

For flow cytometric invasion assay, three types of host cells were used (Table 1). The human endothelial cell line, Ea.hy926 [31], was grown in DMEM (Sigma-Aldrich) supplemented with 10% fetal calf serum (FCS).The fibroblasts (human fetal lung cells) were cultivated as described earlier [31,32]. Another epithelial cell line, the human bladder carcinoma cell line 5637 (ATCC HTB-9™), which secretes several functionally active cytokines, was also used in this study and maintained as described by Quentmeier et al. [33]. Fluorescein-5-isothiocyanate (FITC)-labeling of bacterial cells was performed as described elsewhere [34,35] and labeled cells were used within 24 h. The flow cytometric invasion assay was performed as described previously with minor modifications, such as the addition of propidium iodide just before samples were analyzed on a FacsCALIBUR™ (BD Bioscience) [36]. Results were normalized according to the mean fluorescence intensity of the respective bacterial preparation, as determined by flow cytometry. The invasiveness of the laboratory *S. aureus* strain Cowan I was set as 100% and the results are shown as means ± SEM of three independent experiments performed in duplicates.

2.14. Sortase Inhibitor PVS

5 mL BHI in a glass tube was inoculate with *S. lugdunensis* and grown overnight at 37 °C with 160 rpm. Next day, cultures were diluted to an OD_{595} of 0.01. 180 μL of diluted culture was mixed with 20 μL of 10× concentrated phenyl vinyl sulfone (PVS) in a microtiter plate well. Plates were incubated at 37 °C for 18 h. Growth was determined in a Biorad reader at absorbance 595 nm. The lowest concentration that inhibited the cell growth was considered to be the MIC. For determination of effects of PVS on binding of bacteria to Fn and Fg, microtiter plate wells were coated with Fn and Fg separately. Overnight cultures were diluted to an OD_{595} of 0.01. 180 μL of diluted culture were mixed with 20 μL of concentrated phenyl vinyl sulfone (PVS) to the respective concentrations in a microtiter plate well. Plates were incubated at room temperature for 1 h. Adherence was determined in a Biorad reader at absorbance of 595 nm.

2.15. Hydroxylaminolysis of LPXTG Peptide

Reactions were performed in 260 μL volume containing 50 mM Hepes buffer pH 7.5, 5 mM $CaCl_2$, and 10 μM LPETG fluorescent labeled peptides (DABCYL-LPETG-EDANS). The mixture was heated at 95 °C for 5 min. Then 10 μL of staphylococcal LiCl-extracts or 10 μM recombinant sortase-A, 5 μM of the sortase inhibitor p-hydroxymecuribenzoic acid (pHMB), or 10 mM DTT were added. Reactions were incubated for 1 h at 37 °C and then analyzed fluorometrically at 350 nm for excitations and 495 nm for recordings. Experiments were performed in triplicates.

2.16. Real-Time Reverse-Transcription PCR (qtRT-PCR)

For RNA isolation, *S. lugdunensis* was grown in BHI for 18 h at 37 °C with 160 rpm. After bacterial RNA isolation, real-time amplification and transcripts quantification was done as described earlier [37]. Primer sequences are given in Table 2.

Table 2. Primer sequences used in this study.

Primer	Sequence (5′–3′)	Reference
srtA1FH	AAAAAGCTTTAAGAAAGCTAAAAAAATGACATAGTTG	This study
srtA1RE	AAAGAATTCCTCCAATAATGGTCATCAATTGGTTTGTCC	This study
srtA2FE	AAGAATTCTATTTATAGCAGAACAGATTAAATAATTGTAG	This study
srtA2RB	AAAGGATCCCATCTGAGTCAA GACTACTAGCAAGTGG	This study
Ery-EF,	ATATATCGATTAGGGACCTCTTTAGC	[28]
Ery-ER	ATATATCGATATCATGAGTATTGTCCG	[28]
SrtA1FH	AAAAAGCTTTAAGAAAGCTAAAAAAATGACATAGTTG	This study
SrtA2RB	AAAGGATCCCATCTGAGTCAAGACTACTAGCAAGTGG	This study
srtAF	CTCGGATCCAAACCTCATATTGATAGTTATTTACATGAC	This study
srtAR	CTCGGTACCTTATTTAATCTGTTCTGCTATAAATATTTTACGC	This study
RTFblF	GAAGCAACAACGCAGAACAA	[38]
RTFblR	TGCTTGTGCCTCGCTATTTA	[38]
RT16SF	CAGCTCGTGTCGTGAGATGT	[38]
RT16SR	TAGCACGTGTGTAGCCCAAA	[38]
RTvWbF	GGACCAGGTGAAGGTGATGT	This study
RTvWbR	GCCGCTGATTTTCGTGTAAT	This study

2.17. Statistical Analysis

Using GraphPad Instat3, the statistical significance of the results was analyzed by ANOVA in combination with Bonferroni´s post test (compare all pairs of column or compare selected pairs of column) or Dunnett´s post test (compare all columns vs. control). Differences with p values ≤ 0.05 were considered as significant and are indicated with asterisks: * ($p \leq 0.05$), ** ($p \leq 0.01$), and *** ($p \leq 0.001$).

3. Results

3.1. Sortase A-Dependent Proteins

A comparative bioinformatic analysis of publicly available *S. lugdunensis* genomes identified the presence of only 11 sortase-A dependent proteins (Table 3) [39]. All 11 MSCRAMMs were found to be highly conserved in the *S. lugdunensis* strains as genomic data from strain HKU09-01, M23590, VCU139 and N920143 revealed [39]. However, there exist only minor differences in the number of repeats within the stalk regions that in turns affect the length of mature proteins [39].

Table 3. Known *S. lugdunensis* putative sortase-A mediated cell surface proteins with relevant properties [39].

Genetic Identifiers (GN)	Annotation	Cleavage Motif	Size (aa)	Predicted Protein Size (kDa)	NCBI BLAST Hit (Protein [1], Strain [2], Length [3])
SLUG_00890 SLGD_00061	IsdB	LPATG	690	76.9	Surface protein SasI, HKU09-01
SLUG_00930 SLGD_00065 SasE	IsdJ	LPNTG	646	71.5	Cell surface protein IsdA, HKU09-01 LPXTG cell wall surface anchor protein, M23590
SLUG_02990 SLGD_00301	SlsF	LPASG	659	73.4	Predicted cell-wall-anchored protein SasF, HKU09-01
SLUG_03480 SLGD_00351	SlsA	LPDTG	1930	207.3	Cell wall associated biofilm protein, HKU09-01, 3799
SLUG_03490 SEVCU139_1800 SLGD_00352	SlsD [4]	LPATG	1619	175.8	Putative serine-aspartate repeat protein F, VCU139, 2190 Putative uncharacterized protein, HKU09-01, 1136
SLUG_03850 SLGD_00389 HMPREF0790_1688	Slsc	LPETG	190	21	LPXTG protein, HKU09-01 Cell wall surface anchor family protein, M23590, 196
SLUG_04710 SEVCU139_1680 SLGD_00473	SlsE	LPETG	3459	364	Gram-positive signal peptide protein, VCU139, 2988 Hypothetical membrane protein, HKU09-01, 3232
SLUG_04760 SLGD_00478	SlsB	LPNTG	277	30.6	Putative uncharacterized protein, HKU09-01
SLUG_22400 SLGD_02322 bca PE	SlsG	LPDTG	2079	222.1	Putative uncharacterized protein, HKU09-01, 2886 C protein alpha-antigen, VCU139, 2031

Table 3. *Cont.*

Genetic Identifiers (GN)	Annotation	Cleavage Motif	Size (aa)	Predicted Protein Size (kDa)	NCBI BLAST Hit (Protein [1], Strain [2], Length [3])
SLUG_16350 HMPREF0790_0533 SLGD_01633	Fbl	LPKTG	881	94.2	Clumping factor A, M23590, 857 Clumping factor A (fragment), VCU139, 688 Methicillin-resistant surface protein, HKU09-01, 701
SLUG_23290 SLGD_02429	vWbF	LPETG	1869	209.4	Von Willebrand factor-binding protein, HKU09-01, 2194

[1] Different designation as given in annotation column. [2] Strain different from N920143. Annotation, cleavage motif, size (aa), and predicted protein size (kDa) columns are based on strain N920143. [3] Length if different from size (aa) column. [4] The *slsD* contains a nonsense codon located just 5′ to the region encoding LPXTG.

3.2. Alignment of Sortase A Sequences

In silico analyses identified homologs of *S. aureus srtA* in published genomes of *S. lugdunensis*. A nucleotide NCBI BLAST search identified SLGD_00472 (strain HKU09-01), *srtA* (strain N920143 and M23590), and SEVCU139_1681 (strain VCU139) as homologs of staphylococcal *srtA*. A ClustalW alignment of amino acid sequences was performed for pairwise comparison of SrtA between the four *S. lugdunensis* strains (N920143, HKU09-01, M23590, VCU139) and *S. aureus* SrtA. The SrtA sequences of four strains of *S. lugdunensis* (N920143, HKU09-01, M23590, VCU139) showed an intra-alignment score of 98–100%. The identity of SrtA sequences from four *S. lugdunensis* strains to *S. aureus* Newman was between 76–77% and revealed significant similarities. We identified conserved amino acid residues in the region corresponding to the calcium binding cleft of the *S. aureus* SrtA, such as the three glutamate residues Glu105, Glu108 and Glu166 as well as the aspartate residue Asp112, which is also highly conserved in the SrtA sequences of the *S. lugdunensis* strains (Figure 1) [40].

```
S.lug-N920143   MRKWTNQLMTIIGVILILVAIYLFAKPHIDSYLHDKDNSDKIENYDKTVSKANKNDK--- 57
S.lug-HKU09-01  MRKWTNQLMTIIGVILILVAIYLFAKPHIDSYLHDKDNSDKIENYDKTVSKANKNDK--- 57
S.lug-M23590    MRKWTNQLMTIIGVILILVAIYLFAKPHIDSYLHDKDNSDKIENYDKTVSKANKNDK--- 57
S.lug-VCU139    MKKWTNQLMTIIGVILILVAIYLFAKPHIDSYLHDKDNSDKIENYDKIVSKANKNDK--- 57
S.aur-Newman    MKKWTNRLMTIAGVVLILVAAYLFAKPHIDNYLHDKDKDEKIEQYDKNVKEQASKDKKQQ 60
                *:****:**** **:***** ********** ******:.:***:*** *.:  ..:**

S.lug-N920143   --PTIPKDKAEMAGYLRIPDADINEPVYPGPATPEQLNRGVSFAEEQESLDDQNIAIAGH 115
S.lug-HKU09-01  --PTIPKDKAEMAGYLRIPDADINEPVYPGPATPEQLNRGVSFAEEQESLDDQNIAIAGH 115
S.lug-M23590    --PTIPKDKAEMAGYLRIPDADINEPVYPGPATPEQLNRGVSFAEEQESLDDQNIAIAGH 115
S.lug-VCU139    --PTIPKDKAEMAGYLRIPDADINEPVYPGPATPEQLNRGVSFAEEQESLDDQNIAIAGH 115
S.aur-Newman    AKPQIPKDKSKVAGYIEIPDADIKEPVYPGPATPEQLNRGVSFAEENESLDDQNISIAGH 120
                *  *****::.***:.******:***********************:*******:****

S.lug-N920143   TYIGRPHYQFTNLKAAKKGSKVYFKVGNETREYKMTTIRDVNPDEIDVLDEHRGDKNRLT 175
S.lug-HKU09-01  TYIGRPHYQFTNLKAAKKGSKVYFKVGNETREYKMTTIRDVNPDEIDVLDEHRGDKNRLT 175
S.lug-M23590    TYIGRPHYQFTNLKAAKKGSKVYFKVGNETREYKMTTIRDVNPDDIDVLDEHRGDKNRLT 175
S.lug-VCU139    TYIGRPHYQFTNLKAAKKGSKVYFKVGNETREYKMTTIRDVNPDEIDVLDEHRGDKNRLT 175
S.aur-Newman    TFIDRPNYQFTNLKAAKKGSMVYFKVGNETRKYKMTSIRDVKPTDVGVLDEQKGKDKQLT 180
                *:*.**:*********** ********:****.****:****.*  ::.****::*..::**

S.lug-N920143   LITCDDYNEKTGVWEKRKIFIAEQIK 201
S.lug-HKU09-01  LITCDDYNEKTGVWEKRKIFIAEQIK 201
S.lug-M23590    LITCDDYNEKTGVWEKRKIFIAEQIK 201
S.lug-VCU139    LITCDDYNEKTGVWEKRKIFIAEQIK 201
S.aur-Newman    LITCDDYNEKTGVWEKRKIFVATEVK 206
                *******************:* .:.*
```

Figure 1. Multiple sequence alignments of SrtA sequences of four strains of *S. lugdunensis* (N920143, HKU09-01, M23590, and VCU139) and *S. aureus* Newman using CLUSTAL 2.1. The highly conserved amino acids glutamate (E) and aspartate (D) involved in binding of calcium ions are highlighted in yellow. Stars denote highly conserved amino acids.

3.3. Generation of Sortase-A Mutants

To generate knockout mutants of *srtA*, we used the homologous recombination method leading to an insertion of *ermB* cassette into the *srtA* genes of *S. lugdunensis* strains Sl48 and Sl44 (Figure 2A). The *srtA* deficient mutants Mut7 and Mut47 were analyzed by PCR amplification using primers that anneal to *ermB* or to sequences flanking the *srtA* gene. The *ermB* gene could be amplified from the chromosomal DNA of the *srtA* deficient mutants Mut7 and Mut47, but not from the isogenic parent strains Sl48 and Sl44 respectively. Amplifications with primers specific for sequences flanking *srtA* revealed the insertion of *ermB* into the *srtA* genes of strains Mut7 and Mut47. The final double cross-over *srtA::ermB* allelic replacement in the Δ*srtA* mutant was confirmed by PCR amplification. Figure 2 shows the map of *srtA* as well as the positions of the primer pairs and the expected sizes of the amplified PCR products. The expected PCR product size with primers SrtA1F and SrtA2R was 2.2 Kb for the wild type and is 3.1 kb for the mutant. The observed sizes for the increased PCR products could only be generated from the Δ*srtA* mutants with primers SrtA1F and SrtA2R, but not with DNA of the wild type strains. In addition, amplifications of the *ermB* cassette were only successful with the Δ*srtA* mutant, but not with wild type strain DNAs using primers Ery-FE and Ery-RE which confirmed the allelic replacement of the wild type *srtA* gene by *srtA::ermB* (Figure 2). Results are exemplary shown for the Sl48 parental and Mut47 mutant strain.

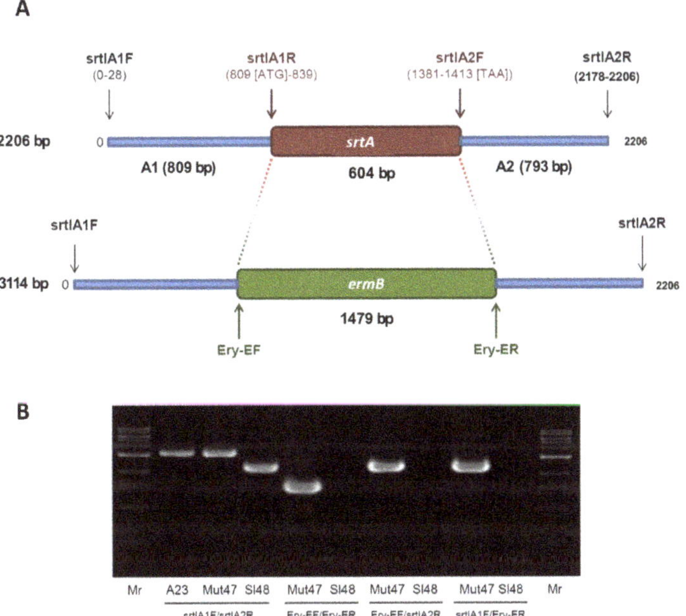

Figure 2. Genetic map of srtA and primer positions for PCRs, and results of PCR confirming the gene depletion of *srtA* in *S. lugdunensis*. (**A**) The gene loci of *srtA* in *S. lugdunensis* are shown before and after the recombination processes. In addition, the gene loci of *srtA* in *S. lugdunensis* are shown before and after the recombination processes and the positions of primer binding during the PCR experiments. The primer sequences are given in Table 3. (**B**) PCR products from genomic DNA of *S. lugdunensis* wild-type strain (Sl48), Δ*srtA* mutant strain (Mut47) and plasmid DNA of pBT*srtA* (A23) were separated by 1% agarose gels. Higher molecular bands (PCR with primer pair srtlA1F and srtlA2R) and the occurrence of corresponding bands to PCR experiments with primer pairs Ery-EF/Ery-ER, Ery-EF/srtlA2R and srtlA1F/Ery-ER confirming the deletion of *srtA* and following substitution by the *ermB* gene in the genome *S. lugdunensis* Δ*srtA* mutant strain Mut47.

3.4. Characterisation of the Surface Proteins by SDS Page and Western Blot Experiments

First, we verified the absence of the sortase A protein in mutant strains and confirmed their complementation with plasmid coded *srtA*. Specific polyclonal antibodies rose in rabbits against recombinant sortase-A lacking the N-terminal signal peptide, revealed sortase-A as a polypeptide of 26 kDa in staphylococcal extracts. Immunoblotting with anti-SrtA antibodies failed to detect the SrtA in the whole cell extract from mutant strains, but complementation of the mutant strain with plasmid pCUSrtA encoded sortase-A restored the appearance of the sortase-A (Figure 3A). As a control, the transformation of the mutant strain with empty vector DNA did not affect the expression of the SrtA (data not shown). Together, these results confirmed the successful depletion of the *srtA* gene from the genome of *S. lugdunensis*. The SrtA is described as a membrane bound protein. Therefore, an attempt was made to localize SrtA within *S. lugdunensis* cells. For this purpose, *S. lugdunensis* cultivated in BHI broth medium was fractionated into the extracellular fraction, cell wall fraction, cytosol fraction and membrane digest. SrtA was detected in the membrane digest (Figure 3B). Strongest signal was observed for the membrane fractions. Only weak signals were found for the other protein fractions. It shows that SrtA of *S. lugdunensis* is also a membrane bound protein.

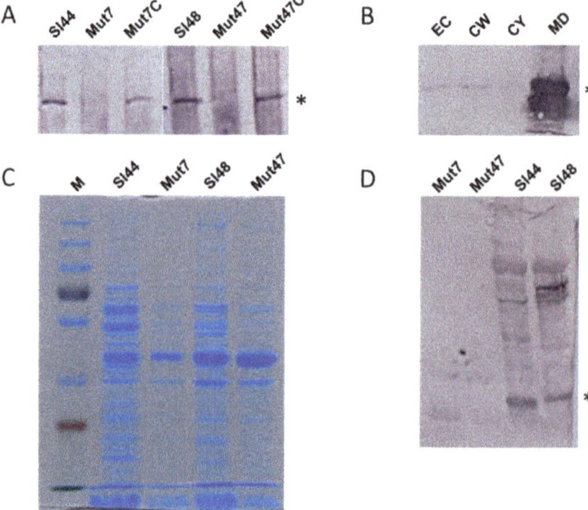

Figure 3. Western blot and SDS page experiments of *S. lugdunensis* wild type, ΔsrtA mutant and complemented mutant strains showing the presence and localization of SrtA and consequences of a *srtA* deletion for the surface proteome pattern. (**A**) Western blots showing the absence of SrtA in the mutant strains but specific signals in wild types and complemented mutants. Whole cell extracts of wild type, mutant and complemented mutant strains were separated via SDS page and blotted. The Western blots were probed with specific polyclonal antibodies rose in rabbits against recombinant SrtA (anti-SrtA). Asterisk mark the specific SrtA band. (**B**) Localization of SrtA within *S. lugdunensis* cells. For this purpose, *S. lugdunensis* were grown in BHI broth medium. Cells were fractionated into the extracellular fraction (EC), cell wall fraction (CW), cytosol fraction (CY), and membrane digest fraction (MD). Proteins were separated by SDS page. After blotting, SrtA was detected in the different fractions with anti-SrtA antibodies. Asterisk mark the specific SrtA band. (**C**) Analyses of cell surface proteins of the wild-type strains and the mutant strains. The cell surface proteins were extracted in 0.1 M hydroxylamine hydrochloride. The protein extracts were separated on SDS-Page and stained with Coomassie Blue. (**D**) Cell surface protein fractions of wild type and mutant strains were separated via SDS page. After blotting, the Western blots were probed with a mixture of antibodies raised in rabbits against formaldehyde fixed whole cells of *S. lugdunensis*. Asterisk mark the specific SrtA band. M—page ruler, Sl44 and Sl48—wild type strains, Mut7 and Mut47—ΔsrtA mutant strains, Mut7C and Mut47C—complemented ΔsrtA mutant strains.

3.5. The Surface Proteomes of the Wild-Type and Mutant Strains Differ in the Absence of the SrtA

The cell surface protein expression patterns were determined by extracting the whole cells in 0.1 M hydroxylamine hydrochloride. In Coomassie Blue-stained SDS-PAGEs, the wild type and the mutant strains showed similar but non-identical patterns of proteins. In general, gels revealed less protein concentrations and protein bands for the mutants strains compared to the wild type isolates. In all strains, lower molecular mass proteins were detected in the both wild type and mutant strains, but higher molecular mass proteins were dominant in the wild type strains (Figure 3C). The results were confirmed by Western immunoblots probed with a mixture of antibodies raised in rabbits against formaldehyde fixed whole cells of *S. lugdunensis*. This experiment revealed a lack of bands for several proteins of the cell surface protein extracts of the Δ*srtA* mutant strains, although these bands were present in wild type strains (Figure 3D).

*3.6. Determination of Growth and Biofilm Formation of the Δ*srtA* Mutants*

The parental strains of *S. lugdunensis* Sl44/Sl48 and Δ*srtA* mutants Mut7/Mut47 showed nearly the same growth rate for 24 h when cultivated in 100 mL of BHI in 500 mL flask (Figure 4A). Furthermore, in little cultivation volumes of only 5 mL in 14 mL glass tubes, no difference in growth could be observed (data not shown). Next, we determined the biofilm formation of all strains. *S. lugdunensis* wild type strains and the Δ*srtA* isogenic mutants were grown in BHI, TSB, and were supplemented with or without additional 0.5% (*w/v*) glucose. The addition of 0.5% glucose to TSB resulted in a strong biofilm formation but it was not observed in BHI with glucose. Mut7 produced 70% less biofilm in TSB plus glucose compared to the wild type strain Sl44. Mut47 showed only a decrease of around 25% biofilm formation capability in TSB plus glucose compared to the parental strain Sl48. In the complemented mutants Mut47C and Mut47C, the biofilm formation was completely restored to the levels of the wild-type strains Sl44 and Sl48 (Figure 4B).

3.7. Recombinant SrtA As Well As Cell Extracts Catalyzes Hydroxylaminolysis

To verify whether our recombinant SrtA as well as native SrtA of cell extracts still had catalytic activities, we incubated them with LPXTG fluorescent peptides (DABCYL-LPETG-EDANS) and CaCl$_2$. An increase in fluorescence is observed when cleavage of DABCYL-LPETG-EDANS peptide separates the fluorophore from Dabcyl which in turn confirmed the well described catalytic mechanism of the sortase. First, we tested the catalytic activity of the recombinant sortase SrtA from *S. lugdunensis* in this experimental setup. We observed an increase in fluorescence caused by SrtA which confirmed the cleavage of the DABCYL-LPETG-EDANS peptide (Figure 5A). As hydroxylaminolysis of LPXTG peptides depends on the sulfhydryl of the sortase, an addition of the known sortase inhibitor pHMB at least strongly abolished enzymatic activity of SrtA. However, the enzymatic activity was restored when DTT was added. We next tested cell lysates containing the native SrtA of the wild type strain Sl48, the mutant Mut47, the complemented mutant Mut47C and compare them with the recombinant SrtA in one experiment (Figure 5B). We observed an increase in fluorescence intensity when cell extracts of Sl48, Mut47C and the recombinant SrtA were incubated with the LPXTG peptide in presence of CaCl$_2$. As expected, only low activity was measured for cell extracts of the mutant strain Mut47 and the control (Figure 5B). Comparable results were found in cell extracts of the Sl44 wild type, the SrtA deficient mutant Mut7, and the Mut7C complemented mutant strains.

*3.8. Agglutination Test and Adherence of Δ*srtA* Mutants to Immobilized Fibronectin (Fn), Fibrinogen (Fg), and Vitronectin (Vn)*

Clumping factor (ClfA) of *S. aureus* is a SrtA substrate and constitutes an important virulence factor. It binds to the C-terminus of the γ-chain of fibrinogen and allowed the adhesion to different eukaryotic cell types. Besides, the ClfA is used as an antigen for rapid diagnostic identification based on latex-agglutination test systems. *S. lugdunensis* also possesses a clumping factor (Fbl) with an amino acid identity of 58% to ClfA of *S. aureus*. Hence, we applied the Pastorex Staph Plus (Sanofi Diagnostic Pasteur, S.A., Marnes la Coquette, France) rapid agglutination test for simultaneous detection of the fibrinogen affinity antigen (clumping factor) combined with the detection of protein A and capsular polysaccharides of *S. aureus*. The WT strains showed positive agglutination reactions, but the Δ*srtA* strains did not reveal any agglutination (Figure 6A). It suggested the absence of Fbl in mutant strains, but we verified this result and measured the adherence of the Δ*srtA* mutants to different ECM proteins via ELISA assays. All strains were investigated for binding to immobilized Fn, Fg, and Vn. Compared to wild-type strain Sl44, the adherence of the Δ*srtA* mutant Mut7 was significantly decreased to immobilized Fg, Fn, and Vn (Figure 6B). The binding of complemented mutant Mut7C was indistinguishable to the parent level binding capacity. For the mutant strain Mut47, the results differed a little bit. We observed only a strong decrease in binding Fg, but the binding to Fn- and Vn-coated surfaces of Mut47 decreased less and not to a significant extent. However, binding of complemented mutant Mut47C to Fg, Fn and Vn was restored to the same extent as the wild type strain Sl48 (Figure 6B).

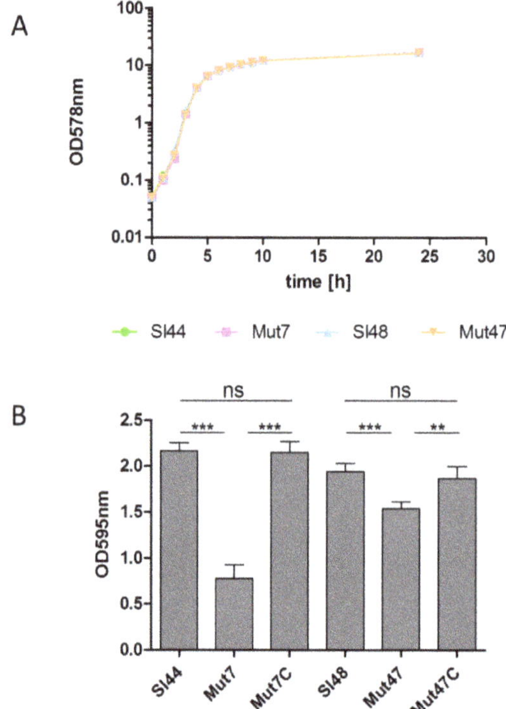

Figure 4. Growth and biofilm formation of *S. lugdunensis* wild-type strains (Sl44, Sl48) and Δ*srtA* mutant strains (Mut7, Mut47). (**A**) Bacterial growth was monitored for 24 h. Bacteria were cultivated in 100 mL BHI in 500 mL flask at 37 °C under permanent agitation. The experiment was done on triplicates. One representative experiment is shown. (**B**) Biofilm-forming capacities of the *S. lugdunensis* wild type (Sl44 and Sl48), Δ*srtA* mutant (Mut7 and Mut47) and complemented mutant (Mut7C and Mut47C) strains were assessed by a quantitative biofilm assay performed in microtiter plates applying crystal violet and determination of the OD$_{595nm}$. Results are shown as the mean of five independent experiments with the standard deviation (SD). Statistical analyses were performed using one-way ANOVA with Bonferroni multiple comparisons posttest (** $p < 0.01$; *** $p < 0.001$). ns—not significant.

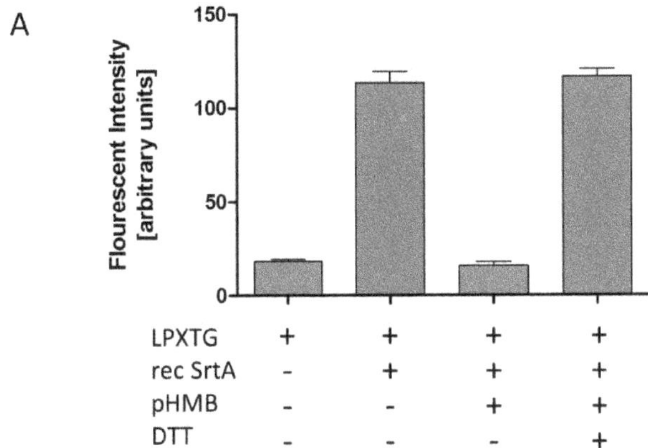

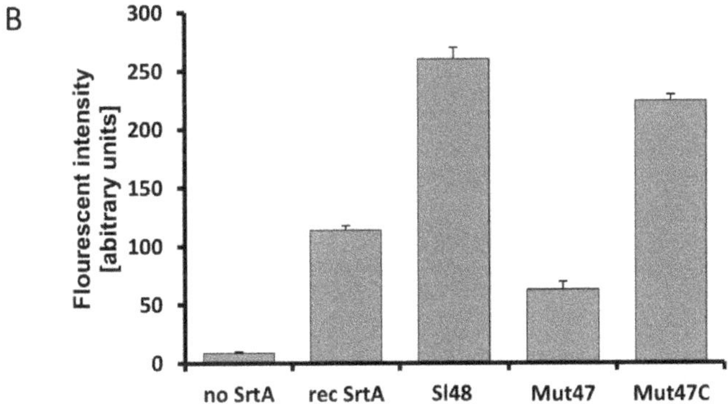

Figure 5. Hydroxylaminolysis of LPXTG peptide by recombinant SrtA (**A**) and whole *S. lugdunensis* cells (**B**). (**A**) Recombinant SrtA was incubated with the sorting substrate Dabcyl-QALPETGEE-Edans (LPXTG), and peptide cleavage was monitored as an increase in fluorescence. The reactions were influenced by the addition of pHMB (inhibitor of sortase activity) and by DTT. Presence and absence of the respective substance is shown as "+" or "-", respectively. (**B**) Same assay using the LPXTG substrate, recombinant SrtA, *S. lugdunensis* cell extracts from the wild-type strain Sl48, Δ*srtA* mutant Mut47 and complemented Mut47C. All results are shown as the mean of three independent experiments with the standard deviation (SD).

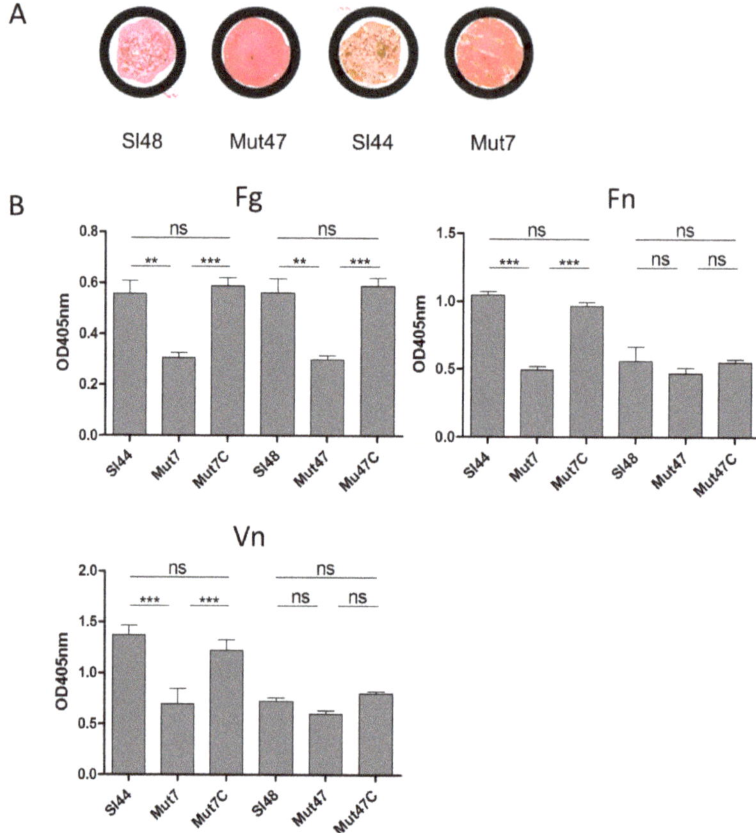

Figure 6. Pastorex staph plus agglutination test and binding of *S. lugdunensis* strains to ECM proteins. (**A**): Pastorex staph plus test of *S. lugdunensis* wild-type strains (Sl44, Sl48) and Δ*srtA* mutant strains (Mut7, Mut47) grown overnight on blood agar plates. Material was mixed in one drop of Pastorex reagent on a Pastorex disposable card. Results were recorded as positive on visual agglutination of wild-type strains and as negative for both mutants showing no agglutination. (**B**): Binding of *S. lugdunensis* wild-type strains (Sl44 and Sl48), Δ*srtA* mutants (Mut7 and Mut47), and complemented mutant strains (Mut7C and Mut47C) to immobilized fibrinogen (Fg), fibronectin (Fn), and vitronectin (Vn) assessed by ELISA adherence assays. Results are shown as the mean of four independent experiments with the standard deviation (SD). Statistical analyses were performed using one-way ANOVA with Bonferroni multiple comparisons posttest (** $p < 0.01$; *** $p < 0.001$). ns—not significant.

3.9. Sortase A Inhibition Resulted in Decreased Biofilm Formation and Binding to Fg and Fn

The function of sortase A could be blocked by different classes of sortase A inhibitors like berberine chloride (BBCl), phenyl vinyl sulfone (PVS) and pHMB as shown before in Section 3.7 [41]. In our study, we used the well-known SrtA blocking reagent PVS. Our experiments with PVS showed MIC's of about 8-12 mM for strains Sl48, Sl44 and the corresponding Δ*srtA* mutants (Figure 7).

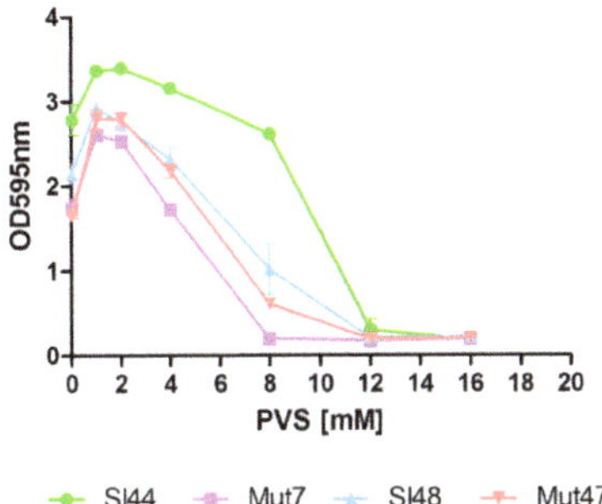

Figure 7. Effect of the sortase A inhibitor phenyl vinyl sulfone (PVS) on biofilm formation. The minimum inhibitory concentrations (MIC) of the sortase inhibitor PVS of *S. lugdunensis* wild type (SL44 and Sl48), Δ*srtA* mutant (Mut7 and Mut47) were found about 12 mM. Data are presented as mean adsorptions of triplicate determinations. Single representative experiments out of three are presented. Error bars show standard deviations (SD).

The sortase A inhibition by PVS appeared to be highly effective because at 8 mM concentration we observed a reduction in biofilm formation of strains Sl48 and Sl44, but also for the corresponding Δ*srtA* mutants. We further determined the effect of PVS on the binding capability of *S. lugdunensis* to Fn- and Fg-coated surfaces (Figure 8). A PVS treatment decreased the ability of *S. lugdunensis* to adhere to Fn- and Fg-coated surfaces. PVS at a concentration of 12 mM abolishes the binding to Fn of the parental strains Sl48 and Sl44 and the Δ*srtA* mutants Mut47 and Mut7. The effective PVS concentration to stop binding to Fg is about 16–20 mM (Figure 8).

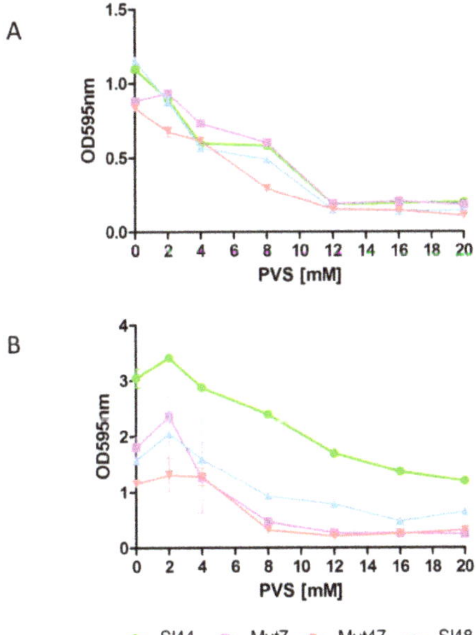

Figure 8. Effect of PVS on binding of *S. lugdunensis* to Fn- (**A**) or Fg- (**B**) coated wells. Overnight cultures were diluted to an OD_{595} of 0.01. 180 μL of diluted culture were mixed with 20 μL of concentrated phenyl vinyl sulfone (PVS) to the respective concentrations in a microtiter plate well. Plates were incubated at room temperature for 1 h. Adherence was determined in a Biorad reader at absorbance of 595 nm. Data are presented as mean adsorptions of triplicate determinations. Single representative experiments out of three are presented. Error bars show standard deviations (SD).

3.10. Hydroxylamine HCl Treatment Decrease Binding to Immobilized Fg and Fn

A hydroxylamine treatment causes a formation of the C-terminal threonine hydroxamate of surface proteins, which are thereby released into the culture medium (17). Therefore, a hydroxylamine exposure of *S. lugdunensis* should result in a decreased adherence to Fg, Fn or other eukaryotic proteins. Here, we incubated different wild type strains (Sl44, Sl48, Sl253 and Sl241) with different concentrations of hydroxalamine HCL and found that 20 mM hydroxylamine HCl tendentially reduced the ability to adhere to Fn (Figure 9A). The reduction of Fg adherence was much lower than that observed for adherence to Fn (Figure 9B). In addition, we observed strain specific differences in binding Fg and Fn.

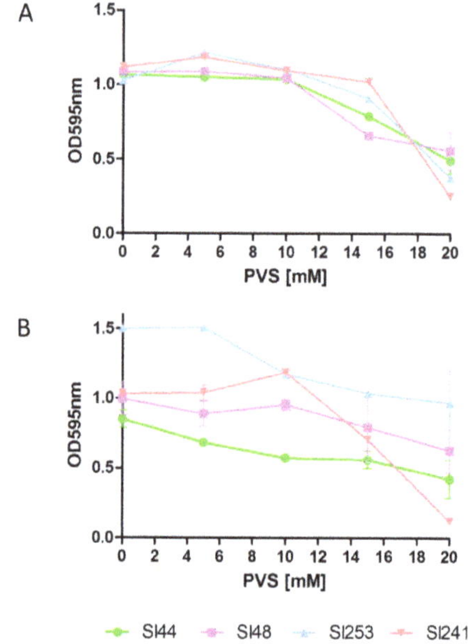

Figure 9. Effect of hydroxylamine HCl on binding of *S. lugdunensis* wild type strains to Fn (**A**) or Fg (**B**) coated wells. Overnight cultures were diluted to an OD_{595} of 0.01. 180 µL of diluted culture were mixed with 20 µL of concentrated hydroxylamine hydrochloride to the respective concentrations in a microtiter plate well. Plates were incubated at room temperature for 1 h. Adherence was determined in a Biorad reader at absorbance of 595 nm. Data are presented as mean adsorptions of triplicate determinations. Single representative experiments out of three are presented. Error bars show standard deviations (SD).

3.11. Adherence and Invasion to Eucaryotic Cell Lines

LPXTG motif cell surface proteins are important for the adherence of staphylococci to eukaryotic cells. As shown above, Δ*srtA* mutants showed decreased adherence to different ECM proteins. Therefore we tested the adherence of the wild type, mutant, and complemented strains to three different eukaryotic cell lines. Surprisingly, the Δ*srtA* mutant Mut47 showed only slight reduced adherence to confluent epithelial cells, fibroblast, and 5637 cells. However, another Δ*srtA* mutant Mut7 bound more to the above mention three cell lines. However, these differences were not statistically significant (Figure 10).

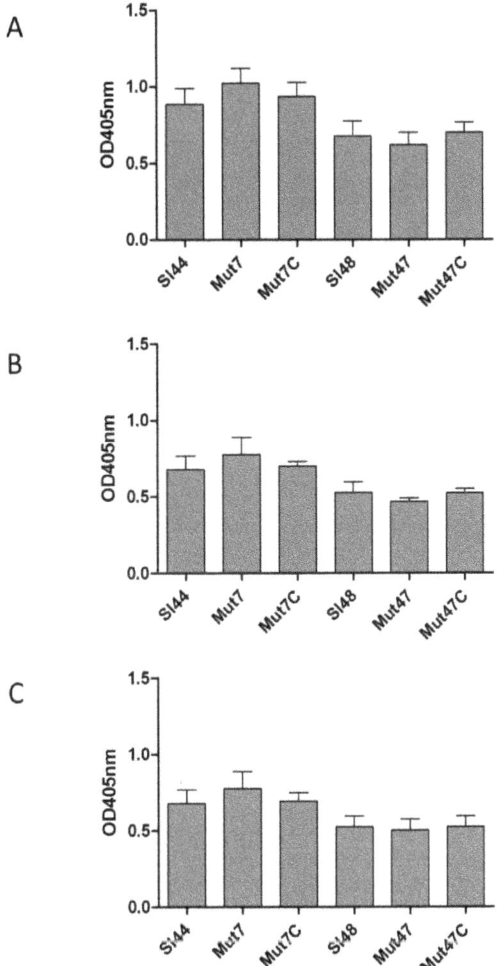

Figure 10. Adherence of *S. lugdunensis* wild-type (Sl44 and Sl48), Δ*srtA* mutant (Mut7 and Mut47) and complemented mutant (Mut7C and Mut47C) strains to immobilized human host cells ((**A**): endothelial cell line EA.hy926, (**B**): fetal lung A549 fibroblasts and (**C**): urinary bladder carcinoma cell line 5637 (ATCC HTB-9™)) assessed by ELISA adherence assays. Results are shown as the mean of at three independent experiments with the standard deviation (SD). Statistical analyses were performed using one-way ANOVA with Bonferroni multiple comparisons posttest, but the differences were not significant.

Next, we analyzed the invasion to the same eukaryotic cell lines. In experiments with FITC-labeled wild type and mutant strains, we observed a significant increase in invasion of both mutant strains to 5637 cells, Ea.hy926, and fibroblast cells (Figure 11). In addition, we observed that the Mut7 strain reached same invasion levels as the other mutant Mut47, but the wild type Sl44 revealed a comparable low capability to invade these cell lines. However, complemented strains showed the same invasion behavior as the respective wild type strains.

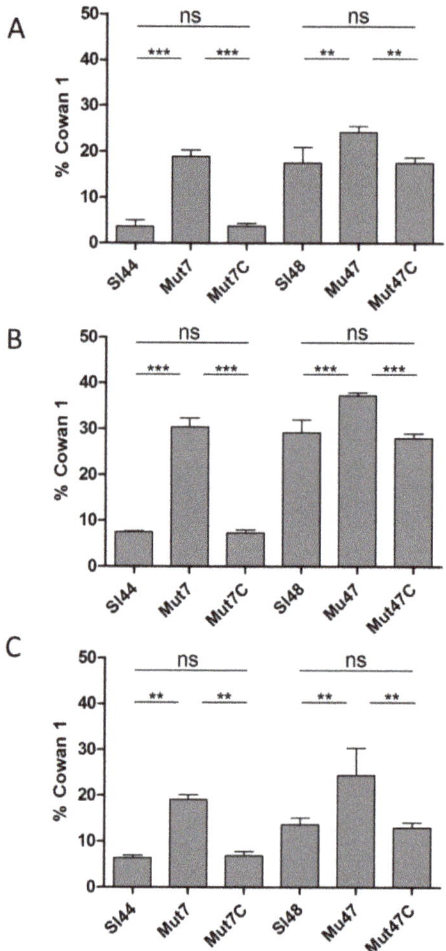

Figure 11. Internalization of FITC-labelled *S. lugdunensis* wild-type (Sl44 and Sl48), Δ*srtA* mutant (Mut7 and Mut47) and complemented mutant (Mut7C and Mut47C) strains by human host cells ((**A**): endothelial cell line EA.hy926, (**B**): fetal lung A549 fibroblasts and (**C**): urinary bladder carcinoma cell line 5637 (ATCC HTB-9™)) assessed by flow cytometry and computed in relation to *S. aureus* Cowan 1. Results are shown as the mean of three independent experiments with the standard deviation (SD). Statistical analyses were performed using one-way ANOVA with Bonferroni multiple comparisons posttest (** $p < 0.01$; *** $p < 0.001$). ns—not significant.

3.12. SrtA Influence the Gene Expression of Further Adhesins

As shown for Δ*srtA* mutants of other species [42], we finally measured the influence of a *srtA* gene deletion on the gene expression of two further LPXTG proteins. Using real-time PCR, expression of specific mRNA for the LPXTG-bearing adhesin proteins Fbl and vWbF were compared in wild type strains Sl44, Sl48 and their respective Δ*srtA* mutants Mut7 and Mut47. The RT-PCR analyses of *fbl* and *vwbF* genes showed a modest up-regulation of about 2 fold in both Δ*srtA* mutant backgrounds (Table 4), verifying results observed for sortase mutants of other species [42].

Table 4. Real-time quantification of *fbl* and *vWbF* genes.

Strain	fbl [a]	vWbF [a]
Mut47 vs. Sl48	1.88 ± 0.12	2.35 ± 0.25
Mut7 vs. Sl44	1.99 ± 0.18	1.68 ± 0.12

[a] Relative levels of *fbl* and *vWbF* specific RNA in *S. lugdunensis srtA* deletion mutants were compared to wild-type strains. The fold changes in gene expression of *fbl* and *vWbF* are shown for the Δ*srtA* mutants Mut47 and Mut7 relative to wild type strains Sl48 and Sl44. Above given values represent mean ± SD of three independent experiments performed in triplicate.

4. Discussion

The LPXTG-anchored proteins are covalently anchored to the cell surface of many bacteria and play key roles as virulence factors for the establishment of bacterial infections [43]. The executive enzyme, the sortase, is essential for the functional assembly of cell surface virulence factors and, hence, important for the pathogenesis of staphylococci and in particular *S. aureus* [22,44]. Mutations in the *srtA* gene result in non-anchoring of 19 surface proteins to the cell envelope [23]. Consequently, *S. aureus* sortase mutants are defective in assembling surface proteins and are highly attenuated in the pathogenesis shown in animal infection studies [23]. However, only a few reports are available on the properties of *S. lugdunensis* in terms of interaction with ECM molecules or host tissues, as well as on its possession of the LPXTG-anchored proteins and secretable expanded repertoire adhesive molecules (SERAMs). *S. lugdunensis* strains were shown to bind to varying extents to collagen, fibronectin, vitronectin, laminin, fibrinogen, thrombospondin, plasminogen, and human IgG [45]. A comparative bioinformatic analysis of publicly available *S. lugdunensis* genomes identified the presence of 11 sortase A dependent proteins (Table 3) [39]. In the published genome sequence of *S. lugdunensis* strain N920143, Heilbronner et al. (2011) identified *fbl*, *vWbF*, *isdB*, *isdJ* and seven more genes coding for proteins with a LPXTG motif [39]. In the present study, we report the characterization of *S. lugdunensis* housekeeping SrtA and its role in LPXTG motif proteins' cell wall anchoring. To address the role of SrtA to adhere to different ECMs and cell lines, stable and defined Δ*srtA* mutants of *S. lugdunensis* were generated by allelic replacement as described before and respective complemented mutant strains were generated to exclude any polar effects [30].

The SrtA from four *S. lugdunensis* strains (N920143, HKU09-01, M23590, and VCU139) showed identity of 76-77% to the SrtA of *S. aureus* strain Newman, and a very high identity of 98-100% among themselves. Specific residues identified in the region corresponding to the described calcium binding cleft in the *S. aureus* SrtA were also highly conserved in the amino acid sequence of the *S. lugdunensis* SrtA. In our study, the recombinant SrtA as well as whole cell extracts of *S. lugdunensis* could catalyze the hydrolysis of a fluorescent quenched polypeptide carrying the LPXTG motif, which is characteristic of sortases of other species (Figure 5) [16,23]. In addition, it was observed that the *S. lugdunensis* SrtA activity also required calcium ions for hydroxyaminolysis of LPXTG peptide, as reported previously for SrtA of *S. aureus* [40]. Therefore, we propose a quite similar catalytic mechanism as shown for SrtA from *S. aureus* [16,17]. These assumption was approved by further phenotypic analyzes showing an essential role for SrtA in anchoring LPXTG proteins, as Δ*srtA* mutants differ in their cell surface proteome from that of the wild type strains, showing the absence of several proteins in the Δ*srtA* mutants (Figure 3). It's therefore not surprising that we found a reduced biofilm formation and a reduced adherence to Fn, Fg, and Vn in the Δ*srtA* mutants (Figures 4B and 6). Similar results were shown for *S. aureus* as the inhibition of SrtA activity caused loss of binding to Fn, Fg, and IgG and reduced biofilm formation [46,47]. Furthermore, these results are consistent with our observations that the Δ*srtA* mutants showed no agglutination in the Pastorex test.

The presence of an active sortase enzyme in *S. aureus* was found essential for the adherence to host eukaryotic cell [46]. In a previous study, the Δ*srtA* mutants of *S. aureus* and *Lactococcus salivarius* showed significant reductions in adhesion to different epithelial cell lines compared to the WT strain [48,49]. Here in this report, the Δ*srtA* mutant Mut47 showed only slightly reduced adherence to confluent epithelial cells, fibroblast, and 5637 cells, whereas another Δ*srtA* mutant Mut7 bound to above mention cell lines (Figure 10). Obviously, the differences were not statistically significant, which was surprising because formerly we observed significant lower adherences to Fn, Fg, and Vn. As shown previously, the autolysin AtlL of *S. lugdunensis* was identified as a major adhesin being crucially involved in the internalization process to cells [47]. Here, we showed a fairly strong imbalance in the proteome of the cell surface compartment in the mutant strains. These probably make AtlL more available to the cell surface, which might compensate for the loss of other LPXTG adhesins in case of Mut47 and eventually enhance, to a limited extent, the binding of Mut7 to eukaryotic cells. On the other hand, we cannot exclude the absence of the respective LPXTG counterpart proteins on the cell surfaces of the eukaryotic cell lines under our experimental conditions. This rather would lead to indistinguishable results between the adherence of wild types and Δ*srtA* mutant to cell lines, similar to what we have observed here.

We further observed that SrtA is not only involved in the covalent binding of LPXTG motif proteins to cell wall, but also influenced the transcriptional regulation of at least two adhesin genes *fbl* and *vWbF*. RT-PCR results showed that expression of *fbl* and *vWbI* genes were upregulated at least 2 fold in the Δ*srtA* mutants relative to their wild type strains (Table 4). This is in agreement with the previous finding that mutations in the *srtA* gene result in the upregulated expression of cell surface proteins in an oral *Streptococcus* strain [42]. Nobbs et al. (2007) reported a significant upregulation of the expression of adhesin genes like *sspA/B*, *cshA/B*, and *fbpA* in Δ*srtA* deficient mutants relative to their wild type strains of *S. gordonii* [50]. Moreover, Hall et al. (2019) recently described a quality control mechanism that monitors the processing of LPXTG adhesins by SrtA via measuring the left C-terminal cleaved LPXTG proteins (C-peps) which stayed membrane located and are recognized by a previously uncharacterized intramembrane two-component system (TCS) [51]. Prevention of C-peps generation de-repressed this TCS and resulted in increased expression of further adhesins compensating for the loss of LXPTG adhesins [51]. Since TCS is conserved in streptococci but not in staphylococci, the same C-pep-driven regulatory circuit is unlikely in *S. lugdunensis*, but similar regulation processes cannot be ruled out and might be an explanation for the upregulation of *fbl* and *vWbF*. Thus, the comparable high adherence to host cells of Δ*srtA* mutants might also be a result of the upregulation of other adhesins in the mutant strains and is object of future research. However, detailed in vitro infection studies are necessary to understand the impact of the single LPXTG proteins of *S. lugdunensis* to the respective adherence mechanisms.

A sudden inhibition of the SrtA activity could disrupt the pathogenesis of bacterial infections [22]. The surface-protein anchoring is sensitive to sulfhydryl-modifying reagents like methanethiosulfonates, berberine chloride, (Z)-3-(2,5-dimethoxyphenyl)-2-(4-methoxyphenyl)acrylonitrile or PVS under in vitro conditions [17,42,52]. Here, the SrtA inhibitor PVS reduced the biofilm formation and also reduced the adherence to Fg and Fn (Figures 7 and 8). Further, a hydroxylamine HCl treatment, releasing LPXTG proteins from the cell wall, was found to greatly reduce the adherence of four wild type strains of *S. lugdunensis* to immobilized Fg and Fn (Figure 9). This clearly showed the important role of the LPXTG proteins in *S. lugdunensis* for the binding to some ECM proteins, and confirmed the data observed for *S. aureus* [23,53]. The inhibition of the sortase activity may provide a novel approach for the treatment of infections with staphylococci and a development of such inhibitors could complement the current dependence on antibiotics. Therefore, SrtA is an attractive target to attenuate virulence and hamper infections. In addition, the Δ*srtA* mutant of *S. lugdunensis* showed significant defects in virulence including reduced bacteremia, reduced bacterial spreading to the kidneys and reduced size/density of endocardial vegetations [24]. This highlights the importance of LPXTG-anchored proteins in the pathogenesis of this germ. These and our results suggest that several surface proteins probably act in concert to promote the adhesion to different host structures and enable a survival in the host [24]. Sortase A mutants of *S. aureus* displayed strongly reduced virulence in various infection models including experimental sepsis, highlighting the impact of properly displayed surface proteins during infection [54,55]. Therefore, the Δ*srtA* mutants were recently discussed as vaccine candidates due to the attenuated but still immune-stimulating phenotype [56].

In conclusions, SrtA was found as the housekeeping sortase of *S. lugdunensis* like in other Gram-positive bacteria and in particular in *S. aureus*. Disrupting the presentation of surface proteins by gene deletion or blocking the activity of SrtA via different SrtA inhibitors could therefore effectively cause a reduction in binding to ECM proteins and could probably disrupt the pathogenesis of bacterial infections and promote bacterial clearance by the host. Although some strain-specific differences may exist, SrtA could be an attractive target to attenuate the virulence of *S. lugdunensis*.

Author Contributions: Conceptualization, M.H. and K.B.; methodology, M.H.; software, M.H.; validation, M.H., C.K. and K.B.; formal analysis, M.H.; investigation, M.H.; resources, M.H.; data curation, M.H.; writing—original draft preparation, M.H., C.K.; writing—review and editing, C.K., K.B.; visualization, M.H., C.K.; supervision, K.B.; project administration, M.H.; funding acquisition, K.B. All authors have read and agreed to the published version of the manuscript.

Funding: This study was funded in part from the IZKF (Be2/023/11).

Acknowledgments: The paper is dedicated to Georg Peters (1951–2018).

Conflicts of Interest: The authors declare no conflict of interest.

References

1. Becker, K.; Heilmann, C.; Peters, G. Coagulase-negative staphylococci. *Clin. Microbiol. Rev.* **2014**, *27*, 870–926. [CrossRef]
2. Sundqvist, M.; Bieber, L.; Smyth, R.; Kahlmeter, G. Detection and identification of *Staphylococcus lugdunensis* are not hampered by use of defibrinated horse blood in blood agar plates. *J. Clin. Microbiol* **2010**, *48*, 1987–1988. [CrossRef]
3. van der Mee-Marquet, N.; Achard, A.; Mereghetti, L.; Danton, A.; Minier, M.; Quentin, R. *Staphylococcus lugdunensis* infections: High frequency of inguinal area carriage. *J. Clin. Microbiol.* **2003**, *41*, 1404–1409. [CrossRef]
4. Kaspar, U.; Kriegeskorte, A.; Schubert, T.; Peters, G.; Rudack, C.; Pieper, D.H.; Wos-Oxley, M.; Becker, K. The culturome of the human nose habitats reveals individual bacterial fingerprint patterns. *Environ. Microbiol.* **2016**, *18*, 2130–2142. [CrossRef]
5. Frank, K.L.; Del Pozo, J.L.; Patel, R. From clinical microbiology to infection pathogenesis: How daring to be different works for *Staphylococcus lugdunensis*. *Clin. Microbiol. Rev.* **2008**, *21*, 111–133. [CrossRef]
6. Anguera, I.; Del, R.A.; Miro, J.M.; Matinez-Lacasa, X.; Marco, F.; Guma, J.R.; Quaglio, G.; Claramonte, X.; Moreno, A.; Mestres, C.A.; et al. *Staphylococcus lugdunensis* infective endocarditis: Description of 10 cases and analysis of native valve, prosthetic valve, and pacemaker lead endocarditis clinical profiles. *Heart* **2005**, *91*, e10. [CrossRef]
7. Hussain, M.; Steinbacher, T.; Peters, G.; Heilmann, C.; Becker, K. The adhesive properties of the *Staphylococcus lugdunensis* multifunctional autolysin AtlL and its role in biofilm formation and internalization. *Int. J. Med. Microbiol.* **2015**, *305*, 129–139. [CrossRef]
8. Marlinghaus, L.; Becker, K.; Korte, M.; Neumann, S.; Gatermann, S.G.; Szabados, F. Construction and characterization of three knockout mutants of the *fbl* gene in *Staphylococcus lugdunensis*. *APMIS* **2012**, *120*, 108–116. [CrossRef]
9. Nilsson, M.; Bjerketorp, J.; Wiebensjo, A.; Ljungh, A.; Frykberg, L.; Guss, B. A von Willebrand factor-binding protein from *Staphylococcus lugdunensis*. *FEMS Microbiol. Lett.* **2004**, *234*, 155–161. [CrossRef]
10. Nilsson, M.; Bjerketorp, J.; Guss, B.; Frykberg, L. A fibrinogen-binding protein of *Staphylococcus lugdunensis*. *FEMS Microbiol. Lett.* **2004**, *241*, 87–93. [CrossRef]
11. Donvito, B.; Etienne, J.; Denoroy, L.; Greenland, T.; Benito, Y.; Vandenesch, F. Synergistic hemolytic activity of *Staphylococcus lugdunensis* is mediated by three peptides encoded by a non-agr genetic locus. *Infect. Immun.* **1997**, *65*, 95–100. [CrossRef]
12. Heilmann, C. Adhesion mechanisms of staphylococci. *Adv. Exp. Med. Biol.* **2011**, *715*, 105–123. [CrossRef]
13. Foster, T.J.; Geoghegan, J.A.; Ganesh, V.K.; Hook, M. Adhesion, invasion and evasion: The many functions of the surface proteins of *Staphylococcus aureus*. *Nat. Rev. Microbiol.* **2014**, *12*, 49–62. [CrossRef]

14. Patti, J.M.; Höök, M. Microbial adhesins recognizing extracellular matrix macromolecules. *Curr. Opin. Cell. Biol.* **1994**, *6*, 752–758. [CrossRef]

15. Navarre, W.W.; Schneewind, O. Proteolytic cleavage and cell wall anchoring at the LPXTG motif of surface proteins in gram-positive bacteria. *Mol. Microbiol.* **1994**, *14*, 115–121. [CrossRef]

16. Mazmanian, S.K.; Liu, G.; Ton-That, H.; Schneewind, O. *Staphylococcus aureus* sortase, an enzyme that anchors surface proteins to the cell wall. *Science* **1999**, *285*, 760–763. [CrossRef]

17. Ton-That, H.; Liu, G.; Mazmanian, S.K.; Faull, K.F.; Schneewind, O. Purification and characterization of sortase, the transpeptidase that cleaves surface proteins of *Staphylococcus aureus* at the LPXTG motif. *Proc. Natl. Acad. Sci. USA* **1999**, *96*, 12424–12429. [CrossRef]

18. Perry, A.M.; Ton-That, H.; Mazmanian, S.K.; Schneewind, O. Anchoring of surface proteins to the cell wall of *Staphylococcus aureus*. III. Lipid II is an in vivo peptidoglycan substrate for sortase-catalyzed surface protein anchoring. *J. Biol. Chem.* **2002**, *277*, 16241–16248. [CrossRef]

19. Ruzin, A.; Severin, A.; Ritacco, F.; Tabei, K.; Singh, G.; Bradford, P.A.; Siegel, M.M.; Projan, S.J.; Shlaes, D.M. Further evidence that a cell wall precursor [C(55)-MurNAc-(peptide)-GlcNAc] serves as an acceptor in a sorting reaction. *J. Bacteriol.* **2002**, *184*, 2141–2147. [CrossRef]

20. Spirig, T.; Weiner, E.M.; Clubb, R.T. Sortase enzymes in Gram-positive bacteria. *Mol. Microbiol.* **2011**, *82*, 1044–1059. [CrossRef]

21. Oshida, T.; Sugai, M.; Komatsuzawa, H.; Hong, Y.M.; Suginaka, H.; Tomasz, A. A *Staphylococcus aureus* autolysin that has an N-acetylmuramoyl-L-alanine amidase domain and an endo-beta-N-acetylglucosaminidase domain: Cloning, sequence analysis, and characterization. *Proc. Natl. Acad. Sci. USA* **1995**, *92*, 285–289. [CrossRef]

22. Mazmanian, S.K.; Liu, G.; Jensen, E.R.; Lenoy, E.; Schneewind, O. *Staphylococcus aureus* sortase mutants defective in the display of surface proteins and in the pathogenesis of animal infections. *Proc. Natl. Acad. Sci. USA* **2000**, *97*, 5510–5515. [CrossRef]

23. Mazmanian, S.K.; Ton-That, H.; Su, K.; Schneewind, O. An iron-regulated sortase anchors a class of surface protein during *Staphylococcus aureus* pathogenesis. *Proc. Natl. Acad. Sci. USA* **2002**, *99*, 2293–2298. [CrossRef]

24. Heilbronner, S.; Hanses, F.; Monk, I.R.; Speziale, P.; Foster, T.J. Sortase A promotes virulence in experimental *Staphylococcus lugdunensis* endocarditis. *Microbiology* **2013**, *159*, 2141–2152. [CrossRef]

25. Edgell, C.J.; McDonald, C.C.; Graham, J.B. Permanent cell line expressing human factor VIII-related antigen established by hybridization. *Proc. Natl. Acad. Sci. USA* **1983**, *80*, 3734–3737. [CrossRef]

26. Giard, D.J.; Aaronson, S.A.; Todaro, G.J.; Arnstein, P.; Kersey, J.H.; Dosik, H.; Parks, W.P. In vitro cultivation of human tumors: Establishment of cell lines derived from a series of solid tumors. *J. Natl. Cancer Inst.* **1973**, *51*, 1417–1423. [CrossRef]

27. Fogh, J.; Wright, W.C.; Loveless, J.D. Absence of HeLa cell contamination in 169 cell lines derived from human tumors. *J. Natl. Cancer Inst.* **1977**, *58*, 209–214. [CrossRef]

28. Khan, S.A.; Novick, R.P. Terminal nucleotide sequences of Tn551, a transposon specifying erythromycin resistance in *Staphylococcus aureus*: Homology with Tn3. *Plasmid* **1980**, *4*, 148–154. [CrossRef]

29. Palma, M.; Nozohoor, S.; Schennings, T.; Heimdahl, A.; Flock, J.I. Lack of the extracellular 19-kilodalton fibrinogen-binding protein from *Staphylococcus aureus* decreases virulence in experimental wound infection. *Infect. Immun.* **1996**, *64*, 5284–5289. [CrossRef]

30. Hussain, M.; Becker, K.; von Eiff, C.; Schrenzel, J.; Peters, G.; Herrmann, M. Identification and characterization of a novel 38.5-kilodalton cell surface protein of *Staphylococcus aureus* with extended-spectrum binding activity for extracellular matrix and plasma proteins. *J. Bacteriol.* **2001**, *183*, 6778–6786. [CrossRef]

31. Tuchscherr, L.; Löffler, B.; Buzzola, F.R.; Sordelli, D.O. *Staphylococcus aureus* adaptation to the host and persistence: Role of loss of capsular polysaccharide expression. *Future Microbiol.* **2010**, *5*, 1823–1832. [CrossRef]

32. Hussain, M.; Haggar, A.; Heilmann, C.; Peters, G.; Flock, J.I.; Herrmann, M. Insertional inactivation of Eap in *Staphylococcus aureus* strain Newman confers reduced staphylococcal binding to fibroblasts. *Infect. Immun.* **2002**, *70*, 2933–2940. [CrossRef]

33. Quentmeier, H.; Zaborski, M.; Drexler, H.G. The human bladder carcinoma cell line 5637 constitutively secretes functional cytokines. *Leuk. Res.* **1997**, *21*, 343–350. [CrossRef]

34. Sinha, B.; Francois, P.; Que, Y.A.; Hussain, M.; Heilmann, C.; Moreillon, P.; Lew, D.; Krause, K.H.; Peters, G.; Herrmann, M. Heterologously expressed *Staphylococcus aureus* fibronectin-binding proteins are sufficient for invasion of host cells. *Infect. Immun.* **2000**, *68*, 6871–6878. [CrossRef]

35. Juuti, K.M.; Sinha, B.; Werbick, C.; Peters, G.; Kuusela, P.I. Reduced adherence and host cell invasion by methicillin-resistant *Staphylococcus aureus* expressing the surface protein Pls. *J. Infect. Dis.* **2004**, *189*, 1574–1584. [CrossRef]

36. Sinha, B.; Francois, P.P.; Nüße, O.; Foti, M.; Hartford, O.M.; Vaudaux, P.; Foster, T.J.; Lew, D.P.; Herrmann, M.; Krause, K.H. Fibronectin-binding protein acts as *Staphylococcus aureus* invasin via fibronectin bridging to integrin $\alpha_5\beta_1$. *Cell. Microbiol.* **1999**, *1*, 101–117. [CrossRef]

37. Hussain, M.; Schafer, D.; Juuti, K.M.; Peters, G.; Haslinger-Loffler, B.; Kuusela, P.I.; Sinha, B. Expression of Pls (plasmin sensitive) in *Staphylococcus aureus* negative for *pls* reduces adherence and cellular invasion and acts by steric hindrance. *J. Infect. Dis.* **2009**, *200*, 107–117. [CrossRef]

38. Pinsky, B.A.; Samson, D.; Ghafghaichi, L.; Baron, E.J.; Banaei, N. Comparison of real-time PCR and conventional biochemical methods for identification of *Staphylococcus lugdunensis*. *J. Clin. Microbiol.* **2009**, *47*, 3472–3477. [CrossRef]

39. Heilbronner, S.; Holden, M.T.; van Tonder, A.; Geoghegan, J.A.; Foster, T.J.; Parkhill, J.; Bentley, S.D. Genome sequence of *Staphylococcus lugdunensis* N920143 allows identification of putative colonization and virulence factors. *FEMS Microbiol. Lett.* **2011**, *322*, 60–67. [CrossRef]

40. Hirakawa, H.; Ishikawa, S.; Nagamune, T. Design of Ca^{2+}-independent *Staphylococcus aureus* sortase A mutants. *Biotechnol. Bioeng.* **2012**, *109*, 2955–2961. [CrossRef]

41. Kim, S.H.; Shin, D.S.; Oh, M.N.; Chung, S.C.; Lee, J.S.; Oh, K.B. Inhibition of the bacterial surface protein anchoring transpeptidase sortase by isoquinoline alkaloids. *Biosci. Biotechnol. Biochem.* **2004**, *68*, 421–424. [CrossRef] [PubMed]

42. Maresso, A.W.; Schneewind, O. Sortase as a target of anti-infective therapy. *Pharmacol. Rev.* **2008**, *60*, 128–141. [CrossRef] [PubMed]

43. Que, Y.A.; Francois, P.; Haefliger, J.A.; Entenza, J.M.; Vaudaux, P.; Moreillon, P. Reassessing the role of *Staphylococcus aureus* clumping factor and fibronectin-binding protein by expression in *Lactococcus lactis*. *Infect. Immun.* **2001**, *69*, 6296–6302. [CrossRef]

44. Jonsson, I.M.; Mazmanian, S.K.; Schneewind, O.; Verdrengh, M.; Bremell, T.; Tarkowski, A. On the role of *Staphylococcus aureus* sortase and sortase-catalyzed surface protein anchoring in murine septic arthritis. *J. Infect. Dis.* **2002**, *185*, 1417–1424. [CrossRef] [PubMed]

45. Paulsson, M.; Petersson, A.C.; Ljungh, A. Serum and tissue protein binding and cell surface properties of *Staphylococcus lugdunensis*. *J. Med. Microbiol.* **1993**, *38*, 96–102. [CrossRef]

46. Oh, K.B.; Oh, M.N.; Kim, J.G.; Shin, D.S.; Shin, J. Inhibition of sortase-mediated *Staphylococcus aureus* adhesion to fibronectin via fibronectin-binding protein by sortase inhibitors. *Appl. Microbiol. Biotechnol.* **2006**, *70*, 102–106. [CrossRef]

47. Sibbald, M.J.; Yang, X.M.; Tsompanidou, E.; Qu, D.; Hecker, M.; Becher, D.; Buist, G.; van Dijl, J.M. Partially overlapping substrate specificities of staphylococcal group A sortases. *Proteomics* **2012**, *12*, 3049–3062. [CrossRef]

48. Weidenmaier, C.; Kokai-Kun, J.F.; Kristian, S.A.; Chanturiya, T.; Kalbacher, H.; Gross, M.; Nicholson, G.; Neumeister, B.; Mond, J.J.; Peschel, A. Role of teichoic acids in *Staphylococcus aureus* nasal colonization, a major risk factor in nosocomial infections. *Nat. Med.* **2004**, *10*, 243–245. [CrossRef]

49. van Pijkeren, J.P.; Canchaya, C.; Ryan, K.A.; Li, Y.; Claesson, M.J.; Sheil, B.; Steidler, L.; O'Mahony, L.; Fitzgerald, G.F.; van Sinderen, S.; et al. Comparative and functional analysis of sortase-dependent proteins in the predicted secretome of *Lactobacillus salivarius* UCC118. *Appl. Environ. Microbiol.* **2006**, *72*, 4143–4153. [CrossRef]

50. Nobbs, A.H.; Vajna, R.M.; Johnson, J.R.; Zhang, Y.; Erlandsen, S.L.; Oli, M.W.; Kreth, J.; Brady, L.J.; Herzberg, M.C. Consequences of a sortase A mutation in *Streptococcus gordonii*. *Microbiology* **2007**, *153*, 4088–4097. [CrossRef]

51. Hall, J.W.; Lima, B.P.; Herbomel, G.G.; Gopinath, T.; McDonald, L.; Shyne, M.T.; Lee, J.K.; Kreth, J.; Ross, K.F.; Veglia, G.; et al. An intramembrane sensory circuit monitors sortase A-mediated processing of streptococcal adhesins. *Sci. Signal.* **2019**, *12*. [CrossRef] [PubMed]

52. Sudheesh, P.S.; Crane, S.; Cain, K.D.; Strom, M.S. Sortase inhibitor phenyl vinyl sulfone inhibits *Renibacterium salmoninarum* adherence and invasion of host cells. *Dis. Aquat. Organ.* **2007**, *78*, 115–127. [CrossRef] [PubMed]

53. Ton-That, H.; Schneewind, O. Anchor structure of staphylococcal surface proteins. IV. Inhibitors of the cell wall sorting reaction. *J. Biol. Chem.* **1999**, *274*, 24316–24320. [CrossRef] [PubMed]

54. Jonsson, I.M.; Mazmanian, S.K.; Schneewind, O.; Bremell, T.; Tarkowski, A. The role of *Staphylococcus aureus* sortase A and sortase B in murine arthritis. *Microbes Infect.* **2003**, *5*, 775–780. [CrossRef]
55. Weiss, W.J.; Lenoy, E.; Murphy, T.; Tardio, L.; Burgio, P.; Projan, S.J.; Schneewind, O.; Alksne, L. Effect of srtA and srtB gene expression on the virulence of *Staphylococcus aureus* in animal models of infection. *J. Antimicrob. Chemother.* **2004**, *53*, 480–486. [CrossRef] [PubMed]
56. Kim, H.K.; Kim, H.Y.; Schneewind, O.; Missiakas, D. Identifying protective antigens of *Staphylococcus aureus*, a pathogen that suppresses host immune responses. *FASEB J.* **2011**, *25*, 3605–3612. [CrossRef]

Publisher's Note: MDPI stays neutral with regard to jurisdictional claims in published maps and institutional affiliations.

 *microorganisms*

Article

Combined Effect of Naturally-Derived Biofilm Inhibitors and Differentiated HL-60 Cells in the Prevention of *Staphylococcus aureus* Biofilm Formation

Inés Reigada [1,*], Clara Guarch-Pérez [2], Jayendra Z. Patel [3], Martijn Riool [2], Kirsi Savijoki [1], Jari Yli-Kauhaluoma [3], Sebastian A. J. Zaat [2] and Adyary Fallarero [1]

[1] Drug Research Program, Division of Pharmaceutical Biosciences, Faculty of Pharmacy, University of Helsinki, FI-00014 Helsinki, Finland; kirsi.savijoki@helsinki.fi (K.S.); adyary.fallarero@helsinki.fi (A.F.)

[2] Department of Medical Microbiology and Infection Prevention, Amsterdam institute for Infection and Immunity, Amsterdam UMC, University of Amsterdam, 1105 AZ Amsterdam, The Netherlands; c.m.guarchperez@amsterdamumc.nl (C.G.-P.); m.riool@amsterdamumc.nl (M.R.); s.a.zaat@amsterdamumc.nl (S.A.J.Z.)

[3] Drug Research Program, Division of Pharmaceutical Chemistry and Technology, Faculty of Pharmacy, University of Helsinki, FI-00014 Helsinki, Finland; jayendra.patel@helsinki.fi (J.Z.P.); jari.yli-kauhaluoma@helsinki.fi (J.Y.-K.)

* Correspondence: ines.reigada@helsinki.fi; Tel.: +358-294159749

Received: 12 October 2020; Accepted: 7 November 2020; Published: 9 November 2020

Abstract: Nosocomial diseases represent a huge health and economic burden. A significant portion is associated with the use of medical devices, with 80% of these infections being caused by a bacterial biofilm. The insertion of a foreign material usually elicits inflammation, which can result in hampered antimicrobial capacity of the host immunity due to the effort of immune cells being directed to degrade the material. The ineffective clearance by immune cells is a perfect opportunity for bacteria to attach and form a biofilm. In this study, we analyzed the antibiofilm capacity of three naturally derived biofilm inhibitors when combined with immune cells in order to assess their applicability in implantable titanium devices and low-density polyethylene (LDPE) endotracheal tubes. To this end, we used a system based on the coculture of HL-60 cells differentiated into polymorphonuclear leukocytes (PMNs) and *Staphylococcus aureus* (laboratory and clinical strains) on titanium, as well as LDPE surfaces. Out of the three inhibitors, the one coded **DHA1** showed the highest potential to be incorporated into implantable devices, as it displayed a combined activity with the immune cells, preventing bacterial attachment on the titanium and LDPE. The other two inhibitors seemed to also be good candidates for incorporation into LDPE endotracheal tubes.

Keywords: *Staphylococcus aureus*; biomaterials; medical devices; HL-60 cells; PMNs; biofilm; endotracheal tube; titanium; implantable devices; nosocomial diseases

---

1. Introduction

Over 2.6 million new cases of healthcare-associated infections are annually reported just in the European Union [1], and over 33,000 result in death [2] due to the increasing number of antimicrobial-resistance cases [3]. At least 25% of these infections are associated with the use of medical devices, and 80% of them are estimated to be caused by bacterial biofilms [4,5]. Biofilms are defined as a community of microorganisms encased within a self-produced matrix that adheres to biological or nonbiological surfaces [6,7]. They are currently regarded as the most important nonspecific mechanism of antimicrobial resistance [8,9].

During the worldwide crisis of SARS-CoV-2, as well as in many other pathological conditions, mechanical ventilation is used to assist or replace spontaneous breathing as a life-saving procedure in intensive care. However, the use of endotracheal intubation also poses major risks in prolonged ventilation. The endotracheal tube provides an ideal opportunity for bacteria to form biofilms on both the outer and luminal surface of the tube, increasing the risk of pulmonary infection by 6 to 10 times [10–12], with *Staphylococcus* spp. and *Pseudomonas aeruginosa* among the most frequent colonizing agents [13]. Colonization by microorganisms and the subsequent formation of a biofilm can happen within hours [13], but these kinds of devices are relatively easy to replace. On the other hand, infection of orthopedic implants is particularly problematic, as these devices remain in the body, often causing chronic and/or recurring infections mediated by biofilms. These infections also frequently require removal of the infected implant, thereby causing implant failure [14–16]. Given the rising number of implantations, the absolute number of complications is inevitably increasing at the same pace, causing not only distress for the patients but also an increasing economic burden [15,16].

The most common causative agents of infection in orthopedic implants are Gram-positive cocci of the genus *Staphylococcus*, e.g., *Staphylococcus aureus* and *Staphylococcus epidermidis* [17]. In the absence of a foreign body, contaminations caused by these opportunistic pathogens are usually cleared by the immune system. In contrast, the placement of an implant per se represents a risk factor for the development of a chronic infection. This is due to the fact that the surgical procedure causes tissue damage resulting in the local generation of damage-associated molecular patterns (DAMPs), endogenous danger molecules that are released from damaged or dying cells and activate the innate immune system by interacting with pattern recognition receptors (PRRs) [18]. This is sensed by host neutrophils, which migrate to the injured tissue sites, activating defense mechanisms, such as the generation of oxygen-derived and nitrogen-derived reactive species as well as phagocytosis, to unsuccessfully attempt to clear the foreign material. These events lead to immune cell exhaustion and death, and tissue damage caused by the triggered inflammation eventually leads to a niche of immune suppression around the implant [19]. Under these specific conditions, the clearance of planktonic bacteria by immune cells becomes impaired [14], which predisposes the implant to microbial colonization and biofilm-mediated infection.

However, it is possible that not only host immune cells and bacterial cells can be present at the moment of implantation, but also the cells of the tissue where the material is being implanted. The "race for the surface" concept describes the competition between bacteria and the host cells of the tissue where the device is implanted to colonize the surface of the material [20]. The rapid integration of the material into the host tissue is a key component in the success of an implant, as the colonization of the surface by the cells of the host not only ensures correct integration, but also prevents bacterial colonization [21]. However, if bacterial adhesion occurs first, the host defense system is unable to prevent the colonization and subsequent biofilm formation [22].

One of the current challenges to prevent and resolve biofilm-mediated infections is the limited repertoire of compounds that are able to act on them at sufficiently low concentrations [23]. Because of this, there is intense ongoing research focused on the search for new biofilm inhibitors by means of chemical screenings. However, for compounds to be truly effective when used for the protection of biomaterials in translational applications, they must be tested in meaningful experimental models. Based on this, we previously optimized an in vitro system based on the coculture of SaOS-2, osteogenic cells, and *S. aureus* (laboratory and clinical strains) on a titanium surface (Reigada; et al. [24]), and studied the effect of three naturally derived biofilm inhibitors (Figure 1). Two of them were dehydroabietic acid (DHA) derivatives, namely, *N*-(abiet-8,11,13-trien-18-oyl)cyclohexyl-L-alanine and *N*-(abiet-8,11,13-trien-18-oyl)-D-tryptophan, coded **DHA1** (Figure 1a) and **DHA2** (Figure 1b), respectively. The third one was a flavan derivative, 6-chloro-4-(6-chloro-7-hydroxy-2,4,4-trimethylchroman-2-yl)benzene-1,3-diol, coded **FLA1** (Figure 1c). All of them were previously reported by our group and demonstrated to display activity in preventing biofilm formation, as well as disrupting preformed *S. aureus* biofilms on 96-well plates [25,26], but the testing in the coculture model with osteogenic cells provided with new insights into their applicability as part of anti-infective implantable devices.

The coculture model developed previously [24] involving SaOS-2 cells and *S. aureus* strains offered an in vitro environment that was closer in terms of host–bacteria interactions and substrate materials to the in vivo scenario of the implanted device. Moreover, in terms of antimicrobial evaluation, it provided information not only regarding the antibiofilm activity but also about the effects on tissue integration. However, this model did not assess the effect of the antimicrobials on immune cells. As mentioned earlier, chronic inflammation not only lowers the antimicrobial efficacy, but also complicates the integration of the implant material as a consequence of maintained inflammation and tissue damage. Therefore, it is essential to assess the effect that antimicrobials might have in the presence of immune cells, particularly for those intended to be used in medical devices. Endotracheal tubes significantly differ from orthopedic implants in material, function, and implantation procedure, but they both cause impairment of the host immune antimicrobial capacity.

Because of this, in this contribution, we move one step forward toward emulating in vivo conditions encountered by medical devices in an in vitro setting. In this case, we introduce immune cells, specifically neutrophils, in a coculture model with bacterial cells. Neutrophils were selected as immune cells as they are the first cell types to migrate toward damaged tissue cells [27]. Our aim was to develop cocultures of *S. aureus* and differentiated HL-60 cells, grown on titanium coupons or low-density polyethylene (LDPE) tubes, to simulate biofilm formation on orthopedic implants or endotracheal tubes, respectively. As a proof-of-concept, we also study the possible antimicrobial effects of naturally derived antibiofilm inhibitors **DHA1**, **DHA2**, and **FLA1**, and their possible applicability as part of medical devices.

2. Materials and Methods

2.1. Compounds

The dehydroabietic acid derivatives coded **DHA1** and **DHA2** (previously coded 11 and 9b, respectively) were synthesized according to [25]. Their spectral data were identical to those reported in [28]. The flavan derivative coded **FLA1** (previously coded 291 in [26]), was purchased from TimTec (product code: ST075672, www.timtec.net). These compounds were selected given their low minimum inhibitory concentrations (MICs) and the low concentrations needed in order to prevent *S. aureus* biofilm formation (Table S1). The control antibiotic rifampicin was purchased from Sigma-Aldrich (CAS number 13292-46-1) and coded **RIF**. Molecular structures are shown in Figure 1.

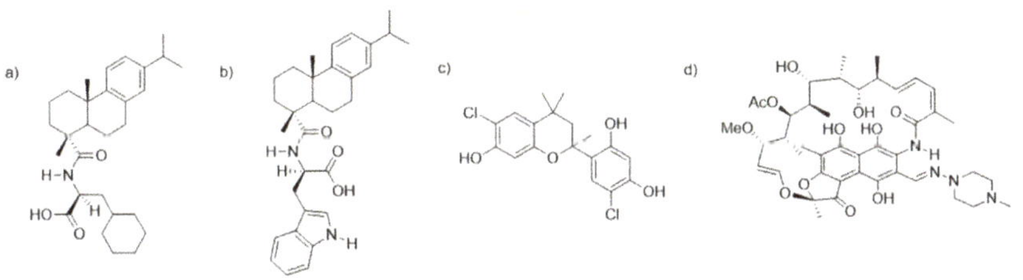

Figure 1. Chemical structures of the two dehydroabietic acid (DHA) derivatives, (**a**) *N*-(abiet-8,11,13-trien-18-oyl)cyclohexyl-L-alanine, (**b**) *N*-(abiet-8,11,13-trien-18-oyl)-D-tryptophan, coded **DHA1** and **DHA2**, the flavan derivative, (**c**) 6-chloro-4-(6-chloro-7-hydroxy-2,4,4-trimethylchroman-2-yl)benzene-1,3-diol, coded **FLA1**, and (**d**) rifampicin, coded **RIF**.

2.2. Bacterial Strains

Bacterial studies were performed using the laboratory strain *S. aureus* ATCC 25923 (American Type Culture Collection, Manassas, VA, USA) and one clinical strain isolated from hip prostheses and osteosynthesis implants at the Hospital Fundación Jiménez Díaz (Madrid, Spain) [29] (coded *S. aureus* P2).

2.3. HL-60 Cell Culture and Differentiation

HL-60 (ATCC CCL-240) cells were grown and maintained in Roswell Park Memorial Institute (RPMI) 1640 Medium (R8758, Sigma Aldrich, St. Louis, MO, USA), supplemented with 20% (*v/v*) heat-inactivated fetal bovine serum (FBS) (Sigma Aldrich, St. Louis, MO, USA) and 1% (*v/v*) penicillin/streptomycin (Sigma Aldrich, St. Louis, MO, USA). Cells were maintained in suspension at a concentration between 10^5–10^6 cells/mL in 72 cm^2 culture flasks (VWR, Radnor, PA, USA) at 37 °C in 5% CO_2 in a humidified incubator. *N,N*–Dimethylformamide (DMF)(Sigma Aldrich, St. Louis, MO, USA) was utilized in order to differentiate the cells into polymorphonuclear-like cells [30]. In order to carry out the differentiation, cells were cultured for 6 days in the maintenance medium at a concentration of 100 mM DMF. The success of the differentiation was assessed visually after Giemsa staining using a Leica DMLS microscope (Leica, Wetzlar, Germany). Differentiated cells were neutrophil-like, with a multilobar nucleus and a fairly clear cytoplasm.

2.4. Biofilm Prevention Efficacy of Differentiated HL-60 Cells against Different Bacterial Concentrations of S. aureus ATCC 25923

2.4.1. Bacterial Inoculum Preparation

Pure colonies (2–3) of *S. aureus* ATCC 25923 were added to 5 mL of tryptic soy broth (TSB, Neogen®, Lansing, MI, USA) and incubated overnight at 37 °C, 120 rpm. After incubation, 10 µL of the preculture was added to 5 mL fresh TSB and incubated for 3–4 h (37 °C, 120 rpm), in order to obtain a midlogarithmic growth phase of bacteria. Bacterial cultures were washed twice with sterile phosphate-buffered saline (PBS; 140 mM NaCl, pH 7.4) and the concentration was adjusted to 2×10^8 in RPMI 1640 based on the optical density of the suspension at 600 nm. From this stock, serial dilutions were made in RPMI 1640 between 2×10^8 and 2×10^2 CFU/mL.

2.4.2. Immune Cell (HL-60) Preparation

Twenty-four hours prior to starting the experiments, the media of the differentiated HL-60 cells were refreshed with nonsupplemented RPMI 1640 in order to remove possible traces of antibiotics from the maintenance media. Differentiated HL-60 cells (after 6 days exposure to a concentration of 100 mM DMF) were counted with a Countess™ II automated cell counter (Thermo Scientific, Waltham, MA, USA) and adjusted to a concentration of 2×10^5 cells/mL. In order to test the influence of activating the cells with phorbol 12-myristate 13-acetate (PMA) (Sigma Aldrich, St. Louis, MO, USA) on the antimicrobial capacity, half of the suspension of differentiated HL-60 cells was incubated for 20 min in RPMI 1640 supplemented with 25 nM PMA. PMA-activated and nonactivated HL-60 cells were separately added (100 µL) to flat-bottomed, 96-well microplates (Nunclon Δ surface, Nunc, Roskilde, Denmark).

2.4.3. Coculture of S. aureus ATCC 25923 and HL-60 Cells

S. aureus ATCC 25923 suspensions at the different concentrations (100 µL) were added to the wells of the 96-well microplate already containing 100 µL of one of the two different HL-60 cell suspensions (activated or nonactivated). In the bacterial control wells, 100 µL of the *S. aureus* ATCC 25923 suspension at different concentrations were added to wells containing 100 µL of RPMI 1640 alone. The wells corresponding to the HL-60 cell control consisting of a suspension of HL-60 cells at a concentration 10^5 cells/mL in RPMI 1640 were used to observe the cell morphology after 18 h of incubation, but no quantitative viability test was carried out. The 96-well microplates were incubated for 18 h at 37 °C and 5% CO_2 in a humidified incubator.

2.4.4. Biofilm Quantification in 96-Well Microplates

After *S. aureus* ATCC 25923 biofilms were grown, the media were carefully removed and each well was washed twice with 125 µL sterile PBS, followed by the addition of 150 µL of TSB. Each 96-well microplate was closed with a plastic seal and parafilm and placed in a plastic bag, which was sealed with heat in order to prevent leakage. The plate was sonicated for 10 min at 35 kHz in an Ultrasonic Cleaner 3800 water bath sonicator (Branson Ultrasonics, Danbury, CT, USA) at 25 °C. This procedure did not affect the viability of the staphylococci (Figure S1).

S. aureus ATCC 25923 was then resuspended in RPMI 1640 using 3 pipetting cycles (up/down). Samples (10 µL) from each tested condition were transfer to 90 µL of TSB and serial dilutions were made from 10^{-1} up to 10^{-7}. Aliquots (10 µL) of all dilutions were transferred to sheep blood agar plates (Amsterdam UMC, Amsterdam, The Netherlands) and incubated at 37 °C overnight. Viable plate colonies were counted the next day.

2.5. Influence of Opsonizing S. aureus ATCC 25923 on the Efficacy of HL-60 Cells in Preventing Bacterial Attachment on Titanium Coupons

S. aureus ATCC 25923 suspensions were prepared as described in Section 2.4.1. The inoculum was adjusted to 2×10^7 CFU/mL and opsonized using 50% (*v/v*) pooled human serum in PBS (pooled from 15 healthy blood donors) by incubating at 37 °C for 30 min with gentle agitation, washing twice with PBS, and resuspending in RPMI 1640 [31]. As a control, a nonopsonized suspension of *S. aureus* ATCC 25923 was used. The two different suspensions (opsonized and nonopsonized), were then added to sterilized titanium coupons (0.4 cm height, 1.27 cm diameter, BioSurface Technologies Corp., Bozeman, MT, USA).

On the other hand, HL-60 cells were prepared as described in Section 2.4.2 and added to the titanium coupons, to which the bacterial suspension was previously added. The final volume covering each coupon was 1 mL, made up of 500 µL of a suspension of 2×10^7 CFU/mL *S. aureus* ATCC 25923, and 500 µL of a suspension of 2×10^5 HL-60 cells/mL in RPMI 1640.

2.6. Effect of the Antimicrobial Compounds on the Prevention of S. aureus ATCC 25923 and S. aureus P2 Adhesion in Coculture with Differentiated HL-60 Cells on Titanium Coupons

2.6.1. Culture of Staphylococci and HL-60 Cells

Bacterial inocula of *S. aureus* ATCC 25923 and *S. aureus* P2 were prepared as described in Section 2.4.1. The concentration was adjusted to 2×10^7 CFU/mL in RPMI 1640. For the differentiated HL-60 cells, the media were refreshed with RPMI 1640 24 h before the experiments started in order to remove possible traces of antibiotics from the maintenance media. On the day of the experiment, cells were counted with a Countess™ II automated cell counter (Thermo Scientific, Waltham, MA, USA) and adjusted to a concentration of 2×10^5 cells/mL in RPMI 1640. Each suspension (bacterial and HL-60 cells suspensions, 500 µL of each) was added to sterilized titanium coupons that were inserted in the different wells of a 24-well plate (Nunclon Δ surface, Nunc, Roskilde, Denmark) and to which the different tested compounds or the control antibiotics were added at a concentration of 50 µM (0.25% DMSO).

The 24-well plates containing the titanium coupons were maintained in cocultures with a solution containing 10^7 CFU of *S. aureus* ATCC 25923 or *S. aureus* P2 and 10^5 HL-60 cells in a total volume of 1 mL of RPMI 1640 for 24 h. Titanium coupons with added **RIF** (50 µM, 0.25% DMSO) were used as positive antibiotic controls. As coculture controls, titanium coupons without the addition of the tested antimicrobial compounds or control antibiotics were exposed simultaneously to both cellular systems (*S. aureus* and HL-60 cells) at the concentrations previously described. In addition, bacterial controls (exposed or not to the antimicrobial compounds in the absence of differentiated HL-60 cells) were also included.

2.6.2. Bacterial Adherence on Titanium Coupons

Titanium coupons were gently washed with TSB to remove remaining adhering planktonic cells and transferred to Falcon tubes containing 1 mL of 0.5% (*w/v*) Tween® 20-TSB solution. Next, the tubes were sonicated in an Ultrasonic Cleaner 3800 water bath sonicator (Branson Ultrasonics, Danbury, CT, USA) at 25 °C for 5 min at 35 kHz. The tubes were vortexed for 20 s prior to and after the sonication step. Serial dilutions of the resulting bacterial suspensions were made from 10^{-1} up to 10^{-7} and plated on tryptic soy agar (Neogen®, Lansing, MI, USA) plates.

2.7. Effect of the Antimicrobial Compounds on the Prevention of S. aureus ATCC 25923 Adhesion in Coculture with Differentiated HL-60 Cells on LDPE Tubes

The assay was carried out as described in Section 2.6 but instead of titanium coupons the materials added to the 24-well plates were 1 cm long sections of a sterilized fine bore LDPE tubing (Smiths Medical ASD, Minneapolis, MN, USA).

2.8. Scanning Electron Microscopy (SEM)

In order to visualize the effect of the antimicrobial compounds on the prevention of *S. aureus* ATCC 25923 adhesion in coculture with immune cells on titanium coupons, *S. aureus* ATCC 25923 was cocultured with differentiated HL-60 cells on titanium coupons, as described in Section 2.6. Prior to SEM, samples were fixed in 4% (*v/v*) paraformaldehyde and 1% (*v/v*) glutaraldehyde (Merck, Kenilworth, NJ, USA) overnight at room temperature. Samples were rinsed twice with distilled water, with the duration of each cycle being 10 min. The dehydration procedure consisted of 2 cycles of incubation for 15 min in 50% (*v/v*) ethanol, 2 cycles of incubation for 20 min in 70% (*v/v*) ethanol, 2 cycles of incubation for 30 min in 80% (*v/v*) ethanol, 2 cycles of incubation for 30 min in 90% (*v/v*) ethanol, and 2 cycles of incubation for 30 min in 100% (*v/v*) ethanol. In order to reduce the sample surface tension, samples were immersed in hexamethyldisilazane (Polysciences Inc., Warrington, FL, USA) overnight and air-dried. Before imaging, samples were mounted on aluminum SEM stubs and sputter-coated with a 4 nm platinum–palladium layer using a Leica EM ACE600 sputter coater (Leica Microsystems, Wetzlar, Germany). Images were acquired at 2 kV using a Zeiss Sigma 300 SEM (Zeiss, Oberkochen, Germany) at the Electron Microscopy Center Amsterdam (Amsterdam UMC).

Of each coupon, 8–10 fields were viewed and photographed at magnifications of 250×, 500×, 1000×, and 3000×. Representative images are shown in the results.

2.9. Statistical Analysis

The quantitative data are reported as the mean and standard deviation (SD) of at least three independent experiments. Data were analyzed using GraphPad Prism 8 for Windows. For statistical comparisons, Tukey's multiple comparison test and Welch's unpaired *t*-test were applied, and $p < 0.05$ was always considered as statistically significant.

3. Results

3.1. Effect of PMA Activation of Differentiated HL-60 Cells on Prevention of S. aureus ATCC 25923 Biofilm Formation

Before performing the experiments with the titanium coupons and the antimicrobial compounds, the initial concentration of *S. aureus* ATCC 25923 was determined where the bacteria were able to form a biofilm in absence of the HL-60 cells, but were prevented from forming a biofilm when cocultured with 10^5 HL-60 cells. At the same time, it was assessed whether the activation of the HL-60 cells with PMA enhanced their bacterial clearance capability (Figure 2). At an initial *S. aureus* ATCC 25923 concentration of 10^8 CFU/mL, HL-60 cells did not significantly affect the bacterial attachment in 96-well microplates. A reduction on the adhered *S. aureus* ATCC 25923 viable cell counts was observed at an *S. aureus* ATCC 25923 concentration of 10^7 CFU/mL and below ($p < 0.001$ in all cases when comparing the bacterial control with bacteria cocultured with HL-60 cells, and $p = 0.001$ for cells activated with PMA and a starting inoculum of 10^3 CFU/Ml), with the exception of the lowest bacterial concentration tested, i.e., 10^2 CFU/mL, where no difference was found. Both PMA-activated and nonactivated

HL-60 cells (gray and green columns, respectively) showed similarly reduced numbers of adherent *S. aureus* ATCC 25923 viable cell counts at 24 h, so it was concluded that activation with PMA does not significantly enhance *S. aureus* ATCC 25923 clearance.

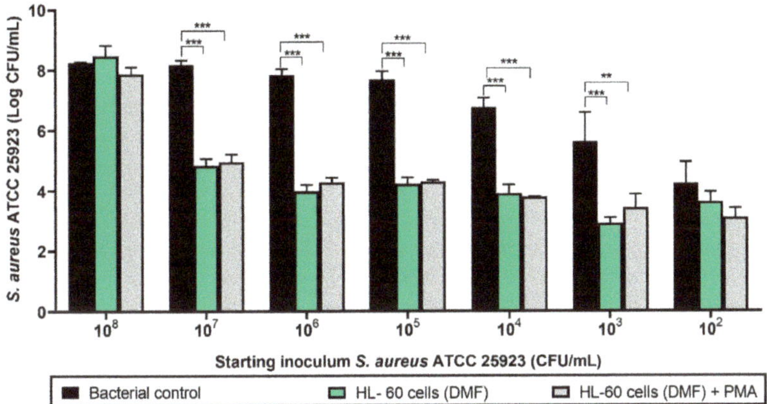

Figure 2. Viable counts of 24-hour-old biofilms formed by different concentrations of *S. aureus* ATCC 25923 cocultured with HL-60 cells on 96-well microplates. Black columns represent the bacterial control (viable attached cells in the absence of HL-60 cells). Green columns show the coculture of *S. aureus* ATCC 25923 with HL-60 cells differentiated with *N,N*-dimethylformamide (DMF). Gray columns show the coculture of *S. aureus* ATCC 25923 with HL-60 cells differentiated with DMF and activated with phorbol 12-myristate 13-acetate (PMA). Values are means and SD of three independent experiments (*** $p < 0.001$; ** $p < 0.01$).

Based upon these results, the optimal initial *S. aureus* ATCC 25923 concentration was found to be 10^7 CFU/mL, which was then selected for the rest of the experiments. Since PMA activation of the HL-60 cells did not influence their phagocytic activity, no PMA stimulation was performed in subsequent experiments.

3.2. Influence of Opsonizing S. aureus ATCC 25923 on the Efficacy of HL-60 Cells in Preventing Bacterial Attachment on Titanium Coupons

One of the most relevant mechanisms of the host defense against *Staphylococcus* spp. is phagocytosis. Given that opsonization of *S. aureus* is important for neutrophils to be able to clear planktonic *S. aureus* [32], the phagocytic efficacy of HL-60 cells, as well as the impact of opsonization of *S. aureus* ATCC 25923, in preventing *S. aureus* ATCC 25923 biofilm formation on titanium coupons was simultaneously explored (Figure 3). The left part of Figure 3 shows how the preventive capability of HL-60 cells was preserved when tested on titanium surfaces ($p < 0.001$, when comparing *S. aureus* + HL-60 (gray column) with the bacterial control (black column)). This bacterial clearance activity was also observed when *S. aureus* ATCC 25923 was opsonized ($p = 0.007$), but no significant differences were found when comparing the effects of the HL-60 cells on opsonized versus nonopsonized *S. aureus* ATCC 25923 ($p = 0.260$). The antibacterial effects of PMA-activated HL-60 cells against opsonized *S. aureus* ATCC 25923 was additionally explored, but no differences were found with the antimicrobial activity of nonactivated HL-60 against nonopsonized *S. aureus* ATCC 25923 (Figure S2). Because opsonization did not enhance the bacterial clearance capacity of HL-60, the use of opsonized *S. aureus* during the rest of the experiments was decided against.

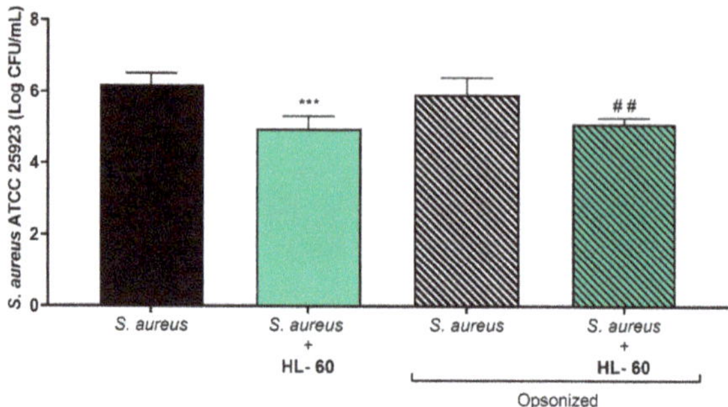

Figure 3. Viable counts of adhered opsonized and nonopsonized *S. aureus* ATCC 25923 on titanium coupons after coculture with HL-60 cells for 24 h. The left half of the graph includes the nonopsonized *S. aureus* ATCC 25923 biofilm formation, in absence (black column) or cocultured with HL-60 cells (green column). The right half includes the opsonized *S. aureus* ATCC 25923 biofilm formation, in absence (white/black column) or cocultured with HL-60 cells (green/black column). "*" indicates statistical differences with the nonopsonized *S. aureus* ATCC 25923 in monoculture, while "#" represents statistical differences with the opsonized *S. aureus* ATCC 25923 in monoculture with Welch's unpaired *t*-test (*** $p < 0.001$; ## $p < 0.01$). Values are means and SD of three independent experiments.

3.3. Effect of the Antimicrobial Compounds on the Prevention of S. aureus Adhesion in Coculture with Differentiated HL-60 Cells on Titanium Coupons

The effects of the three biofilm inhibitors **DHA1**, **DHA2**, and **FLA1** and one control antibiotic, **RIF**, on the prevention of *S. aureus* attachment on titanium coupons were investigated using two strains, the laboratory strain ATCC 25923 (Figure 4a) and the clinical isolate P2 (Figure 4b), either in the absence (gray bars) or presence of HL-60 cells (green bars). In the absence of HL-60 cells, all of the tested antimicrobial compounds, as well as the control antibiotic, significantly reduced the attachment of *S. aureus* ATCC 25923 ($p < 0.001$, $p = 0.040$, $p < 0.001$, and $p < 0.001$, for **DHA1**, **DHA2**, **FLA1**, and **RIF** versus the control, respectively). In the case of the clinical strain (Figure 4b), all the antimicrobial compounds, except **DHA2**, also showed antimicrobial activity in the absence of HL-60 cells ($p = 0.0067$, $p < 0.001$, and $p = 0.0087$ versus control for **DHA1**, **FLA1**, and **RIF**, respectively).

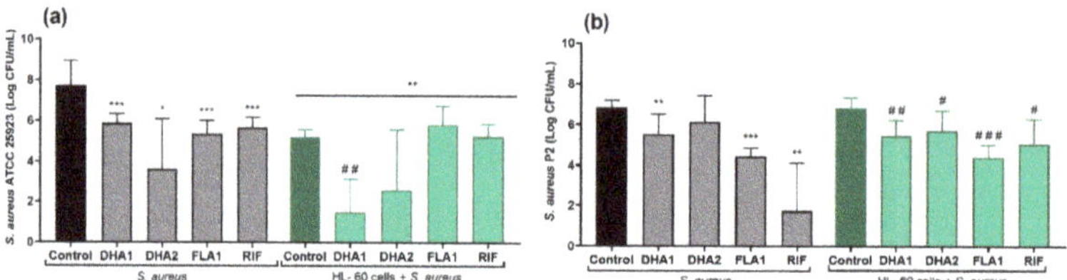

Figure 4. Viable counts of adhered *S. aureus* ATCC 25923 (**a**) and *S. aureus* P2 (**b**) on titanium coupons exposed to different antimicrobial compounds (tested at 50 μM) after 24 h of incubation. The gray bars show the attached viable bacteria when exposed just to the antimicrobial compounds, while the green bars show the results when *S. aureus* strains were cocultured with HL-60 cells. "*" indicates statistical differences with the control in monoculture while "#" represents statistical differences with the control in cocultured controls with Welch's unpaired *t*-test (* $p < 0.05$; ** $p < 0.01$; *** $p < 0.001$)/ # $p < 0.05$; ## $p < 0.01$; ### $p < 0.001$). Values are means and SD of three independent experiments.

The right part of Figure 4a (green bars) corresponds to the same experiments performed in the presence of HL-60 cells. Under such conditions, there was also a significant reduction of the attached viable *S. aureus* ATCC 25923 when compared with the material incubated without HL-60 cells (green control bar versus black control bar, $p < 0.001$). Moreover, this reduction was further increased by compound **DHA1** ($p = 0.025$ when comparing the coculture control column with the **DHA1** column in the HL-60 cells + *S. aureus* ATCC 25923 section of Figure 4, and $p < 0.001$ when comparing the **DHA1** gray column with the **DHA1** green column). The same tendency was observed for **DHA2**, but no statistical differences were found when comparing this to the control (dark green control: HL-60 cells + *S. aureus* ATCC 25923 section of Figure 4). Despite the fact that **FLA1** and the model antibiotic (**RIF**) successfully prevented the adhesion of *S. aureus* ATCC 25923, they did not cause a further increase in the bacterial clearance activity of the HL-60 cells. In contrast, the mere presence of HL-60 cells did not result in a significant reduction of *S. aureus* P2 attachment (Figure 4b), since no differences were found between the *S. aureus* P2 control in monoculture and in coculture with HL-60 cells. In this case, no differences were found between the antibacterial effects of the compounds in monoculture when compared with the same treatment in coculture with HL-60 cells (**DHA1**, **DHA2**, **FLA1**, and **RIF** gray columns versus their corresponding green columns).

Using SEM imaging, it was visually confirmed that **DHA1** reduced the number of *S. aureus* ATCC 25923 adherent to the titanium surface (Figure 5). In fact, almost no cocci were observed on the **DHA1**-treated titanium across the entire coupon. In addition, the presence of HL-60 cells also reduced the bacterial attachment, which was further enhanced by the treatment of **DHA1**. Adherent HL-60 cells were observed, as seen in the last row of images.

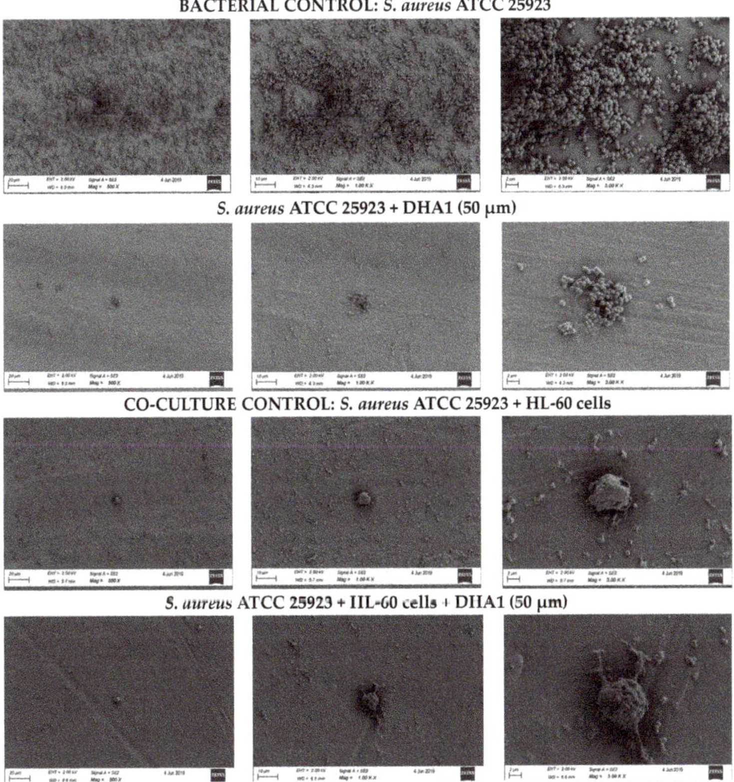

Figure 5. Representative images acquired by SEM. From left to right the same section of the titanium coupon is shown with different magnification, 500, 1000 and 3000×. Upper row of images, bacterial control, i.e., coupons incubated with *S. aureus* ATCC 25923 only. Second row, titanium coupons incubated with *S. aureus* ATCC 25923 and treated with **DHA1**. Third row, cocultured control, titanium coupons incubated with *S. aureus* ATCC 25923 and HL-60 cells simultaneously. Fourth row, titanium coupons cocultured with *S. aureus* ATCC 25923 and HL-60 and treated with **DHA1**.

3.4. Effect of the Antimicrobial Compounds on the Prevention of S. aureus ATCC 25923 Adhesion in Coculture with Differentiated HL-60 Cells on LDPE tubes

The adhesive capacity of *S. aureus*, as well as the antimicrobial effect of different compounds, is known to significantly differ depending on the material [33]. For this reason, the applicability of the compounds as part of endotracheal tubes was tested on a clinically relevant material, LDPE. Figure 6 shows the effects of the tested compounds and the control antibiotic (**RIF**) on the prevention of *S. aureus* ATCC 25923 attachment on LDPE tubes. The left part of the figure shows that all the antimicrobial compounds significantly reduced the numbers of attached viable *S. aureus* ATCC 25923 cells, in the absence of HL-60 ($p < 0.001$, $p = 0.0035$, $p < 0.001$, and $p < 0.001$, when comparing, respectively, **DHA1**, **DHA2**, **FLA1**, and **RIF** with the control). Adding HL-60 cells resulted in significant prevention of *S. aureus* ATCC 25923 attachment to LDPE tubes ($p < 0.001$ when comparing the coculture control (dark green column) with the bacterial control (black column)). In this case, all the antimicrobial compounds looked to be able to further potentiate this bacterial clearance capability ($p < 0.001$, $p = 0.023$, $p = 0.002$, and $p < 0.001$, when comparing, respectively, **DHA1**, **DHA2**, **FLA1**, and **RIF** light green columns with the control dark green column). However, similarly to what was observed on the titanium model, it was only the **DHA1** treatment that further potentiated the action of HL-60 cells against *S. aureus* ATCC 25923 ($p = 0.015$, when comparing the **DHA1** gray column with the **DHA1** green column). No differences were found between the viable cells (CFU/mL) attached on the LDPE exposed to *S. aureus* ATCC 25923 and treated with **DHA2**, **FLA1**, and **RIF** in monoculture and those exposed to both *S. aureus* ATCC 25923 and HL-60 with the same treatments (gray columns versus green columns). Full bacterial clearance was detected in LDPE tubes in the presence of **DHA1** and **RIF** (i.e., no viable bacterial counts measured), where cocultures of *S. aureus* ATCC 25923 with differentiated HL-60 cells were formed. These findings further highlight the relevance of **DHA1** as an antimicrobial candidate for incorporation into medical devices.

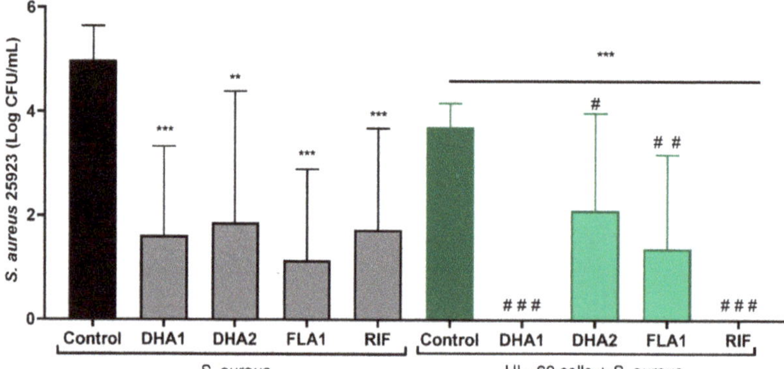

Figure 6. Viable counts of adhered *S. aureus* ATCC 25923 on LDPE tubes exposed to different antimicrobial compounds (tested at 50 μM) after 24 h of incubation. The gray bars show the attached viable bacteria when exposed just to the antimicrobial compounds, while the green bars show the results when *S. aureus* strains were cocultured with HL-60 cells. "*" indicates statistical differences with the control in monoculture, while "#" represents statistical differences with the control in cocultured controls with Welch's unpaired *t*-test (** $p < 0.01$; *** $p < 0.001$)/(# $p < 0.05$; ## $p < 0.01$; ### $p < 0.001$). Values are means and SD of three independent experiments.

4. Discussion

In this study, we explored the potential of incorporation of three previously identified naturally derived biofilm inhibitors into medical devices, particularly for titanium implantable devices and LDPE endotracheal tubes. From the three tested antimicrobial compounds, in line with previous findings [24], **DHA1** appeared to be the best candidate for incorporation as part of implantable medical devices. All of the compounds proved to be interesting candidates to include into anti-infective endotracheal tubes, but it was **DHA1** that again showed itself to be the most promising candidate, as it was the only one that significantly increased the bacterial clearance capacity of HL-60 cells against *S. aureus* ATCC 25923.

For compounds to be truly effective when used for protection of biomaterials in translational applications, they must be tested in meaningful experimental models. The insertion of any device provokes an acute inflammatory response that may cause ineffectiveness of innate immune cells such as neutrophils in cleaning planktonic bacteria, since these cells are directed to degrade the material. For this reason, it is vitally important to study the effects of antimicrobial compounds on neutrophils to assess their suitability as part of medical devices.

Before this investigation, no published reports existed on the effect of the two DHA derivatives (**DHA1** and **DHA2**) on neutrophils, however, the parent compound (dehydroabietic acid, DHA) was previously reported to have slight toxicity toward this cell type [34]. This prior knowledge further justified the need for an assessment of the effects of **DHA1** and **DHA2** on the bacterial clearance capacity of neutrophils. Similarly, no data existed on effects of **FLA1** on neutrophils, but other flavan derivatives were studied. Out of 10 different flavan-3-ol derivatives tested on human neutrophils, only two presented a slight toxic effect toward the neutrophils, but all of them reduced reactive oxygen species (ROS) and interleukin-8 production [35]. On the other hand, flavan-3-ol derivatives extracted from *Bistorta officinalis* (Delarbre) were reported to inhibit tumor necrosis factor-α (TNF-α) release from neutrophils [36]. These earlier findings were promising in terms of applying the flavan derivative **FLA1** as part of medical devices, since its antimicrobial capacity in combination with its anti-inflammatory effects could result in prevention of infection while providing ideal cues toward material integration and resolution of inflammation.

In this study, we utilized HL-60 cells differentiated to polymorphonuclear-like cells in order to study the effects of the antimicrobial compounds in the presence of neutrophils. Alternatively, freshly extracted neutrophils could also have been used, but in such case differences would be encountered between individual donors in terms of reproducibility, the total number of cells that can be harvested, their short lifespan, or the disturbances in their physiology due to isolation procedures [37,38].

The three tested antimicrobial compounds and the control antibiotic (**RIF**) reduced *S. aureus* ATCC 25923 adhesion to titanium in the absence of HL-60 at the tested concentration of 50 μM. In the case of the clinical *S. aureus* strain, all compounds, except **DHA2**, significantly reduced *S. aureus* P2 biofilm formation on titanium in the absence of HL-60. The prevention of biofilm formation by **DHA1** and **FLA1** was as expected, as the compounds were used at concentrations higher than their MIC values [25,26]. Compound **DHA2** showed some prevention activity, despite being tested at a concentration slightly below its MIC (i.e., 60 μM) [26]. As with most antimicrobials, cytotoxicity is a concern, and **DHA2** was shown to reduce viability of HL cells (originating from the human respiratory tract) at a concentration of 100 μM [25].

The mere presence of HL-60 cells significantly reduced the bacterial attachment of *S. aureus* ATCC 25923. In contrast, this effect was not observed for the clinical *S. aureus* strain P2. This was also observed in our previous study, with the results obtained with the laboratory strain *S. aureus* ATCC 25923 significantly differing from the ones obtained with *S. aureus* P2 under the same experimental conditions. The latter has a key relevance in assessing the applicability of biofilm inhibitors as part of titanium implantable devices, as it was isolated from patients with orthopedic device-related infections. For this reason, the additional measurement of the preventive capacity of the biofilm inhibitors against the clinical *S. aureus* P2 strain is of great relevance.

Compound **DHA1** was shown to further potentiate *S. aureus* ATCC 25923 clearance caused by HL-60, and this effect was further confirmed by SEM. This compound also managed to effectively prevent the adherence of the clinical strain (*S. aureus* P2), but in this case it was difficult to establish if the reduction in bacterial attachment caused by **DHA1** (as well as **FLA1** and **RIF**) was due to a combined antimicrobial effect with the HL-60 cells or if it was due to their intrinsic antimicrobial capacity, as their effects in monoculture were equal to those observed in coculture with the HL-60 cells. These results emphasized the importance of not limiting the in vitro experimentation to laboratory strains, especially in cases aimed at finding compounds effective against medical device-associated infections, as the results obtained with laboratory strains may overestimate the efficacy of the compound.

These results demonstrate that **DHA1** seems to further increase *S. aureus* ATCC 25923 clearance by HL-60, while none of the other compounds negatively affect the antimicrobial effect of these immune cells. This is also of high relevance, as adverse effects on the immune response could be detrimental, and even increase the risk of infection. As an example, Croes et al. [39] biofunctionalized the surface of titanium implants with chitosan-based coatings that were incorporated with different concentrations of silver nanoparticles. Despite the good antimicrobial results obtained in the in vitro tests, these coatings did not demonstrate antibacterial effects *in vivo*. Due to the toxicity of the silver nanoparticles on the immune cells, these coatings aggravated infection-mediated bone remodeling, including increased osteoclast formation and inflammation-induced new bone formation.

Similar results were obtained in LDPE tubes, with the bacterial clearance capacity of the HL-60 cells against *S. aureus* ATCC 25923 also observed, and **DHA1** further potentiated this activity. The prevention of *S. aureus* adherence on the surface of endotracheal tubes may have potential to significantly reduce the rates of ventilator-associated pneumonia caused by these bacteria [40]. Additionally, by preventing the attachment of *S. aureus*, the attachment of *P. aeruginosa* may also be hampered, as several infection models demonstrated how early colonization by *S. aureus* facilitated subsequent *P. aeruginosa* colonization [41,42]. The development of a dual-species biofilm is expected to not only strongly worsen the pathology but significantly complicate the treatment [43].

In this study, and in concordance with our previous findings, **DHA1** was identified as the best candidate to be incorporated into implantable devices. This is because, in addition to its intrinsic antibiofilm capacity against both laboratory and clinical strains of *S. aureus*, it also seems to be able to enhance *S. aureus* ATCC 25923 clearance by HL-60 cells. Further mechanistic studies should be performed in order to elucidate if **DHA1** has a direct effect on the antimicrobial activity of PMNs. In the near future, we plan to assess the effects of **DHA1** on phagocytosis, ROS production, and formation of neutrophil extracellular traps (NETs).

Given the promising results that **DHA1** showed, both in our previous publications and in the current one, this biofilm inhibitor is involved in plans to be integrated as part of a titanium coating by means of 3D printing, with the coating formulation consisting of **DHA1**-loaded poly(lactic-co-glycolic acid) micro particles that suspended in a gelatine–methacrylate gel inkjet-printed onto titanium coupons [44]. The printing procedure is already validated and the prototype materials are currently being tested. In the case of endotracheal tubes, our current results suggest that all of the tested antimicrobial compounds would be beneficial for the prevention of *S. aureus* adhesion and subsequent biofilm formation, but **DHA1** appears to be the best candidate.

To the best of our knowledge, this is one of the first studies showing a positive effect of novel antimicrobials on the antibiofilm-clearing capacity of immune cells. This is of particular relevance because it does not only provide new alternatives to fight against the immense burden of bacterial biofilms, but it also sets the basis for a new in vitro system to accelerate the drug discovery process, thereby enabling better selection of antimicrobial incorporation into medical devices.

5. Conclusions

We showed the suitability of a coculture of *S. aureus* and differentiated HL-60 cells as an in vitro assay to assess the applicability of antimicrobial compounds for the protection of medical devices. As a proof-of-concept, we tested three antimicrobials, concluding that the DHA derivative **DHA1** is the best candidate for incorporation into implantable devices, as it does not only prevent biofilm formation on titanium but also seems to enhance the antibacterial capability of immune cells. On the other hand, according to our results, all of the antimicrobial compounds studied here, i.e., the two DHA derivatives, **DHA1** and **DHA2,** and the flavan derivative, **FLA1**, can tentatively be regarded as promising candidates to form part of anti-infective endotracheal tubes.

Supplementary Materials: Supplementary materials can be found at http://www.mdpi.com/2076-2607/8/11/1757/s1.

Author Contributions: Conceptualization, I.R., A.F. and S.A.J.Z.; methodology, J.Z.P., C.G.-P., M.R. and I.R.; software, I.R.; validation, I.R. and A.F.; formal analysis, I.R.; investigation, I.R.; resources, J.Y.-K., J.Z.P., K.S., A.F. and S.A.J.Z.; data curation, I.R.; writing—original draft preparation, I.R.; writing—review and editing, A.F., M.R., S.A.J.Z., C.G.-P. and J.Y.-K.; visualization, I.R., and A.F.; supervision, A.F., K.S. and S.A.J.Z.; project administration, A.F.; funding acquisition, A.F., S.A.J.Z. and J.Y.-K. All authors read and agreed to the published version of the manuscript.

Funding: This project has received funding from the European Union's Horizon 2020 research and innovation program under the Marie Skłodowska—Curie grant agreement No. 722467 (PRINT-AID consortium). We also acknowledge the funding received from the Jane and Aatos Erkko Foundation.

Acknowledgments: We thank Teemu J. Kinnari and Ramón Pérez-Tanoira for their support during experimentation with the clinical strain. We thank Firas Hamdan for his help obtaining the human serum.

Conflicts of Interest: The authors declare no conflict of interest.

References

1. Friedrich, A.W. Control of hospital acquired infections and antimicrobial resistance in Europe: The way to go. *Wien. Med. Wochenschr.* **2019**, *169*, 25–30. [CrossRef]

2. Cassini, A.; Högberg, L.D.; Plachouras, D.; Quattrocchi, A.; Hoxha, A.; Simonsen, G.S.; Colomb-Cotinat, M.; Kretzschmar, M.E.; Devleesschauwer, B.; Cecchini, M.; et al. Attributable deaths and disability-adjusted life-years caused by infections with antibiotic-resistant bacteria in the EU and the European Economic Area in 2015: A population-level modelling analysis. *Lancet Infect. Dis.* **2019**, *19*, 56–66. [CrossRef]

3. Ferri, M.; Ranucci, E.; Romagnoli, P.; Giaccone, V. Antimicrobial resistance: A global emerging threat to public health systems. *Crit. Rev. Food Sci. Nutr.* **2017**, *57*, 2857–2876. [CrossRef] [PubMed]

4. Percival, S.L.; Suleman, L.; Vuotto, C.; Donelli, G. Healthcare-associated infections, medical devices and biofilms: Risk, tolerance and control. *J. Med. Microbiol.* **2015**, *64*, 323–334. [CrossRef] [PubMed]

5. Khan, H.A.; Baig, F.K.; Mehboob, R. Nosocomial infections: Epidemiology, prevention, control and surveillance. *Asian Pac. J. Trop. Biomed.* **2017**, *7*, 478–482. [CrossRef]

6. Paharik, A.E.; Horswill, A.R. The Staphylococcal Biofilm: Adhesins, Regulation, and Host Response. *Microbiol. Spectr.* **2016**, *4*. [CrossRef]

7. Hall-Stoodley, L.; Costerton, J.W.; Stoodley, P. Bacterial biofilms: From the Natural environment to infectious diseases. *Nat. Rev. Microbiol* **2004**, *2*, 95. [CrossRef]

8. Kumar, A.; Alam, A.; Rani, M.; Ehtesham, N.Z.; Hasnain, S.E. Biofilms: Survival and defense strategy for pathogens. *Int. J. Med. Microbiol.* **2017**, *307*, 481–489. [CrossRef]

9. Singh, S.; Singh, S.K.; Chowdhury, I.; Singh, R. Understanding the Mechanism of Bacterial Biofilms Resistance to Antimicrobial Agents. *Open Microbiol. J.* **2017**, *11*, 53–62. [CrossRef]

10. Ferreira Tde, O.; Koto, R.Y.; Leite, G.F.; Klautau, G.B.; Nigro, S.; Silva, C.B.; Souza, A.P.; Mimica, M.J.; Cesar, R.G.; Salles, M.J. Microbial investigation of biofilms recovered from endotracheal tubes using sonication in intensive care unit pediatric patients. *Braz. J. Infect. Dis. Off. Publ. Braz. Soc. Infect. Dis.* **2016**, *20*, 468–475. [CrossRef] [PubMed]

11. Bardes, J.M.; Waters, C.; Motlagh, H.; Wilson, A. The Prevalence of Oral Flora in the Biofilm Microbiota of the Endotracheal Tube. *Am. Surg.* **2016**, *82*, 403–406. [CrossRef]

12. Li, H.; Song, C.; Liu, D.; Ai, Q.; Yu, J. Molecular analysis of biofilms on the surface of neonatal endotracheal tubes based on 16S rRNA PCR-DGGE and species-specific PCR. *Int. J. Clin. Exp. Med.* **2015**, *8*, 11075–11084. [PubMed]

13. Vandecandelaere, I.; Matthijs, N.; Van Nieuwerburgh, F.; Deforce, D.; Vosters, P.; De Bus, L.; Nelis, H.J.; Depuydt, P.; Coenye, T. Assessment of microbial diversity in biofilms recovered from endotracheal tubes using culture dependent and independent approaches. *PLoS ONE* **2012**, *7*, e38401. [CrossRef] [PubMed]

14. Arciola, C.R.; Campoccia, D.; Montanaro, L. Implant infections: Adhesion, biofilm formation and immune evasion. *Nat. Rev. Microbiol.* **2018**, *16*, 397–409. [CrossRef]

15. Kurtz, S.M.; Lau, E.; Watson, H.; Schmier, J.K.; Parvizi, J. Economic burden of periprosthetic joint infection in the United States. *J. Arthroplast.* **2012**, *27*, 61–65.e1. [CrossRef] [PubMed]

16. Kaufman, M.G.; Meaike, J.D.; Izaddoost, S.A. Orthopedic Prosthetic Infections: Diagnosis and Orthopedic Salvage. *Semin. Plast. Surg.* **2016**, *30*, 66–72. [CrossRef]

17. Arciola, C.R.; Campoccia, D.; Ehrlich, G.D.; Montanaro, L. Biofilm-based implant infections in orthopaedics. *Adv. Exp. Med. Biol.* **2015**, *830*, 29–46. [CrossRef]

18. Roh, J.S.; Sohn, D.H. Damage-Associated Molecular Patterns in Inflammatory Diseases. *Immune Netw.* **2018**, *18*, e27. [CrossRef]

19. Franz, S.; Rammelt, S.; Scharnweber, D.; Simon, J.C. Immune responses to implants—A review of the implications for the design of immunomodulatory biomaterials. *Biomaterials* **2011**, *32*, 6692–6709. [CrossRef]

20. Gristina, A.G.; Naylor, P.T.; Myrvik, Q. The Race for the Surface: Microbes, Tissue Cells, and Biomaterials. In *Molecular Mechanisms of Microbial Adhesion*; Springer: New York, NY, USA, 1989; pp. 177–211.

21. Perez-Tanoira, R.; Han, X.; Soininen, A.; Aarnisalo, A.A.; Tiainen, V.M.; Eklund, K.K.; Esteban, J.; Kinnari, T.J. Competitive colonization of prosthetic surfaces by *Staphylococcus aureus* and human cells. *J. Biomed. Mater. Res. Part A* **2017**, *105*, 62–72. [CrossRef] [PubMed]

22. Stones, D.H.; Krachler, A.M. Against the tide: The role of bacterial adhesion in host colonization. *Biochem. Soc. Trans.* **2016**, *44*, 1571–1580. [CrossRef] [PubMed]

23. Jaskiewicz, M.; Janczura, A.; Nowicka, J.; Kamysz, W. Methods Used for the Eradication of Staphylococcal Biofilms. *Antibiotics* **2019**, *8*, 174. [CrossRef]

24. Reigada, I.; Perez-Tanoira, R.; Patel, J.Z.; Savijoki, K.; Yli-Kauhaluoma, J.; Kinnari, T.J.; Fallarero, A. Strategies to Prevent Biofilm Infections on Biomaterials: Effect of Novel Naturally-Derived Biofilm Inhibitors on a Competitive Colonization Model of Titanium by *Staphylococcus aureus* and SaOS-2 Cells. *Microorganisms* **2020**, *8*, 345. [CrossRef] [PubMed]

25. Manner, S.; Vahermo, M.; Skogman, M.E.; Krogerus, S.; Vuorela, P.M.; Yli-Kauhaluoma, J.; Fallarero, A.; Moreira, V.M. New derivatives of dehydroabietic acid target planktonic and biofilm bacteria in *Staphylococcus aureus* and effectively disrupt bacterial membrane integrity. *Eur. J. Med. Chem.* **2015**, *102*, 68–79. [CrossRef] [PubMed]

26. Manner, S.; Skogman, M.; Goeres, D.; Vuorela, P.; Fallarero, A. Systematic Exploration of Natural and Synthetic Flavonoids for the Inhibition of *Staphylococcus aureus* Biofilms. *Int. J. Mol. Sci.* **2013**, *14*, 19434–19451. [CrossRef]

27. Jhunjhunwala, S. Neutrophils at the Biological–Material Interface. *ACS Biomater. Sci. Eng.* **2018**, *4*, 1128–1136. [CrossRef]

28. Hochbaum, A.I.; Kolodkin-Gal, I.; Foulston, L.; Kolter, R.; Aizenberg, J.; Losick, R. Inhibitory effects of D-amino acids on *Staphylococcus aureus* biofilm development. *J. Bacteriol.* **2011**, *193*, 5616–5622. [CrossRef]

29. Esteban, J.; Gomez-Barrena, E.; Cordero, J.; Martin-de-Hijas, N.Z.; Kinnari, T.J.; Fernandez-Roblas, R. Evaluation of quantitative analysis of cultures from sonicated retrieved orthopedic implants in diagnosis of orthopedic infection. *J. Clin. Microbiol.* **2008**, *46*, 488–492. [CrossRef]

30. Collins, S.J.; Ruscetti, F.W.; Gallagher, R.E.; Gallo, R.C. Terminal differentiation of human promyelocytic leukemia cells induced by dimethyl sulfoxide and other polar compounds. *Proc. Natl. Acad. Sci. USA* **1978**, *75*, 2458–2462. [CrossRef]

31. Lu, T.; Porter, A.R.; Kennedy, A.D.; Kobayashi, S.D.; DeLeo, F.R. Phagocytosis and killing of *Staphylococcus aureus* by human neutrophils. *J. Innate Immun.* **2014**, *6*, 639–649. [CrossRef]

32. van Kessel, K.P.M.; Bestebroer, J.; van Strijp, J.A.G. Neutrophil-Mediated Phagocytosis of *Staphylococcus aureus*. *Front. Immunol.* **2014**, *5*, 467. [CrossRef] [PubMed]

33. Hiltunen, A.K.; Savijoki, K.; Nyman, T.A.; Miettinen, I.; Ihalainen, P.; Peltonen, J.; Fallarero, A. Structural and Functional Dynamics of *Staphylococcus aureus* Biofilms and Biofilm Matrix Proteins on Different Clinical Materials. *Microorganisms* **2019**, *7*, 584. [CrossRef]

34. Sunzel, B.; Söderberg, T.A.; Reuterving, C.O.; Hallmans, G.; Holm, S.E.; Hänström, L. Neutralizing effect of zinc oxide on dehydroabietic acid-induced toxicity on human polymorphonuclear leukocytes. *Biol. Trace Elem.* **1991**, *30*, 257–266. [CrossRef]

35. Czerwińska, M.E.; Dudek, M.K.; Pawłowska, K.A.; Pruś, A.; Ziaja, M.; Granica, S. The influence of procyanidins isolated from small-leaved lime flowers (Tilia cordata Mill.) on human neutrophils. *Fitoterapia* **2018**, *127*, 115–122. [CrossRef] [PubMed]

36. Pawłowska, K.A.; Hałasa, R.; Dudek, M.K.; Majdan, M.; Jankowska, K.; Granica, S. Antibacterial and anti-inflammatory activity of bistort (*Bistorta officinalis*) aqueous extract and its major components. Justification of the usage of the medicinal plant material as a traditional topical agent. *J. Ethnopharmacol.* **2020**, *260*, 113077. [CrossRef]

37. Manda-Handzlik, A.; Bystrzycka, W.; Wachowska, M.; Sieczkowska, S.; Stelmaszczyk-Emmel, A.; Demkow, U.; Ciepiela, O. The influence of agents differentiating HL-60 cells toward granulocyte-like cells on their ability to release neutrophil extracellular traps. *Immunol. Cell Biol.* **2018**, *96*, 413–425. [CrossRef]

38. Yaseen, R.; Blodkamp, S.; Luthje, P.; Reuner, F.; Vollger, L.; Naim, H.Y.; von Kockritz-Blickwede, M. Antimicrobial activity of HL-60 cells compared to primary blood-derived neutrophils against *Staphylococcus aureus*. *J. Negat. Results Biomed.* **2017**, *16*, 7. [CrossRef] [PubMed]

39. Croes, M.; Bakhshandeh, S.; van Hengel, I.A.J.; Lietaert, K.; van Kessel, K.P.M.; Pouran, B.; van der Wal, B.C.H.; Vogely, H.C.; Van Hecke, W.; Fluit, A.C.; et al. Antibacterial and immunogenic behavior of silver coatings on additively manufactured porous titanium. *Acta Biomater.* **2018**, *81*, 315–327. [CrossRef]

40. Seitz, A.P.; Schumacher, F.; Baker, J.; Soddemann, M.; Wilker, B.; Caldwell, C.C.; Gobble, R.M.; Kamler, M.; Becker, K.A.; Beck, S.; et al. Sphingosine-coating of plastic surfaces prevents ventilator-associated pneumonia. *J. Mol. Med.* **2019**, *97*, 1195–1211. [CrossRef]

41. Lyczak, J.B.; Cannon, C.L.; Pier, G.B. Lung infections associated with cystic fibrosis. *Clin. Microbiol. Rev.* **2002**, *15*, 194–222. [CrossRef]

42. Alves, P.M.; Al-Badi, E.; Withycombe, C.; Jones, P.M.; Purdy, K.J.; Maddocks, S.E. Interaction between *Staphylococcus aureus* and *Pseudomonas aeruginosa* is beneficial for colonisation and pathogenicity in a mixed biofilm. *Pathog. Dis.* **2018**, *76*. [CrossRef]

43. Beaudoin, T.; Yau, Y.C.W.; Stapleton, P.J.; Gong, Y.; Wang, P.W.; Guttman, D.S.; Waters, V. *Staphylococcus aureus* interaction with *Pseudomonas aeruginosa* biofilm enhances tobramycin resistance. *NPJ Biofilms Microbiomes* **2017**, *3*, 25. [CrossRef] [PubMed]

44. Yang, Y.; Chu, L.; Yang, S.; Zhang, H.; Qin, L.; Guillaume, O.; Eglin, D.; Richards, R.G.; Tang, T. Dual-functional 3D-printed composite scaffold for inhibiting bacterial infection and promoting bone regeneration in infected bone defect models. *Acta Biomater.* **2018**, *79*, 265–275. [CrossRef]

 *microorganisms*

Article

The Effects of Silver Sulfadiazine on Methicillin-Resistant *Staphylococcus aureus* Biofilms

Yutaka Ueda [1], Motoyasu Miyazaki [2], Kota Mashima [1], Satoshi Takagi [3], Shuuji Hara [4], Hidetoshi Kamimura [1] and Shiro Jimi [5,*]

[1] Department of Pharmacy, Fukuoka University Hospital, Fukuoka 814-0180, Japan; ueday@fukuoka-u.ac.jp (Y.U.); mashimakota210@fukuoka-u.ac.jp (K.M.); kamisan@fukuoka-u.ac.jp (H.K.)
[2] Department of Pharmacy, Fukuoka University Chikushi Hospital, Fukuoka 818-8502, Japan; motoyasu@fukuoka-u.ac.jp
[3] Departments of Plastic, Reconstructive and Aesthetic Surgery, Faculty of Medicine, Fukuoka University, Fukuoka 814-0180, Japan; stakagi@fukuoka-u.ac.jp
[4] Department of Drug Informatics, Faculty of Pharmaceutical Sciences, Fukuoka University, Fukuoka 814-0180, Japan; harashu@fukuoka-u.ac.jp
[5] Central Lab for Pathology and Morphology, Faculty of Medicine, Fukuoka University, Fukuoka 814-0180, Japan
* Correspondence: sjimi@fukuoka-u.ac.jp; Tel.: +81-92-801-1011

Received: 23 September 2020; Accepted: 5 October 2020; Published: 8 October 2020

Abstract: Methicillin-resistant *Staphylococcus aureus* (MRSA), the most commonly detected drug-resistant microbe in hospitals, adheres to substrates and forms biofilms that are resistant to immunological responses and antimicrobial drugs. Currently, there is a need to develop alternative approaches for treating infections caused by biofilms to prevent delays in wound healing. Silver has long been used as a disinfectant, which is non-specific and has relatively low cytotoxicity. Silver sulfadiazine (SSD) is a chemical complex clinically used for the prevention of wound infections after injury. However, its effects on biofilms are still unclear. In this study, we aimed to analyze the mechanisms underlying SSD action on biofilms formed by MRSA. The antibacterial effects of SSD were a result of silver ions and not sulfadiazine. Ionized silver from SSD in culture media was lower than that from silver nitrate; however, SSD, rather than silver nitrate, eradicated mature biofilms by bacterial killing. In SSD, sulfadiazine selectively bound to biofilms, and silver ions were then liberated. Consequently, the addition of an ion-chelator reduced the bactericidal effects of SSD on biofilms. These results indicate that SSD is an effective compound for the eradication of biofilms; thus, SSD should be used for the removal of biofilms formed on wounds.

Keywords: biofilm; MRSA; silver ion; silver sulfadiazine; wound infections

1. Introduction

Biofilms (BFs) are a cause of chronic infections.   Several types of symbiotic bacteria, such as *Staphylococcus aureus* and *Pseudomonas aeruginosa*, colonize our body and form BFs [1–3]. Methicillin-resistant *S. aureus* (MRSA) causes soft-tissue infections, indwelling catheter-associated infections, bacteremia, endocarditis, and osteomyelitis. Approximately 80% of chronic wound infections are attributed to bacteria or BFs [1]. BFs produce a subpopulation of drug-resistant cells called persister cells [4–7]. The BF matrix predominantly contains extracellular polysaccharides [8,9], and interacts with other molecules, including quorum-sensing signaling molecules/autoinducers, polypeptides, lectins, lipids, and extracellular DNA [10–12]. Specific molecules targeting BFs and effective drugs for BF eradication have not been identified yet. Therefore, the BF itself becomes a serious exacerbation factor in antimicrobial resistance.

Silver sulfadiazine (SSD) has been used as an exogenous antimicrobial agent since the 1960s [13,14], and is used on partial and full thickness burns to prevent infection [15,16]. It is registered on the World Health Organization's List of Essential Medicines [17]. SSD possesses broad-spectrum antibacterial activity, reacting nonspecifically to Gram-negative and Gram-positive bacteria, causing distortion of the cell membrane and inhibition of DNA replication [14,18–20]. Nevertheless, common side effects of SSD include pruritus and pain at the site of administration [21], as well as decreased white blood cell counts, allergic reactions, bluish-gray skin discoloration, and liver inflammation [22–24]. However, the Cochrane systematic review (2010) [25] did not recommend the use of SSD because of insufficient evidence on whether silver-containing dressings or topical agents promote wound healing or prevent wound infection. In addition, specific antibiotics are currently obtainable.

To overcome these shortcomings, new nanomedicine technologies using SSD have been adopted to enhance its antibacterial activity [26]. Unfortunately, SSD cytotoxicity was also enhanced [27]. SSD is poorly soluble and has limited penetration through intact skin [28,29]. When it comes in contact with body fluids, free sulfadiazine (SD) can be absorbed systemically, metabolized in the liver [30], and excreted in urine [31].

Notably, the effects of SSD on BFs have been widely studied [32,33]; however, their mechanisms of action have not been investigated yet. In this study, we investigated the essential mechanisms underlying the antibacterial action of SSD, especially against BFs, using SSD elements, silver and SD, and analyzed the effects of the compounds on a clinical strain of BF-forming MRSA.

2. Materials and Methods

2.1. Preparation of BF Chips

ATCC BAA-2856 (OJ-1) [34–36], a high-BF-forming strain of MRSA, was used. BF chips were prepared as previously described [36,37]. In brief, one colony grown on tryptic soy agar (TSA) was inoculated in tryptic soy broth (TSB) and grown at 37 °C until the optical density (OD) = 0.57 (λ = 578 nm). A 1000-fold diluted bacterial solution was used for BF formation in the culture, and a plastic sheet for an overhead projector (3M Japan Ltd., Tokyo, Japan) was used as the substrate. Sterilized plastic sheets (1 × 8 cm) were immersed in bacterial solution and incubated to obtain a mature and uniform BF. After incubation, the sheets were washed 3 times with 10 mL of 0.01 M phosphate buffered saline solution (pH 7.4) to remove planktonic cells. Plastic sheets cut into 1 × 1 cm pieces termed "BF chips" were primarily used in the experiments unless otherwise stated.

2.2. Determination of BF Mass and Number of Live Bacteria

BF mass using crystal violet (CV) stain and live bacteria number by colony forming unit (CFU) counts were determined using a BF sheet (1 × 8 cm) in 10 mL media. CV-positive BF mass was measured according to a previously reported method [38] with some modifications. In brief, BFs stained with CV (Sigma-Aldrich, Tokyo, Japan) were eluted in 1 mL of 30% acetic acid in an RIA tube (Eiken Chemical Co., Ltd., Tokyo, Japan). Absorbance (λ = 570 nm) was measured using a spectrometer (GENESYS 10S VIS, Thermo Scientific, LMS, Tokyo, Japan). Bacteria on BF sheets were dissociated using an ultrasonic generator (Sonifier 250; Branson Ultrasonics, Emerson Japan, Ltd., Kanagawa, Japan), and colony-forming units (CFUs) were assessed.

2.3. Antibacterial Effects of Compounds on BFs

Ethylene oxide-gas sterilized silver nitrate (FUJIFILM Wako Pure Chemical Corporation, Osaka, Japan), SSD (ALCARE Co., Ltd., Tokyo, Japan), and SD (FUJIFILM Wako Pure Chemical Corporation) were freshly prepared in TSB at a maximum concentration of 11,200 μM and serially diluted to 43.75 μM. After dipping the BF chip in 5-mL TSB-containing tubes in the presence of different concentrations of $AgNO_3$, SSD, and SD, the tubes were incubated at 37 °C for 24 h. Because SSD and SD were insoluble in liquid, cultures were continuously and gently mixed on a horizontal-rotation shaker (G10: New Brunswick Scientific, New Brunswick, NJ, USA) at 120 rpm.

The minimum antibacterial concentrations of compounds for bacteria derived from BF were determined by methods described previously [37] with some modifications (Figure S1). After incubating the bacteria for 24 h with different compounds at different concentrations, the culture tubes were kept at 4 °C for 1 h in order for the compounds to sediment, before the turbidity in the supernatant was measured (λ = 578 nm). The values were used to assess the minimum inhibitory concentration for planktonic cells derived from BFs (bMIC). Next, 10 µL of medium from each previously used tube for the bMIC analysis, was blotted on a 1 × 1 cm filter paper on TSA, and then they were incubated overnight at 37 °C. The colonies that grew around the paper were used to assess the minimum bactericidal concentration for planktonic cells derived from BF (bMBC). BF chips used for the bMIC analysis were directly placed on TSA and incubated overnight at 37 °C, and the minimum BF eradication concentration (MBEC) was determined.

In some studies, ethylenediamine-*N,N,N′,N′*-tetraacetic acid (EDTA) (FUJIFILM Wako Pure Chemical Corporation) (pH 7.4) at a concentration of 680 µM was used to chelate ions including silver.

2.4. Measurement of Liberated Silver Ions in the Media

After BF chips were incubated in TSB at 37 °C for 24 h in the presence of different concentrations of $AgNO_3$ and SSD, media were centrifuged (EX-136: TOMY SEIKO Co. Ltd., Tokyo, Japan) at 3000 rpm for 10 min, and filtered with a 0.45-µm filter unit (Merck Millipore, Darmstadt, Germany) to remove bacteria and SSD aggregates. Ionized silver in the media was diluted 4 times with water and was quantified with AGT-131 using a NI-Ag kit (range: 0.01–0.25 ppm Ag-ion/mg (mL), Japan Ion Co., Tokyo, Japan).

2.5. SSD Inducing Direct/Indirect Bactericidal Effects on BFs

To avoid direct contact of insoluble SSD with BFs, a closed chamber with a semipermeable membrane (Intercell S well chamber: KURABO INDUSTRIES LTD, Osaka, Japan) was used along with a syringe gasket to close the lid. Different concentrations of SSD in 500 µL TSB were injected into the chamber, which rested on 4.5 mL of TSB with the BF chip in a tube. The tube was cultured at 120 rpm at 37 °C for 24 h. For the control, 500 µL of SSD was directly added into TSB with the BF chip, and the chamber was filled with 500 µL TSB floated on the medium. After incubation, bMIC, bMBC, and MBEC were determined.

2.6. Quantification of the SD Attachment on BFs

The SD amount was quantified using the diazo-coupling reaction with *N,N*-diethyl-*N′*-1-naphthylethylenediamine oxalate, also known as Tsuda's reagent (FUJIFILM Wako Pure Chemical Corporation) [39]. BFs formed on the surface of plastic tubes after 24 h incubation at 37 °C were washed 3 times with phosphate buffered saline. Freshly prepared SD was added at concentrations between 43.75 and 11,200 µM in 3 mL of TSB in a tube and was incubated at 37 °C for 1 h. After incubation, BFs were washed 3 times to remove any unbound SD on the BFs, and 500 µL of 1 N HCl was added to the tubes, from which 70 µL of the solution was mixed with 20 µL of 10% sodium nitrite (FUJIFILM Wako Pure Chemical Corporation) and reacted for 2 min on ice. Then, 100 µL of 2.5% ammonium amidosulfate (FUJIFILM Wako Pure Chemical Corporation) was added and reacted for 1 min, and 100 µL of Tsuda's reagent was added to the solution and mixed. The solutions in a 96-well plate (Becton, Dickinson and Company, Franklin Lakes, NJ, USA) were measured (λ = 550 nm) using a microplate reader (Model680; Bio-Rad Laboratories, Inc., Hercules, CA, USA). The SD calibration curve was also prepared (Figure S2).

2.7. Morphological Analysis of the Effects of AgNO₃, SSD, and SD on BFs

Plastic sheets in TSB with or without bacteria were incubated at 37 °C for 24 h. After incubation, the sheets were placed in media containing $AgNO_3$, SSD, and SD at a concentration of 2800 μM and incubated at 37 °C for 24 h. After incubation, the sheets were fixed in 5% formalin (pH 7.4), and stained with crystal violet. Another set of unstained sheets was placed in the air for more than one week to be turned black due to the sulfur-oxidized silver properties.

2.8. Ethics Approval

All methods involving bacterial handling were performed in accordance with the relevant guidelines and regulations under the Fukuoka University's experiment regulations.

2.9. Data and Statistical Analysis

Results from two different experimental groups initially underwent a distribution analysis using the F-test, before the Student's *t*-test or Mann–Whitney *U* test were performed. *p* values < 0.05 were considered to denote statistical significance. Sample numbers and repeated experiments are indicated in the legends of the figures and tables. Data are expressed as mean ± standard error.

3. Results

3.1. Antibacterial Effects of AgNO₃, SSD, and SD

Antibacterial effects of $AgNO_3$, SSD, and SD on BF chips were obtained (Table 1): bMIC for planktonic bacteria from BFs: 700 μM, 700 μM, >5600 μM; bMBC for planktonic bacteria from BFs 2800 μM, 1400 μM, >5600 μM; and MBEC for bacteria in BFs: >5600 μM, 2800 μM, >5600 μM, respectively. Results showed that SD had no antibacterial effects, but SSD alone was effective in targeting BFs.

Table 1. Minimum antimicrobial concentration (bMIC) and minimum bactericidal concentration (bMBC) for planktonic bacteria from BFs and bacteria in BFs (minimum BF eradication concentration (MBEC)).

Compounds		0	175	350	700	1400	2800	5600 (μM)
	bMIC	+	+	+	−	−	−	−
AgNO₃	bMBC	+	+	+	+	+/−	−	−
	MBEC	+	+	+	+	+	+	+
	bMIC	+	+	+	−	−	−	−
SSD	bMBC	+	+	+	−/+	−	−	−
	MBEC	+	+	+	+/−	−/+	−	−
	bMIC	+	+	+	+	+	+	+
SD	bMBC	+	+	+	+	+	+	+
	MBEC	+	+	+	+	+	+	+

The antimicrobial effects of different concentrations of the compounds were determined using the BF chip method described in the Materials and Methods. The results of independent replicated examinations were evaluated: "−": 0/6 cases, "−/+":2/6 cases, "+/−":4/6 cases, "+": 6/6 cases. N size = 6; number of experimental replicates = 2.

3.2. Effects of AgNO₃, SSD, and SD on Viable Cells in BFs

Viable cell numbers on the BF chip were analyzed (Figure 1) using a CFU assay technique after collecting all bacteria from the BF chip. After 24-h incubation with $AgNO_3$ in different concentrations, bacterial growth was all arrested to some extent, but a dose-dependent suppression effect was not found. In contrast, SSD suppressed CFU values in a dose-dependent manner, especially at concentrations higher than 2800 μM, and the values reached approximately 1/100 of the initial density (0 h) and approximately 1/100,000 of cell density after 24 h of incubation. However, no effects on cell growth were found in any of the SD exposures.

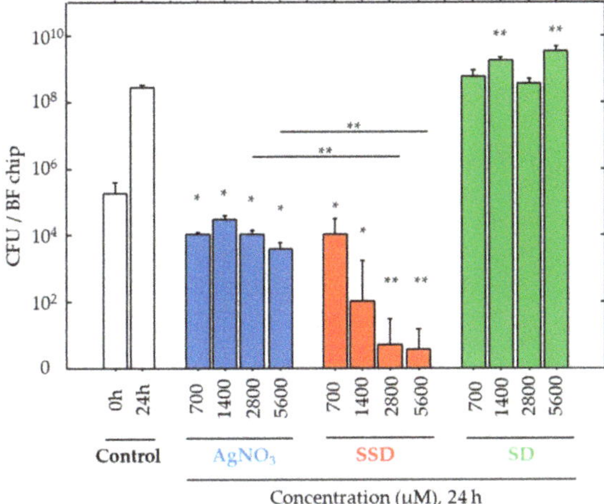

Figure 1. Colony forming unit (CFU) values for bacteria in biofilms (BFs). CFU values before (0 h) and after 24 h incubation at 37 °C in the vehicle alone (Control) and with AgNO$_3$, silver sulfadiazine (SSD), and sulfadiazine (SD) at different concentrations. Data are presented as the mean ± SE. N size = 5; number of experimental replicates = 2. *: $p < 0.05$, **: $p < 0.01$ vs. control after 24-h incubation with different compounds.

3.3. Silver Ion Release in the Culture with SSD

Antibacterial effects of SSD could be due to the ionized silver's action. As such, silver ions liberated in the cultures with AgNO$_3$ and SSD were measured. In both cases, liberated silver ions increased dose-dependently (Figure 2). Silver ions in the culture with AgNO$_3$ demonstrated a linear increase pattern, while those in the culture with SSD showed a logarithmic increase pattern. Significant differences were noted in the samples with concentrations greater than 700 µM (Figure 2). In the cultures with AgNO$_3$ and SSD at the concentration of 700 µM, silver ions increased to 5.91 µM and 4.64 µM, respectively. At the maximum concentration (11,200 µM), the silver ion concentration in the culture with AgNO$_3$ was three times higher than that in SSD culture (Figure 2).

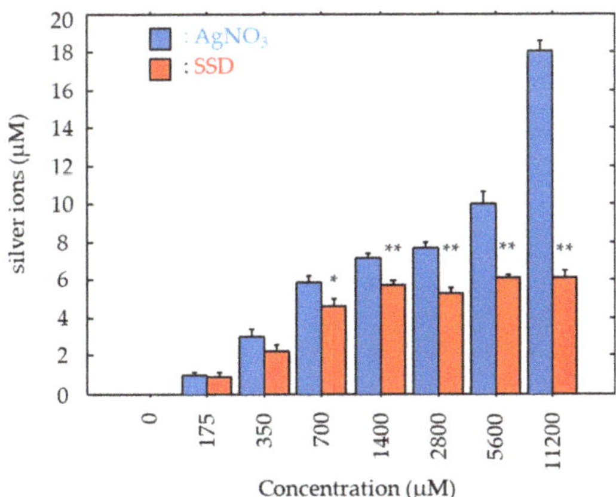

Figure 2. Medial ionized silver concentrations. After 24-h incubation at 37 °C in the presence of BF chips with AgNO$_3$ and SSD, medial ionized silver concentration was determined. Data are presented as the mean ± SE. N size = 3; number of experimental replicates = 2. *: $p < 0.05$ and **: $p < 0.01$ vs. AgNO$_3$.

3.4. Direct/Indirect Bactericidal Effect of SSDs on BFs

We hypothesized that the antibacterial effects of SSD could be due to its direct attachment to the BFs. To examine this mechanism, SSD was confined to a chamber, and a similar culture study was carried out. The bMIC in the culture with the chambered SSD was 1400 μM, which was two times greater than that in the culture with direct addition of SSD (Table 2). In terms of the bactericidal effects, the bMBC increased to >5600 μM, which was greater in the chambered SSD condition compared to that by direct SSD addition. Similarly, the MBEC value also increased to >5600 μM, which was greater in the chambered SSD compared to that in the direct SSD condition (Table 2). Therefore, the direct attachment of SSD to the BFs was necessary for initiating its antibacterial effects.

Table 2. Summary of antimicrobial threshold values.

	bMIC	(+EDTA)	bMBC	(+EDTA)	MBEC	(+EDTA)
<Direct>						
$AgNO_3$ (μM)	700	(700)	2800	(2800)	>5600	(>5600)
SSD (μM)	700	(1400)	1400	(2800)	2800	(>5600)
SSD chambered (μM)	1400	(ND)	>5600	(ND)	>5600	(ND)

ND: not determined.

3.5. Effects of EDTA on $AgNO_3$ and SSD-Induced Antibacterial Activity

EDTA is a chelator for silver ions [40]. To analyze the involvement of silver ions in the antibacterial effects, EDTA at a concentration of 640 μM was used. In the cultures with $AgNO_3$, EDTA did not affect bMIC, bMBC, and MBEC values (700 μM, 2800 μM, and >5600 μM, respectively, Table 2). However, all the values in the cultures with SSD were increased by the addition of EDTA (700 μM → 1400 μM, 1400 μM → 2800 μM, and 2800 μM → >5600 μM, respectively, Table 2).

3.6. Effects of Compounds on BF Eradication

The BF eradication effects of the compounds were analyzed at a dose of 2800 μM, an effective concentration for SSD (Table 2). The results of BF mass, CFU and morphological analysis are shown in Figure 3a–c, respectively. The BF eradication effects were shown to be SSD > $AgNO_3$ > SD, and SD had no effect. Silver-containing compounds, including $AgNO_3$ and SSD, displayed BF eradication activities, where SSD showed the greatest effects.

Given that SSD is chemically synthesized with SD and silver, we examined the BF eradication effects of a mixture of SD and $AgNO_3$ at a concentration of 2800 μM (Figure 3a,c). The values of BF mass and live bacteria number in the mixture were lower than those in SD alone, equivalent to those in $AgNO_3$ alone, but were still significantly higher than those in SSD, suggesting that SSD was still the most effective at eradicating BFs.

3.7. SD Deposition to BFs

In the control without BFs, the addition of any concentration of SD to the media only resulted in a consistently low level of SD being deposited on the plastic surface (Figure 3d). However, in the BF tubes, SD was deposited on the BFs in a dose-dependent manner, and prominent deposition started once concentrations of SD were greater than 1400 μM (Figure 3d).

3.8. BF Morphology

After exposure to the different compounds, the BFs stained with crystal violet are shown in the left panels of Figure 3b, where the overall purple staining intensity among the groups follows the next scheme: Control > SD > $AgNO_3$ > SSD. String-like structures (i.e., thickened BFs) found in the

Control group remained in the SD and AgNO$_3$ groups, and such structures were scarcely detected in the SSD group.

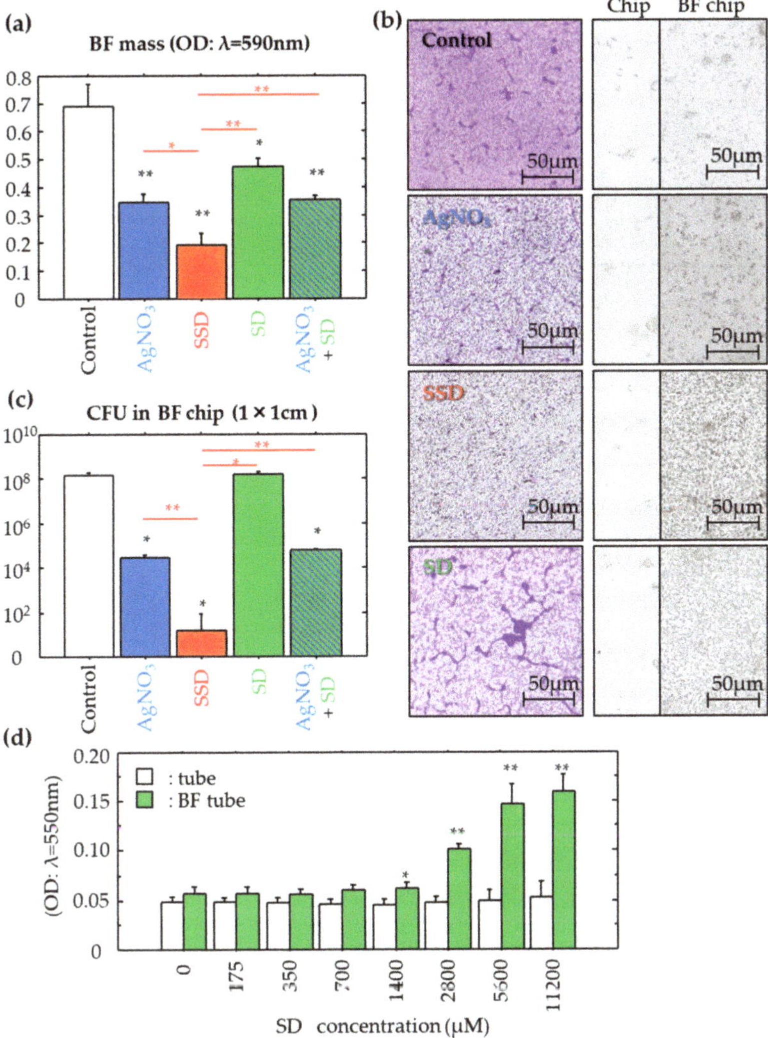

Figure 3. Effects of the compounds on mature BFs formed on plastic chips and SD deposition on BF chips. The effects of the compounds (2800 µM) after 24-h incubation at 37 °C on BF masses (**a**), morphology (**b**) and CFU values (**c**) were determined. Data are presented as the mean ± SE. N size = 5; number of experimental replicates = 2. *: $p < 0.05$ and **: $p < 0.01$ vs. negative control shown in white, and vs. SSD shown in red. In morphology, silver was deposited on the plastic chips with/without matured BFs after 24 h incubation at 37 °C with/without the different compounds. Sulfurized silvers in air are shown as black elements. Objective lens: ×20. In SD quantification, SD deposited on the BFs formed in plastic tubes after 24 h incubation at 37 °C was determined according to the method described in the Materials and Methods (**d**). Data are presented as the mean ± SE. N size = 5; number of experimental replicates = 2. *: $p < 0.05$ and **: $p < 0.01$ vs. BF-free control.

3.9. Localization of Silver on BFs

Oxidized silver turns black in color. No black spots were found on the plastic substrate even after exposure to all compounds, nor were there spots on the BFs following SD exposure (Figure 3b, right panels). In contrast, fine small black spots accumulated on the BFs after exposure to $AgNO_3$ and SSD. Moreover, SSD induced denser accumulation of black spots.

4. Discussion

BFs are formed by bacteria, which contain massive extracellular polysaccharides/exopolymeric substances, and acquire an antibiotic-tolerant nature, which is explained by the decreased drug penetration [41] and the appearance of dormant cells (i.e., persister cells) [42]. Thus, the eradication of BFs has become a difficult task. In this study, we examined the effect of SSD on MRSA-formed BFs. The SSD concentration in the medical drugs, namely Silvadene, etc., is of 1% (10,000 µg/mL), which could be diluted with tissue fluid exuded from wounds after application. The highest concentration used in this study was 4000 µg/mL (0.4%: 11,200 µM), which is slightly lower than that of the medical drug.

The bactericidal actions of silver can be explained through three different mechanisms: (1) the production of dissolved oxygen-derived reactive oxygen species by its catalytic activity [43,44]; (2) the cross-linkage with silver at the sites of hydrogen bonding between the double strands of DNA [44,45]; and (3) the inhibition of enzyme activities by intracellular silver ions [44,46]. SD, a sulfonamide, inhibits intracellular folate metabolism in bacteria, resulting in proliferation arrest. During the investigation of silver-containing agents, SD has been selected amongst different compounds due to its high and broad bactericidal effects [14,18–20]. We showed here that SSD is effective against MRSA, especially in the BF state, but SD itself had no effect.

Among the compounds tested, SSD significantly decreased BF mass and live bacteria number in BFs, which was a considerably greater difference than that of $AgNO_3$. This phenomenon was supported by silver deposition on the BF chip. We also examined the efficacy of the simultaneous addition of $AgNO_3$ and SD instead of the SSD. These effects are equivalent to those of $AgNO_3$ alone, but are significantly lower than SSD, indicating that the molecular form of SSD is important to induce a silver-related BF eradication. In contrast, it was reported that SSD was ineffective for MRSA BFs formed on a polycarbonate filter using a novel in vitro model (colony/drip-flow reactor) [32]. However, its exposure time was quite short (15 min) to induce an eradication reaction as compared with the present study (24 h).

In contrast, SSD and $AgNO_3$ induced a similar antibacterial potency in bMIC and bMBC, both of which were detected in the planktonic state derived from the BFs. In accordance with their threshold levels, silver ion concentrations reached more than 5 µM. In any case, such concentration could be necessary for growth inhibition and the killing of planktonic bacteria. However, in the range of effective doses, silver ions were always generated at higher levels in $AgNO_3$ conditions compared to those in SSD conditions. This reflects the contradiction of the bactericidal effects of SSD. The results suggest that some factor(s) other than simple diffusion of silver ion may be involved.

The MBEC level is a threshold for bacterial killing concentration in BFs. It was detectable only in SSD rather than $AgNO_3$ or SD, which was also confirmed by the CFU assay. It clearly showed that $AgNO_3$ could not efficiently kill the bacteria in BFs, and its effect remained within the level of growth inhibition. On the other hand, SSD strongly and dose-dependently depressed the live cell number in matured BFs, in which the live cell density at a concentration of 700 µM (the equivalent level of bMIC) reached 1/10,000 of the control after 24 h of incubation, and, at a concentration of 2800 µM (the equivalent level of MBEC), the number was 1/300 of the initial cell density and 1/300,000 of the control after 24-h incubation. These results suggest that the medial silver ion concentration is not a direct influencing factor for the SSD.

To clarify its mechanism of action, SSD was confined in a sealed chamber, by which SSD was constrained and bacteria could not penetrate beyond the membrane, but ionized silver could pass freely. As a result, silver ion concentrations in the chambered condition were lower at 700 and 1400 µM SSD, as compared with non-chambered conditions, but, in higher concentrations more than 2800 µM SSD, they became similar levels (Figure S3). Upon confinement, SSD-induced bMBC and MBEC levels were no longer seen. Our study also showed that SD itself had the property to bind to the BFs. Although, the exact binding mechanisms of SSD on BFs are still unknown, the direct attachment of SSD to BFs is crucial to induce bactericidal effects. Therefore, if the SSD binding site on BFs is identified in the near future, this would lead to the discovery of target molecules of the BFs formed by MRSA.

To validate the silver ion's role in the antibacterial properties of SSD, EDTA was used at a concentration of 680 µM, which was about 40 to 100 times greater than that of the liberated silver ions in the media with $AgNO_3$ and SSD. With the addition of EDTA, no changes in bMIC, bMBC, and MBEC were found in the culture with $AgNO_3$. In contrast, the addition of EDTA increased in all of the values in the culture with SSD. The mechanisms of antibacterial action between $AgNO_3$ and SSD are unknown; however, their mode of reaction might be different. Silver ions from $AgNO_3$ could be easily bound by negatively charged components in media as compared to SSD, and a bound–liberation cycle may be repeated. Moreover, the silver holding capacity in SSD may be greater than that in $AgNO_3$, by which SSD could reach the BFs. This mechanism may act especially on the BFs due to the selective adhesion of SD on BFs, which was confirmed by the greater deposition of silver on the BF chip incubated with SSD, whereas sites of the deposition on BFs could not be identified. It therefore seems likely that SSD may act on the bacteria settled in the BFs. After deposition of SSD, silver ions could be released in a micro environment in BFs, by which the opportunities for bacteria to kill could be increased, resulting in severe distortions of BF structure (Figure 3b).

As a limitation of the present study, one MRSA strain was only used. In future study, we will use different clinical strains, such as low-BF formers and high-BF formers in our collection [47].

5. Conclusions

SSD is a reliable anti-bacterial agent, and thus has recently been used as a coating material for indwelling catheters [48,49]. However, the mechanisms of action of SSD on BFs are still unclear. This study is the first to elucidate the mechanisms behind the efficacy of SSD on BFs. Therefore, it is possible that SSD preferably binds to BFs, and then it releases silver ions, by which bacteria settled in the BFs under a micro-environment are killed (Figure 4). In the future, further molecular levels of investigation are needed to identify the SSD binding site on BFs. Additionally, a systematic investigation of the use of SSD in BF-infected wounds should be performed in clinical practices rather than for infection prevention.

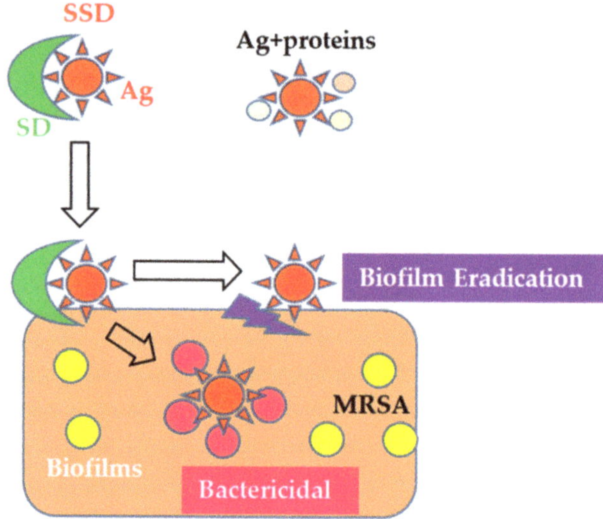

Figure 4. Working hypothesis of the action of SSD on biofilms formed by MRSA.

Supplementary Materials: The following are available online at http://www.mdpi.com/2076-2607/8/10/1551/s1, Figure S1: Summarized procedure of the BF chip method to determine bMIC, bMBC, and MBEC, Figure S2: Standard curve of SD, Figure S3: Filtered SSD and ionized silver concentration.

Author Contributions: All experimental designs were created by S.J. and Y.U. Bacterial experiments were carried out by Y.U., M.M., K.M., and S.H. Statistical analyses were performed by Y.U., S.T., S.H., and H.K. The manuscript was prepared by Y.U., M.M., and S.J. and revised by S.J., K.M., S.T., S.H., and H.K. All authors have read and agreed to the published version of the manuscript.

Funding: This study was partially supported by KAKENHI, a Grant-in-Aid for Scientific Research © (Grant Number 17K10038), Research-support-grant form ALCARE Co., Ltd., Tokyo Japan, Grant of The Clinical Research Promotion Foundation (2019), and the Fukuoka University Research Grand for Science Progression.

Acknowledgments: The authors thank Yui Ohshiro, Arisa Nita, Kaede Munekata, and Yuka Kawamura (Faculty of Pharmaceutical Science, Fukuoka University, Fukuoka, Japan) for their excellent technical assistance. We are also thankful for the research support and the valuable discussion with Takashi Kubo, ALCARE Co., Ltd., Tokyo, Japan.

Conflicts of Interest: The authors declare no conflict of interest.

References

1. Kiedrowski, M.R.; Horswill, A.R. New approaches for treating staphylococcal biofilm infections. *Ann. N. Y. Acad. Sci.* **2011**, *1241*, 104–121. [CrossRef] [PubMed]

2. Otto, M. Staphylococcal infections: Mechanisms of biofilm maturation and detachment as critical determinants of pathogenicity. *Annu. Rev. Med.* **2013**, *64*, 175–188. [CrossRef] [PubMed]

3. Abdulhaq, N.; Nawaz, Z.; Zahoor, M.A.; Siddique, A.B. Association of biofilm formation with multi drug resistance in clinical isolates of. *EXCLI J.* **2020**, *19*, 201–208. [CrossRef] [PubMed]

4. Evans, R.C.; Holmes, C.J. Effect of vancomycin hydrochloride on *Staphylococcus epidermidis* biofilm associated with silicone elastomer. *Antimicrob. Agents Chemother.* **1987**, *31*, 889–894. [CrossRef]

5. Stewart, P.S. Mechanisms of antibiotic resistance in bacterial biofilms. *Int. J. Med. Microbiol.* **2002**, *292*, 107–113. [CrossRef]

6. Singh, R.; Ray, P.; Das, A.; Sharma, M. Role of persisters and small-colony variants in antibiotic resistance of planktonic and biofilm-associated Staphylococcus aureus: An in vitro study. *J. Med. Microbiol.* **2009**, *58*, 1067–1073. [CrossRef]

7. Garcia, L.G.; Lemaire, S.; Kahl, B.C.; Becker, K.; Proctor, R.A.; Denis, O.; Tulkens, P.M.; Van Bambeke, F. Antibiotic activity against small-colony variants of *Staphylococcus aureus*: Review of in vitro, animal and clinical data. *J. Antimicrob. Chemother.* **2013**, *68*, 1455–1464. [CrossRef]

8. Flemming, H.C. The perfect slime. *Colloids Surf. B: Biointerfaces* **2011**, *86*, 251–259. [CrossRef]

9. Dragoš, A.; Kovács, Á. The Peculiar Functions of the Bacterial Extracellular Matrix. *Trends Microbiol.* **2017**, *25*, 257–266. [CrossRef]

10. Roche, F.M.; Downer, R.; Keane, F.; Speziale, P.; Park, P.W.; Foster, T.J. The N-terminal A domain of fibronectin-binding proteins A and B promotes adhesion of Staphylococcus aureus to elastin. *J. Biol. Chem.* **2004**, *279*, 38433–38440. [CrossRef]

11. Kline, K.A.; Fälker, S.; Dahlberg, S.; Normark, S.; Henriques-Normark, B. Bacterial adhesins in host-microbe interactions. *Cell Host Microbe* **2009**, *5*, 580–592. [CrossRef] [PubMed]

12. Lee, J.; Zhang, L. The hierarchy quorum sensing network in *Pseudomonas aeruginosa*. *Protein Cell* **2015**, *6*, 26–41. [CrossRef] [PubMed]

13. Carr, H.S.; Wlodkowski, T.J.; Rosenkranz, H.S. Silver sulfadiazine: In vitro antibacterial activity. *Antimicrob. Agents Chemother.* **1973**, *4*, 585–587. [CrossRef] [PubMed]

14. Fox, C.L.; Modak, S.M. Mechanism of silver sulfadiazine action on burn wound infections. *Antimicrob. Agents Chemother.* **1974**, *5*, 582–588. [CrossRef]

15. Percival, S.L.; Bowler, P.G.; Russell, D. Bacterial resistance to silver in wound care. *J. Hosp. Infect.* **2005**, *60*, 1–7. [CrossRef]

16. Heyneman, A.; Hoeksema, H.; Vandekerckhove, D.; Pirayesh, A.; Monstrey, S. The role of silver sulphadiazine in the conservative treatment of partial thickness burn wounds: A systematic review. *Burns* **2016**, *42*, 1377–1386. [CrossRef]

17. WHO. *World Health Organization Model List of Essential Medicines*; WHO: Geneva, Switzerland, 2019.

18. Hoffmann, S. Silver sulfadiazine: An antibacterial agent for topical use in burns. A review of the literature. *Scand. J. Plast. Reconstr. Surg.* **1984**, *18*, 119–126. [CrossRef]

19. Fuller, F.W.; Parrish, M.; Nance, F.C. A review of the dosimetry of 1% silver sulfadiazine cream in burn wound treatment. *J. Burn Care Rehabil.* **1994**, *15*, 213–223. [CrossRef]

20. Miller, A.C.; Rashid, R.M.; Falzon, L.; Elamin, E.M.; Zehtabchi, S. Silver sulfadiazine for the treatment of partial-thickness burns and venous stasis ulcers. *J. Am. Acad. Dermatol.* **2012**, *66*, e159–e165. [CrossRef]

21. Fuller, F.W. The side effects of silver sulfadiazine. *J. Burn Care Res.* **2009**, *30*, 464–470. [CrossRef]

22. Muller, M.J.; Hollyoak, M.A.; Moaveni, Z.; Brown, T.L.; Herndon, D.N.; Heggers, J.P. Retardation of wound healing by silver sulfadiazine is reversed by Aloe vera and nystatin. *Burns* **2003**, *29*, 834–836. [CrossRef]

23. Chaby, G.; Viseux, V.; Poulain, J.F.; De Cagny, B.; Denoeux, J.P.; Lok, C. Topical silver sulfadiazine-induced acute renal failure. *Ann. Dermatol. Vénéréologie* **2005**, *132*, 891–893. [CrossRef]

24. Abedini, F.; Ahmadi, A.; Yavari, A.; Hosseini, V.; Mousavi, S. Comparison of silver nylon wound dressing and silver sulfadiazine in partial burn wound therapy. *Int. Wound J.* **2013**, *10*, 573–578. [CrossRef]

25. Carter, M.J.; Tingley-Kelley, K.; Warriner, R.A. Silver treatments and silver-impregnated dressings for the healing of leg wounds and ulcers: A systematic review and meta-analysis. *J. Am. Acad. Dermatol.* **2010**, *63*, 668–679. [CrossRef] [PubMed]

26. Adhya, A.; Bain, J.; Ray, O.; Hazra, A.; Adhikari, S.; Dutta, G.; Ray, S.; Majumdar, B.K. Healing of burn wounds by topical treatment: A randomized controlled comparison between silver sulfadiazine and nano-crystalline silver. *J. Basic Clin. Pharm.* **2014**, *6*, 29–34. [CrossRef]

27. Liu, X.; Gan, H.; Hu, C.; Sun, W.; Zhu, X.; Meng, Z.; Gu, R.; Wu, Z.; Dou, G. Silver sulfadiazine nanosuspension-loaded thermosensitive hydrogel as a topical antibacterial agent. *Int. J. Nanomed.* **2019**, *14*, 289–300. [CrossRef] [PubMed]

28. Dellera, E.; Bonferoni, M.C.; Sandri, G.; Rossi, S.; Ferrari, F.; Del Fante, C.; Perotti, C.; Grisoli, P.; Caramella, C. Development of chitosan oleate ionic micelles loaded with silver sulfadiazine to be associated with platelet lysate for application in wound healing. *Eur. J. Pharm. Biopharm.* **2014**, *88*, 643–650. [CrossRef]

29. Kumar, P.M.; Ghosh, A. Development and evaluation of silver sulfadiazine loaded microsponge based gel for partial thickness (second degree) burn wounds. *Eur. J. Pharm. Sci.* **2017**, *96*, 243–254. [CrossRef]

30. Winter, H.R.; Unadkat, J.D. Identification of cytochrome P450 and arylamine N-acetyltransferase isoforms involved in sulfadiazine metabolism. *Drug Metab. Dispos.* **2005**, *33*, 969–976. [CrossRef]

31. Nouws, J.F.; Firth, E.C.; Vree, T.B.; Baakman, M. Pharmacokinetics and renal clearance of sulfamethazine, sulfamerazine, and sulfadiazine and their N4-acetyl and hydroxy metabolites in horses. *Am. J. Vet. Res.* **1987**, *48*, 392–402.

32. Agostinho, A.M.; Hartman, A.; Lipp, C.; Parker, A.E.; Stewart, P.S.; James, G.A. An in vitro model for the growth and analysis of chronic wound MRSA biofilms. *J. Appl. Microbiol.* **2011**, *111*, 1275–1282. [CrossRef] [PubMed]

33. Lemire, J.A.; Kalan, L.; Bradu, A.; Turner, R.J. Silver oxynitrate, an unexplored silver compound with antimicrobial and antibiofilm activity. *Antimicrob. Agents Chemother.* **2015**, *59*, 4031–4039. [CrossRef] [PubMed]

34. Makino, T.; Jimi, S.; Oyama, T.; Nakano, Y.; Hamamoto, K.; Mamishin, K.; Yahiro, T.; Hara, S.; Takata, T.; Ohjimi, H. Infection mechanism of biofilm-forming Staphylococcus aureus on indwelling foreign materials in mice. *Int. Wound J.* **2015**, *12*, 122–131. [CrossRef] [PubMed]

35. Haraga, I.; Abe, S.; Jimi, S.; Kiyomi, F.; Yamaura, K. Increased biofilm formation ability and accelerated transport of *Staphylococcus aureus* along a catheter during reciprocal movements. *J. Microbiol. Methods* **2017**, *132*, 63–68. [CrossRef]

36. Ueda, Y.; Mashima, K.; Miyazaki, M.; Hara, S.; Takata, T.; Kamimura, H.; Takagi, S.; Jimi, S. Inhibitory effects of polysorbate 80 on MRSA biofilm formed on different substrates including dermal tissue. *Sci. Rep.* **2019**, *9*, 3128. [CrossRef]

37. Jimi, S.; Miyazaki, M.; Takata, T.; Ohjimi, H.; Akita, S.; Hara, S. Increased drug resistance of meticillin-resistant *Staphylococcus aureus* biofilms formed on a mouse dermal chip model. *J. Med. Microbiol.* **2017**, *66*, 542–550. [CrossRef]

38. Bendouah, Z.; Barbeau, J.; Hamad, W.A.; Desrosiers, M. Use of an in vitro assay for determination of biofilm-forming capacity of bacteria in chronic rhinosinusitis. *Am. J. Rhinol.* **2006**, *20*, 434–438. [CrossRef]

39. Negoro, H. Colorimetric Estimation of PAS Using Tsuda Reagent. *Yakugaku Zasshi-J. Pharm. Soc. Jpn.* **1951**, *71*, 209–210. [CrossRef]

40. Saran, L.; Cavalheiro, E.; Neves, E.A. New aspects of the reaction of silver(I) cations with the ethylenediaminetetraacetate ion. *Talanta* **1995**, *42*, 2027–2032. [CrossRef]

41. Flemming, H.C.; Wingender, J.; Szewzyk, U.; Steinberg, P.; Rice, S.A.; Kjelleberg, S. Biofilms: An emergent form of bacterial life. *Nat. Rev. Microbiol.* **2016**, *14*, 563–575. [CrossRef]

42. Harms, A.; Maisonneuve, E.; Gerdes, K. Mechanisms of bacterial persistence during stress and antibiotic exposure. *Science* **2016**, *354*. [CrossRef] [PubMed]

43. Lansdown, A.B. Silver I: Its antibacterial properties and mechanism of action. *J. Wound Care* **2002**, *11*, 125–130. [CrossRef] [PubMed]

44. Marx, D.E.; Barillo, D.J. Silver in medicine: The basic science. *Burns* **2014**, *40*, S9–S18. [CrossRef]

45. Russell, A.D.; Hugo, W.B. Antimicrobial activity and action of silver. *Prog. Med. Chem.* **1994**, *31*, 351–370. [CrossRef] [PubMed]

46. Slawson, R.M.; Lee, H.; Trevors, J.T. Bacterial interactions with silver. *Biol. Met.* **1990**, *3*, 151–154. [CrossRef]

47. Oyama, T.; Miyazaki, M.; Yoshimura, M.; Takata, T.; Ohjimi, H.; Jimi, S. Biofilm-Forming Methicillin-Resistant Staphylococcus aureus Survive in Kupffer Cells and Exhibit High Virulence in Mice. *Toxins* **2016**, *8*, 198. [CrossRef]

48. de Sousa, J.K.T.; Haddad, J.P.A.; de Oliveira, A.C.; Vieira, C.D.; Dos Santos, S.G. In vitro activity of antimicrobial-impregnated catheters against biofilms formed by KPC-producing Klebsiella pneumoniae. *J. Appl. Microbiol.* **2019**, *127*, 1018–1027. [CrossRef]

49. Mohseni, M.; Shamloo, A.; Aghababaie, Z.; Afjoul, H.; Abdi, S.; Moravvej, H.; Vossoughi, M. A comparative study of wound dressings loaded with silver sulfadiazine and silver nanoparticles: In vitro and in vivo evaluation. *Int. J. Pharm.* **2019**, *564*, 350–358. [CrossRef]

microorganisms

Methicillin-Resistant *Staphylococcus epidermidis* Lineages in the Nasal and Skin Microbiota of Patients Planned for Arthroplasty Surgery

Emeli Månsson [1,2,*], Staffan Tevell [1,3], Åsa Nilsdotter-Augustinsson [4], Thor Bech Johannesen [5], Martin Sundqvist [6], Marc Stegger [1,5,†] and Bo Söderquist [1,6,†]

1 School of Medical Sciences, Faculty of Medicine and Health, Örebro University, SE-701 82 Örebro, Sweden; staffan.tevell@regionvarmland.se (S.T.); mtg@ssi.dk (M.S.); bo.soderquist@regionorebrolan.se (B.S.)
2 Centre for Clinical Research, Region Västmanland—Uppsala University, Hospital of Västmanland, Västerås, SE-721 89 Västerås, Sweden
3 Department of Infectious Diseases, Karlstad Hospital and Centre for Clinical Research and Education, County Council of Värmland, SE-651 82 Karlstad, Sweden
4 Department of Infectious Diseases, and Department of Clinical and Experimental Medicine, Linköping University, SE-60182 Norrköping, Sweden; asa.nilsdotter-augustinsson@liu.se
5 Department of Bacteria, Parasites and Fungi, Statens Serum Institut, 2300 Copenhagen, Denmark; thej@ssi.dk
6 Department of Laboratory Medicine, Clinical Microbiology, Faculty of Medicine and Health, Örebro University, SE-701 82 Örebro, Sweden; martin.sundqvist@regionorebrolan.se
* Correspondence: emeli.mansson@regionvastmanland.se
† These authors contributed equally to this work.

Citation: Månsson, E.; Tevell, S.; Nilsdotter-Augustinsson, s.; Johannesen, T.B.; Sundqvist, M.; Stegger, M.; Söderquist, B. Methicillin-Resistant *Staphylococcus epidermidis* Lineages in the Nasal and Skin Microbiota of Patients Planned for Arthroplasty Surgery. *Microorganisms* **2021**, *9*, 265. https://doi.org/10.3390/microorganisms9020265

Academic Editor: Rajan P. Adhikari
Received: 23 December 2020
Accepted: 26 January 2021
Published: 28 January 2021

Publisher's Note: MDPI stays neutral with regard to jurisdictional claims in published maps and institutional affiliations.

Abstract: *Staphylococcus epidermidis*, ubiquitous in the human nasal and skin microbiota, is a common causative microorganism in prosthetic joint infections (PJIs). A high proportion of PJI isolates have been shown to harbor genetic traits associated with resistance to/tolerance of agents used for antimicrobial prophylaxis in joint arthroplasties. These traits were found within multidrug-resistant *S. epidermidis* (MDRSE) lineages of multiple genetic backgrounds. In this study, the aim was to study whether MDRSE lineages previously associated with PJIs are present in the nasal and skin microbiota of patients planned for arthroplasty surgery but before hospitalization. We cultured samples from nares, inguinal creases, and skin over the hip or knee (dependent on the planned procedure) taken two weeks (median) prior to admittance to the hospital for total joint arthroplasty from 66 patients on agar plates selecting for methicillin resistance. *S. epidermidis* colonies were identified and tested for the presence of *mecA*. Methicillin-resistant *S. epidermidis* (MRSE) were characterized by Illumina-based whole-genome sequencing. Using this method, we found that 30/66 (45%) of patients were colonized with MRSE at 1–3 body sites. A subset of patients, 10/66 (15%), were colonized with MDRSE lineages associated with PJIs. The *qacA* gene was identified in MRSE isolates from 19/30 (63%) of MRSE colonized patients, whereas genes associated with aminoglycoside resistance were less common, found in 11/30 (37%). We found that MDRSE lineages previously associated with PJIs were present in a subset of patients' pre-admission microbiota, plausibly in low relative abundance, and may be selected for by the current prophylaxis regimen comprising whole-body cleansing with chlorhexidine-gluconate containing soap. To further lower the rate of *S. epidermidis* PJIs, the current prophylaxis may need to be modified, but it is important for possible perioperative MDRSE transmission events and specific risk factors for MDRSE PJIs to be investigated before reevaluating antimicrobial prophylaxis.

Keywords: *Staphylococcus epidermidis*; microbiota; multidrug resistance; genome sequencing; phylogenetic analyses; arthroplasty surgery

1. Introduction

Staphylococcus epidermidis is ubiquitous in the human microbiota of the skin and mucosal membranes. *S. epidermidis* is in several ways beneficial to the human host, involved in the regulation of wound healing and defense against virulent pathogens such as *Staphylococcus aureus* [1], and is an important causative microorganism in healthcare-associated infections (HAI) [2], such as prosthetic joint infections (PJIs). PJIs are classified as acute hematogenous, early postoperative, or chronic [3] and are associated with significant morbidity [4] and mortality [5,6]. The international incidence of PJIs is estimated at 1% for total hip arthroplasties (THAs) and 1–2% for total knee arthroplasties (TKAs) [7]. Due to the rising number of arthroplasty surgeries performed annually, the number of patients with PJIs is increasing [8–11]. In early postoperative and chronic PJIs, coagulase-negative staphylococci (CoNS) are the most common causative microorganisms, reported in 25–29% of infections [12–15], and among CoNS, *S. epidermidis* is the most frequently reported species (61–85%) [16–18].

The pathogenesis of *S. epidermidis* PJIs is not fully understood, but bacteria may be introduced either intraoperatively during surgery or in the immediate postoperative period before the wound is completely healed [14]. Preventive measures to reduce the rate of PJIs include the optimization of modifiable host risk factors [19], the operating room environment [20], and systemic antimicrobial prophylaxis [21]. In Sweden, the national guideline for prevention of infections in prosthetic joint surgery (PRISS, https://lof.se/patientsakerhet/vara-projekt/priss/rekommendationer) recommends systemic antimicrobial prophylaxis with cloxacillin and surgery performed in operating theatres with ultra-clean air (<5 colony forming units/m^3). In addition, preoperative skin cleansing with chlorhexidine-gluconate (CHG)-containing soap should be performed twice before surgery. Aminoglycoside-loaded bone cement of both components is also used in 60% of all primary THAs in Sweden [data from the Swedish Hip Arthroplasty Register (SHAR), www.shpr.se].

Previous molecular epidemiology studies have demonstrated that multidrug-resistant (MDR) *S. epidermidis* (MDRSE) lineages, such as sequence type (ST) 2, ST5, ST59, and ST215, are predominant in PJIs [22–25]. This was verified in a recent comparative genomics study in which our group, using genome-wide association methods on approximately 300 sequenced isolates, demonstrated that *S. epidermidis* from PJIs differed from isolates retrieved from nasal mucosa by harboring genetic traits associated with resistance to/tolerance of compounds used for infection prevention in prosthetic joint surgery, such as betalactams, aminoglycosides, and chlorhexidine [26]. This is worrisome, as methicillin-resistant *S. epidermidis* (MRSE) and MDRSE are associated with worse treatment outcomes in PJIs than susceptible *S. epidermidis* [27,28].

In contrast to the predominance of a limited number of MDRSE lineages found in PJIs, metagenomic sequencing has demonstrated considerable strain heterogeneity of *S. epidermidis* in the human microbiome [29]. Culture-based studies have demonstrated low rates (4–25%) of methicillin-resistance in CoNS from patients sampled prior to or on admission to orthopedic wards [30–33]. The majority of CoNS isolates in these studies were presumably *S. epidermidis*, but identification at the species level was generally not performed. Selective media were only used in two of these studies [31,33], and it is possible that colonization with MRSE (and MDRSE) before admission (in low relative abundance) is more common than previously thought.

To further improve the prevention of *S. epidermidis* PJIs, it is important to investigate whether PJI-associated MDRSE lineages are present in the microbiota of patients scheduled for prosthetic joint surgery before hospital admission or whether these lineages are acquired during hospitalization. The hypothesis in this study was that MDRSE associated with PJIs is present in the pre-admission microbiota of arthroplasty surgery patients but in low relative abundance and thus is largely undetected by nonselective culturing methods. To test this, we sub-cultured samples from the nares, inguinal creases, and skin over the hip/knee on selective agar plates to optimize the retrieval of MRSE isolates, which were further characterized by genomic analyses. The aim was to study whether MDRSE lineages associated with PJIs are present in the nasal and skin microbiota of patients scheduled for prosthetic joint surgery before hospital admission.

2. Materials and Methods

One hundred patients scheduled for prosthetic hip- or knee joint replacement surgery were sampled from the nares, inguinal crease, and skin over the hip/knee (depending on the joint to be replaced) two weeks (median) before admission to a hospital using flocked swabs (ESwab, Copan Italia Spa, Brescia, Italy). Sampling was performed at three orthopedic outpatient clinics in central Sweden: A) Karlstad (regional hospital, n = 40; Jan 2013 to Jan 2014), B; Västerås (regional hospital, n = 37; Dec 2012 to May 2014), and C) Linköping (university hospital, n = 23; Mar 2013 to Jan 2015) (Figure S1), following informed consent. The number of primary THAs and TKAs performed annually at these hospitals was A, 457 (2013); B, 732 (2013); and C, 67 (2014) (data from SHAR and the Swedish Knee Arthroplasty Register). The study protocol was approved by the Regional Ethical Review Board of Uppsala, Sweden (project identification code 2012/092, 18/04/2012).

2.1. Culture, DNA Extraction, and Illumina Sequencing

All samples were sent to the Department of Laboratory Medicine, Clinical Microbiology, Örebro University Hospital and subsequently plated on blood agar [Columbia II Agar 3.9% *w/v* (Oxoid, Basingstoke, Hampshire, UK)] supplemented with 6% defibrinated horse blood (SVA, Uppsala, Sweden)). All growths on the plates were harvested and stored frozen (-80 °C) in a preservation medium (trypticase soy broth, BD Diagnostic Systems, Sparks, MD, USA), supplemented with 0.3% *w/v* yeast extract (BD Diagnostic Systems) and 29% horse serum (SVA, Uppsala, Sweden).

In the present study, we included samples from the nasal mucosa, inguinal crease, and skin over either the hip or knee from 66 patients in whom *S. epidermidis* was identified in nasal samples after culturing on nonselective agar plates (Figure 1). From the frozen bacterial suspensions of the original subculture, 10 µL was cultured on selective media (Mueller-Hinton II agar 3.8% *w/v* (BD Diagnostic Systems) supplemented with 5 mg/L cefoxitin (Sigma Aldrich, Darmstadt, Germany) and 10 µL on a chromogenic medium primarily used for detecting methicillin-resistant *S. aureus* (MRSA) (CHROMID MRSA, bioMérieux, Marcy-l'Étoile, France). After 48 h of incubation in an aerobic atmosphere at 36 °C, five colonies (when available) with macroscopic colony morphology consistent with staphylococci were randomly chosen and identified to the species level using MALDI-TOF MS (Microflex LT, Bruker Daltonik, Bremen, Germany) with Biotyper 3.1 DB7311, DB7854, and DB8468 (Bruker Daltonik). The isolates identified as *S. epidermidis* were analyzed for the presence of *mecA* using PCR [34] or, in selected cases, loop-mediated isothermal amplification (LAMP, Genie II, Amplex Diagnostics GmbH, München, Germany and easyplex MRSAplus, Amplex).

The DNA was purified using the Roche MagNA Pure 96 (F. Hoffman-La Roche Ltd., Basel, Switzerland) system after incubation overnight in 36 °C on blood agar plates (SSI Diagnostica, Denmark). The extracted DNA was quantified using the Qubit fluorometer (Invitrogen, Waltham, MA, USA), followed by library preparation using the Nextera XT DNA Library Prep Kit (Illumina Inc., San Diego, CA, USA), according to the manufacturer's protocol. Sequencing was performed on a NextSeq 550 platform (Illumina Inc., San Diego, CA, USA) to obtain paired-end 151 bp reads. The generated sequencing data were subjected to quality control using bifrost (https://github.com/ssi-dk/bifrost) to ensure adequate sequencing depth of all isolates. Intraspecies contamination was checked using NASP v1.0.0 [35]. All genome sequences were archived at the European Nucleotide Archive under project ID PRJEB34788 with accession numbers ERR3585469–ERR3585625.

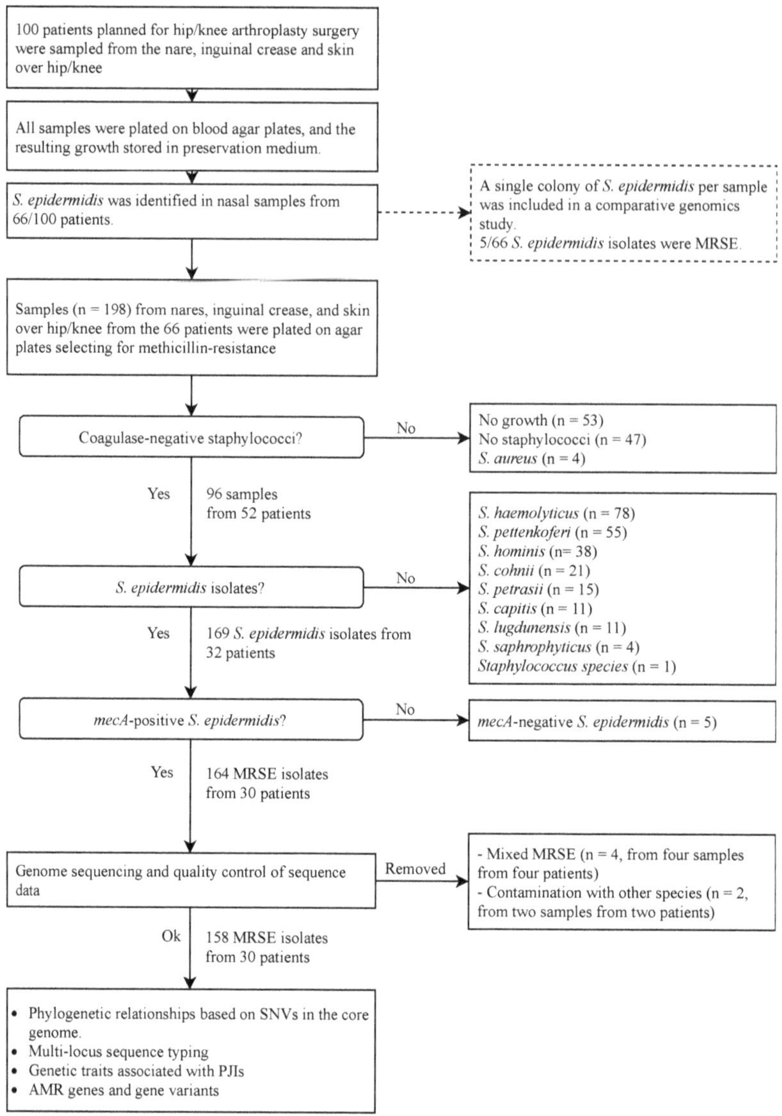

Figure 1. Origin of the isolates sequenced within this study and overview of analyses performed: one-hundred patients planned for arthroplasty surgery were sampled from their nares, inguinal creases, and skin over their hip/knee joints. *S. epidermidis* was retrieved from nasal samples plated on Müller–Hinton (MH) agar plates from 66 patients. A single colony of S. epidermidis was randomly chosen from the nasal sample of these 66 patients. Methicillin-resistant *S. epidermidis* (MRSE) was identified in samples from five out of the 66 patients. To investigate to what extent this underestimated the true colonization rate with MRSE, we plated samples from nares, inguinal creases, and skin over the hip/knee joints for these 66 patients on agar plates selecting for methicillin resistance. Coagulase-negative staphylococci were identified by colony morphology and species-determined using MALDI-TOF. The isolates identified as *S. epidermidis* were analyzed for the presence of *mecA*, and *mecA*-positive *S. epidermidis* isolates were further characterized by Illumina sequencing to determine the phylogenetic relationships, multilocus sequence types, genes associated with prosthetic joint infections (PJIs), and antimicrobial resistance (AMR) genes and gene variants.

2.2. Phylogenetic Analysis

Sequencing data were aligned to the *S. epidermidis* RP62a reference chromosome (GenBank accession number NC_002976.3) using NASP v1.0.0 [35] with BWA-MEM v0.7.12 [36] and subsequent single nucleotide polymorphism (SNP) calling using GATK [37]. If a variant was present in <90% of the base calls per site per individual isolate or a minimum coverage of 10 was not met, the position was excluded across the collection to retain only high-quality variant callings. Phylogenies were obtained with RAxML v8.2.1 [38] using the GTRCAT model and 100 replicates for bootstrapping. The sequence data from four *S. epidermidis* PJI isolates representing the major PJI-associated lineages ST2a, ST2b, ST5, and ST215 were included in the phylogenetic analysis comprising all isolates. Within-sample isolates with ≤3 single nucleotide variant (SNV) distances were regarded as representative multiple isolates of the same strain [39,40]. A single isolate per strain and sample (the sequence with the highest base-pair coverage vs. the reference) was retained, and a phylogenetic tree depicting diversity of MRSE strains retrieved from the nares, inguinal crease, and skin area over the hip or knee before hospitalization was inferred (Figure 2).

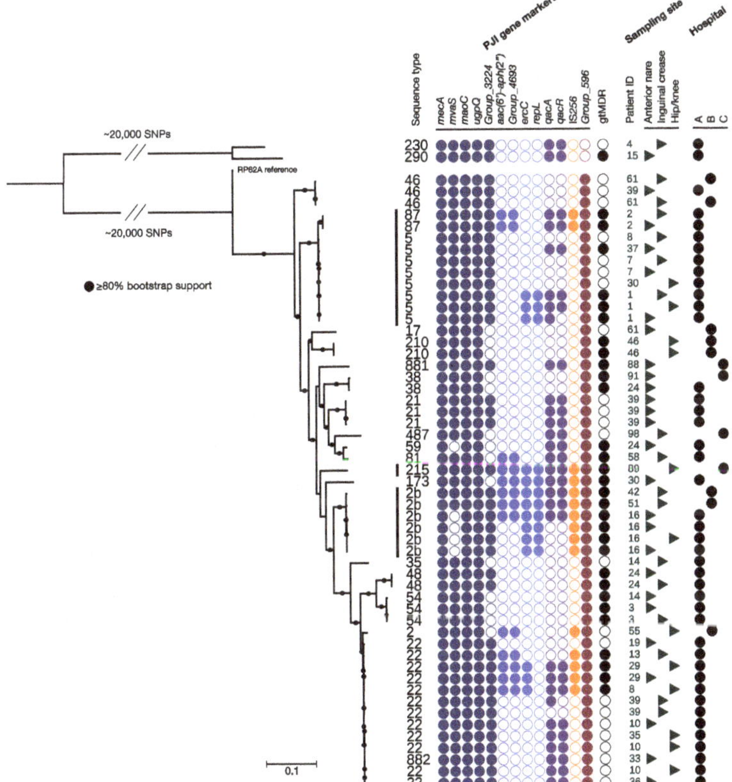

Figure 2. Phylogenetic relatedness of 55 *mecA*-positive *S. epidermidis* isolates retrieved from sampling of 66 patients prior to hospitalization for prosthetic joint surgery: the lineages previously associated with PJIs in Sweden (ST5, ST215, and ST2b) are indicated with a solid bar. Columns from left to right: sequence type (ST), genes previously associated with PJI origin (genes associated with SCC*mec* [dark blue], aminoglycoside resistance [medium blue], macrolide-lincosamide resistance [light blue], chlorhexidine tolerance [purple], IS*256* [orange], and genes coding for proteins of unknown function [dark red]), genotypic multidrug-resistant (black), patient running number, sample origin (body area), and hospital. The mid-point rooted phylogeny was based on 66,152 SNPs, and the scale bar indicates substitutions per site.

2.3. Identification of Genes and Gene Variants Associated with Antimicrobial Resistance

After assembly using SPAdes [41], the genes and their location in the assemblies were identified using Prokka [42]. ABRicate (https://github.com/tseemann/abricate) was used to search the assembled genomes for genes associated with betalactam (*mecA*), aminoglycoside (*aac(6′)-aph(2″)* and *aadD*), fusidic acid (*fusB* and *fusC*), and macrolide-lincosamide (*erm*(A), *erm*(C), *lnu*(A), *mph*(C), *msr*(A), *vat*(B), *vga*(A), and *vga*(B)) resistance present in the ResFinder database [43] accessed on 17 June 2019. The gene presences were determined based on a >80% hit length and >90% sequence identity. Previously identified gene variants associated with fusidic acid (*fusA*) [44], rifampicin (*rpoB*) [45,46], fluoroquinolone (*gyrA*, *grlA*, and *grlB*) [47,48], and trimethoprim/sulfamethoxazole (*dfrG* and F99Y variants of *folA*) [49] resistance were investigated using BLASTN searches in Biomatters Geneious Prime v2019.2.1 (Biomatters Ltd., Auckland, NewZealand) as previously described [26]. Isolates with the presence of genes and/or point mutations associated with resistance towards ≥3 of the following antimicrobial categories were defined as multidrug-resistant (MDR): methicillin, fusidic acid, macrolide-lincosamide, rifampicin, aminoglycosides, fluoroquinolones, and trimethoprim/sulfamethoxazole.

2.4. Identification of STs

Allelic profiles and sequence types were assigned using *mlst* (https://github.com/tseemann/mlst) based on the PubMLST typing scheme for *S. epidermidis* (https://pubmlst.org/sepidermidis/) [50]. ST2 was divided into ST2a and ST2b lineages based on previous findings [26].

2.5. Identification of Genetic Traits in S. epidermidis Associated with PJIs

Genes in *S. epidermidis* previously associated with PJIs [26] were determined using Mykrobe [51] on raw sequencing data as previously described [26]. Genes were identified using 90% sequence length and a minimum coverage of five as the cutoffs.

2.6. Statistics

The Chi-square test and Student's *t*-test were used to test the differences in the MRSE colonization rates and in mean age between the patients recruited between the three hospitals. Fisher's exact test was used to test the differences in genotypic antimicrobial resistance in MRSE from different body areas. A *p*-value (two-sided) < 0.05 was considered statistically significant. All statistical analysis was performed using IBM SPSS Statistics v25 except for the normal approximation confidence interval for the percentage of patients colonized with MDRSE, which was calculated using www.openepi.com v3.01.

3. Results

A total of 198 samples from 66 patients (33 women and 33 men) sampled at a median of 15 [interquartile range (IQR) 10.0–41.0] days before arthroplasty surgery (46 THAs and 20 TKAs) were included in the study. The mean age, which was 67 years (SD 10.4) for the entire study population, differed between hospitals: patients recruited from regional hospital A were older [mean age 70.6 (SD 10.8)] than patients recruited from university hospital C [mean age 59.4 (SD 7.7)] (*p* = 0.0004).

In total, 403 CoNS, originating from 96 samples from 52 patients, were identified on any of the methicillin-resistance selective agar plates. No growth was found on any of the agars for 53 (27%) samples [nares, *n* = 20 (30%); inguinal crease, *n* = 8 (12%); and hip/knee, *n* = 25 (38%)], and no colonies with the macroscopic appearance of staphylococci were found in the 47 samples (Figure 1). The most frequently identified CoNS species was *S. epidermidis* (*n* = 169, 33 patients), followed by *S. haemolyticus* (*n* = 78, 18 patients), *S. pettenkoferi* (*n* = 55, 10 patients), *S. hominis* (*n* = 31, six patients), *S. cohnii* (*n* = 21, four patients), *S. petrasii* (*n* = 15, five patients), *S. capitis* (*n* = 11, four patients), and *S. lugdunensis* (*n* = 11, four patients) (Figure 1). One isolate, identified at the genus level as a *Staphylococcus* species by DB7311 and DB7854, was not available for identification with DB8468 as it had succumbed in the freezer. *S. aureus* was retrieved from two samples from two patients; none of these isolates was *mecA/C* positive.

Out of 169 *S. epidermidis* isolates, 164 isolates originating from 44 samples from 30 patients (30/66, 45% of patients, 95% C.I. 33–57%) were verified to be *mecA*-positive and were further characterized by genome sequencing. The colonization rates with MRSE varied between hospitals: A, 20/29 (69%); B, 6/19 (32%); and C, 4/18 (22%) ($p = 0.003$, df = 2). MRSE was retrieved from a single body site in 17 of the 30 patients colonized with MRSE: seven patients in the nasal samples, six patients in the inguinal crease samples, and four patients from the skin over the joint that was to be replaced (Table 1). In samples from 11 patients, MRSE was retrieved from the nasal mucosa and either the inguinal crease ($n = 7$) or site of planned surgery ($n = 4$); in one MRSE-colonized patient, MRSE was retrieved from the inguinal crease and the site of planned surgery but not from the nasal sample; and in one patient, MRSE was retrieved in all three samples (Table 1).

Table 1. Sequence-types (STs) of methicillin-resistant *S. epidermidis* (MRSE) per patient and sample (nasal/inguinal crease/skin over hip/knee): number in bold indicates multidrug-resistant MRSE. Intra-patient single nucleotide variant (SNV) distances between samples (far right) highlight that MRSE with identical STs differ by 1–40 SNVs between two samples from the same patient.

				STs of MRSE Strains per Sample			
Patient	Age	Sex	Hospital	Nasal Mucosa	Inguinal Crease	Hip/Knee	SNV Distance between Samples [1]
1	84	M	A	**5**	**5**	**5**	6–40
2	83	M	A	87	87	–	2
3	80	F	A	54	54	–	12
4	79	F	A	–	230	–	
7	76	F	A	5	5	–	1
8	75	M	A	–	5	**22**	3917
10	75	F	A	22	–	22, 22	13–20
13	73	F	A	–	**22**	–	
14	71	M	A	54	35	–	3917
15	71	F	A	290	–	–	
16	70	F	A	**2, 2, 2**	–	**2**	9–10
19	61	M	A	22	–	–	
24	54	F	A	48, 38, 59	48	–	3–8466
29	43	F	A	**22**	–	**22**	19
30	73	F	A	**173**	–	5	5842
33	78	F	A	882	–	–	
35	73	F	A	–	–	22	
36	60	F	A	22	–	–	
37	83	F	A	5	–	–	
39	74	F	A	21, 21, 21, 46	22, 22	–	5641–6381
42	77	M	B	–	**2**	–	
46	70	M	B	–	–	**210, 210**	
51	61	F	B	–	**2**	–	
55	60	M	B	–	–	2	
58	81	M	B	–	81	–	
61	57	F	B	17	46, 46	–	5411–5414
88	46	M	C	**881**	–	–	
89	64	M	C	–	–	215	
91	49	M	C	**38**	–	–	
98	53	F	C	–	487	–	

[1] Range when MRSE was found in more than one sample.

Following quality control of the sequence data, six isolates were excluded, four due to signs of intraspecies contamination based on the failed proportion filter using NASP and two due to contamination with other species using Kraken as implemented in bifrost. In total, 157 MRSEs were included in an initial phylogenetic analysis. Based on a conserved core genome of 1.8 Mbp (69.7%), 52 MRSE strains (isolates from the same patient separated by 0–3 SNVs) were identified in the 44 samples [1–4 MRSE strain(s) per sample; Table 1]. As presented above, 13 patients were colonized with MRSE at multiple sites (Table 1). A detailed phylogenetic analysis demonstrated that the same MRSE strain (separated by 0–3 SNVs) was retrieved from two body sites in three patients and that similar strains (here, defined as separated by $\leq$40 SNVs) were found at multiple sites in five patients. Six patients were colonized with MRSE that differed more than 40 SNVs between the sampled sites (range 3917–8466 SNVs); see Table 1. Twenty-three STs were identified among the 52 MRSE strains, including two novel STs: ST881 and ST882.

To visualize the overall relatedness of MRSE strains retrieved from pre-admission sampling, a second phylogeny including only a single isolate per MRSE strain per sample (n = 55; identical strains were found in two samples for three patients) was constructed (Figure 2).

The MRSE belonging to PJI-associated lineages were identified in samples from ten patients, ST2b lineage, n = 3; ST5 lineage, n = 6; and ST215 lineage, n = 1, while no isolate belonging to the ST2a lineage was found. Nine patients were colonized with MRSE isolates belonging to the ST22 lineage, and one patient was colonized with the ST2 isolate that was closely related to the ST22 lineage (previously referred to as ST2c [26]).

The genes previously associated with PJI in *S. epidermidis* [26] are presented in Figure 2. The *aac(6')-aph(2'')* gene (associated with aminoglycoside resistance) was present in MRSE isolates from 11/30 MRSE-colonized patients (37%), *ermC* (associated with macrolide-lincosamide resistance) was present in eight (27%) patients, and *qacA* (associated with chlorhexidine tolerance) was present in 19 (63%) patients. The transposable element IS*256* was present in MRSE isolates from 10 patients (33%). The *group_596* gene was found in 50 out of the 52 MRSE strains (28/30 MRSE colonized patients).

The genes and gene variants associated with resistance towards fluoroquinolones, fusidic acid, macrolides-lincosamides, rifampicin, and trimethoprim-sulfamethoxazole were determined in the MRSE isolates. The isolates within a strain displayed the same pattern of AMR gene absence/presence in all but two cases. The frequency of genes and/or gene variants associated with antimicrobial resistance is presented per MRSE positive sample by antimicrobial category in Table 2. No mutations associated with rifampicin-resistance was found in any of the MRSE isolates. No statistically significant difference in resistance towards different antimicrobial agents was found between samples from the nares, inguinal crease, and skin over the hip/knee. Overall, MDRSE was retrieved from 19 patients (29%).

Table 2. Co-resistance (based on the presence of gene(s)/gene variants) to other antimicrobial agents in methicillin-resistant *S. epidermidis* (MRSE) retrieved from sampling of patients planned for prosthetic joint surgery prior to hospital admission (data are presented as the number (percentage) of samples with resistant MRSE). n.s = not significant.

Antimicrobial Agent	Nasal Mucosa (n = 19)	Inguinal Crease (n = 15)	Hip/Knee (n = 10)	Total	
Aminoglycosides	6 (32%)	5 (33%)	5 (50%)	16 (36%)	n.s
Fluoroquinolones	7 (37%)	6 (40%)	5 (50%)	18 (41%)	n.s
Fusidic acid	6 (32%)	2 (13%)	1 (10%)	9 (20%)	n.s
MLS [1]	12 (63%)	9 (60%)	7 (70%)	28 (63%)	n.s
Rifampicin	0	0	0	0	
TMP/SMX [2]	9 (63%)	5 (33%)	4 (40%)	18 (41%)	n.s
Multidrug-resistant	12 (63%)	8 (53%)	6 (60%)	26 (59%)	n.s

[1] macrolide-lincosamide-streptogramin; [2] trimethoprim-sulfamethoxazole.

4. Discussion

In this study, the aim was to investigate whether MDRSE lineages associated with PJIs are present in the nasal and skin microbiota of patients scheduled for prosthetic joint surgery before hospital admission. HAI with MDRSE is a growing concern, but unlike MRSA [52], there are no guidelines recommending the surveillance of MRSE and/or MDRSE [53]. The possible transmission events of MDRSE still remain unclear, and epidemiological studies of the prevalence of MDRSE in patients at risk of *S. epidermidis* infections has been called for [54]. To our knowledge, this is the first study using whole-genome sequencing to characterize MRSE isolates in the microbiota of patients scheduled for prosthetic joint surgery. We show that a subset of patients (10/66) were colonized with MDRSE lineages associated with PJIs [22,26] and that a number of patients were colonized with MDRSE lineage ST22 (a single locus variant of ST2), represented in publicly available international *S. epidermidis* sequences from PJIs [26]. All ST22-colonized patients were recruited from hospital A, where no data on the molecular epidemiology of *S. epidermidis* in PJIs was available. Overall, MRSEs were retrieved from almost half of patients in this study.

The majority of MRSE-colonized patients (19/30) were colonized with MRSE in the nares, but just over one third (11/30, including 4/10 patients colonized with lineages that predominate in PJIs) were not, suggesting that nasal screening alone is insufficient to detect individuals colonized with MRSE before surgery. Furthermore, we found that colonization with MRSE is likely to be vastly underestimated when a culture-based method using nonselective media is used. In the present study, we used samples from patients from whom *S. epidermidis* had already been identified from nasal samples after culturing on nonselective media. MRSE was only identified in 8% (5/66) of nasal samples using nonselective media, whereas by using selective media, we detected MRSE in nasal samples from an additional 14 patients (in total 29%, 19/66) and from an additional 11 patients from other body areas, summing up to a MRSE colonization rate of 45% (30/66 patients). This indicates that the vast majority of patients colonized with MRSE are missed when a single *S. epidermidis* colony is randomly chosen from a culture of a nasal swab plated on nonselective media.

The proportion of patients colonized with MRSE varied between the three hospitals. This difference could plausibly be explained by differences in prior hospitalization, as colonization rates of methicillin-resistant CoNS has been found to increase from 4–25% to 25–81% during hospitalization for orthopedic surgery [30–32], and almost twice as high carriage rates of methicillin-resistant CoNS have been found in patients undergoing revision THA (46%) compared with primary THA (24%) [55]. In this study, we had no access to data regarding prior admissions, but patients recruited from hospital A (with the highest MRSE colonization rate) were significantly older than those recruited from hospital C (with the lowest MRSE colonization rate), supporting the possibility that prior hospitalization rates may have differed between hospitals.

Preoperative whole-body cleansing with CHG-containing soap at home is recommended in national guidelines for prosthetic joint surgery in Sweden and is supported by the International Consensus Meeting on Musculoskeletal Infections despite a lack of robust evidence [56]. The increased tolerance of *S. epidermidis* to CHG is associated with the gene *qacA* that encodes an efflux pump [57], and *qacA* was one of the genetic traits associated with *S. epidermidis* isolated from PJIs in our previous study [26]. Here, where samples were obtained before whole-body cleansing with CHG-containing soap, almost one third of patients (19/66, 29%) were colonized with MRSE strains harboring *qacA*, possibly in low relative abundance. As *qacA* was found to be rare in methicillin-susceptible colonizing isolates [26], it can be speculated that the relative proportion of MDRSE in the microbiota after preoperative whole-body cleansing with CHG-containing soap may increase. This could subsequently increase the risk of perioperative contamination of the wound and subsequently the prosthetic joint with *S. epidermidis* resistant to standard systemic and locally administered antimicrobial prophylaxis. However, this remains to be explored, and most patients were not colonized with *qacA*-positive MRSE before hospital admission.

The prevalence of the remaining genes previously associated with *S. epidermidis* from PJIs varied. IS*256* and *ermC* were rare in general and almost exclusively found in MDR isolates, whereas *group_596*, encoding a hypothetical protein, was frequently present. The *aac(6′)-aph(2″)* gene encoding aminoglycoside resistance was rare among MRSE isolates other than those belonging to the ST2b, ST5, ST22, and ST215 lineages. This suggests that gentamicin in bone cement is effective for most MRSE strains that patients harbor before prosthetic joint surgery and, thus, can be an important supplement to systemically administered antimicrobial prophylaxis. The reported adjusted risk of revision is lower for cemented than uncemented THA in Sweden [58], but in a previous large registry-based study from Norway, the use of bone cement was associated with increased infections rates, as the rate of revision due to infection was increased for cemented procedures in which no antibiotics was admixed into the bone cement compared to uncemented procedures. However, an equal risk of revision for infection was found for uncemented and cemented hip arthroplasties with antibiotic-loaded cement [59].

The present study has several limitations. We excluded patients from whom *S. epidermidis* was not retrieved in nasal samples cultured on unselective media. However, the colonization rate of MRSE may have been underestimated from (i) MRSE outgrown by other species and/or methicillin-sensitive *Staphylococcus epidermidis* in the first culture on standard media, (ii) the use of a relatively small sample volume for selective culturing, (iii) the choice of cefoxitin concentration in the supplemented plates, and (iv) restricting sampling to five colonies with CoNS appearance per plate. Metagenomic sequencing of the samples could have been used to overcome these limitations related to culturing but would not have been able to give the same strain-level resolution and link mobile genetic elements to specific lineages. A targeted genomic approach would, however, be interesting and useful for further studies. Another limitation is that no data on previous hospital admissions or stays in long-term care facilities was available.

S. epidermidis has been called an "accidental pathogen" as its virulence determinants are also important for colonization, and *S. epidermidis* infections have been regarded as resulting from haphazard contamination by isolates from the microbiota [60]. In this study, we show that a significant proportion of patients scheduled for prosthetic joint surgery was colonized with heterogeneous MRSE strains. Most of the MRSE strains lacked genes encoding resistance to gentamicin and/or tolerance of CHG, compounds used for the prevention of PJIs in Sweden, but a subset of patients were colonized with PJI-associated MDRSE lineages not covered by current prophylaxis guidelines. To lower the rate of PJIs, broader antimicrobial prophylaxis with a dual regimen of a betalactam and a glycopeptide to cover methicillin-resistant isolates has been suggested [61]. However, as prosthetic joint surgery is a high-volume surgery, broader antimicrobial prophylaxis would risk further selection of resistance traits in MDRSE circulating within and between hospitals.

If MDRSE-colonized patients can be demonstrated to have a higher risk of *S. epidermidis* PJIs, screening for MDRSE before surgery could be of interest to limit the use of broader antimicrobial prophylaxis to these patients only [33]. However, we also acknowledge that hospital-acquired re-colonization with MDRSE after preoperative whole-body cleansing with CHG-containing soap can be important and that possible transmission of MDRSE between patients, the hospital environment, and healthcare workers is an important issue for future research.

5. Conclusions

We found that MDRSE lineages previously associated with PJIs are present in a subset of patients' pre-admission microbiota, plausibly in low relative abundance, and could be selected for by the current prophylaxis regimens. To further lower the rate of *S. epidermidis* PJIs, the current prophylaxis may need to be modified, but it is important for possible perioperative MDRSE transmission events as well as specific risk factors for MDRSE PJIs to be investigated before reevaluation of antimicrobial prophylaxis and/or screening and eradication of MDRSE/MRSE.

Supplementary Materials: The following are available online at https://www.mdpi.com/2076-260 7/9/2/265/s1, Figure S1: Map of central Sweden; patients were recruited from Karlstad (regional hospital), Västerås (regional hospital), and Linköping (university hospital).

Author Contributions: Conceptualization, E.M., M.S. (Martin Sundqvist), M.S. (Marc Stegger), and B.S.; methodology, E.M, M.S. (Martin Sundqvist), T.B.J., M.S. (Marc Stegger), and B.S.; software, T.B.J.; formal analysis, E.M. and M.S. (Marc Stegger); investigation, E.M., S.T., Å.N.-A., T.B.J., M.S. (Marc Stegger), and B.S.; resources, E.M, Å.N.-A., and B.S.; data curation, E.M and T.B.J.; writing—original draft preparation, E.M.; writing—review and editing, E.M., S.T., M.S., M.S. (Marc Stegger), and B.S; visualization, E.M and M.S. (Marc Stegger); supervision, M.S. (Marc Stegger) and B.S.; project administration, E.M.; funding acquisition, E.M, Å.N.-A., and B.S. All authors have read and agreed to the published version of the manuscript.

Funding: This research was funded by the Research Committee at Region Örebro County, grant number OLL-767591; Nyckelfonden at Örebro University Hospital, grant number OLL-502241, OLL-502241; Regionala Forskningsrådet I Uppsala-Örebroregionen, grant number RFR-228551; the Research Committee of Östergötland County Council; and Region Västmanland—Uppsala University, Centre for Clinical Research, Hospital of Västmanland.

Institutional Review Board Statement: The study was conducted according to the guidelines of the Declaration of Helsinki and was approved by the Regional Ethical Review Board of Uppsala, Sweden (protocol code 2012/092, 18/04/2012).

Informed Consent Statement: Informed consent was obtained from all subjects involved in the study.

Data Availability Statement: The data presented in this study are openly available in the European Nucleotide Archive under project ID PRJEB34788 with accession numbers ERR3585469–ERR3585625.

Conflicts of Interest: The authors declare no conflict of interest. The funders had no role in the design of the study; in the collection, analyses, or interpretation of data; in the writing of the manuscript; or in the decision to publish the results.

References

1. Stacy, A.; Belkaid, Y. Microbial guardians of skin health. *Science* **2019**, *363*, 227–228. [CrossRef] [PubMed]
2. Becker, K.; Heilmann, C.; Peters, G. Coagulase-negative staphylococci. *Clin. Microbiol. Rev.* **2014**, *27*, 870–926. [CrossRef] [PubMed]
3. Zimmerli, W. Clinical presentation and treatment of orthopaedic implant-associated infection. *J. Intern. Med.* **2014**, *276*, 111–119. [CrossRef] [PubMed]
4. Cahill, J.L.; Shadbolt, B.; Scarvell, J.M.; Smith, P.N. Quality of life after infection in total joint replacement. *J. Orthop. Surg.* **2008**, *16*, 58–65. [CrossRef] [PubMed]
5. Zmistowski, B.; Karam, J.A.; Durinka, J.B.; Casper, D.S.; Parvizi, J. Periprosthetic joint infection increases the risk of one-year mortality. *J. Bone Jt. Surg. Am.* **2013**, *95*, 2177–2184. [CrossRef]
6. Gundtoft, P.H.; Pedersen, A.B.; Varnum, C.; Overgaard, S. Increased Mortality After Prosthetic Joint Infection in Primary THA. *Clin. Orthop. Relat. Res.* **2017**, *475*, 2623–2631. [CrossRef]
7. Kapadia, B.H.; Berg, R.A.; Daley, J.A.; Fritz, J.; Bhave, A.; Mont, M.A. Periprosthetic joint infection. *Lancet* **2016**, *387*, 386–394. [CrossRef]
8. Inacio, M.C.S.; Graves, S.E.; Pratt, N.L.; Roughead, E.E.; Nemes, S. Increase in Total Joint Arthroplasty Projected from 2014 to 2046 in Australia: A Conservative Local Model with International Implications. *Clin. Orthop. Relat. Res.* **2017**, *475*, 2130–2137. [CrossRef]
9. Nemes, S.; Rolfson, O.; W-Dahl, A.; Garellick, G.; Sundberg, M.; Karrholm, J.; Robertsson, O. Historical view and future demand for knee arthroplasty in Sweden. *Acta Orthop.* **2015**, *86*, 426–431. [CrossRef]
10. Kurtz, S.M.; Lau, E.C.; Son, M.S.; Chang, E.T.; Zimmerli, W.; Parvizi, J. Are We Winning or Losing the Battle with Periprosthetic Joint Infection: Trends in Periprosthetic Joint Infection and Mortality Risk for the Medicare Population. *J. Arthroplast.* **2018**, *33*, 3238–3245. [CrossRef]
11. Nemes, S.; Gordon, M.; Rogmark, C.; Rolfson, O. Projections of total hip replacement in Sweden from 2013 to 2030. *Acta Orthop.* **2014**, *85*, 238–243. [CrossRef] [PubMed]
12. Stefansdottir, A.; Johansson, D.; Knutson, K.; Lidgren, L.; Robertsson, O. Microbiology of the infected knee arthroplasty: Report from the Swedish Knee Arthroplasty Register on 426 surgically revised cases. *Scand. J. Infect. Dis.* **2009**, *41*, 831–840. [CrossRef] [PubMed]
13. Lindgren, V.; Gordon, M.; Wretenberg, P.; Karrholm, J.; Garellick, G. Deep infection after total hip replacement: A method for national incidence surveillance. *Infect. Control Hosp. Epidemiol. Off. J. Soc. Hosp. Epidemiol. Am.* **2014**, *35*, 1491–1496. [CrossRef] [PubMed]
14. Tande, A.J.; Patel, R. Prosthetic Joint Infection. *Clin. Microbiol. Rev.* **2014**, *27*, 302–345. [CrossRef] [PubMed]

15. Benito, N.; Mur, I.; Ribera, A.; Soriano, A.; Rodriguez-Pardo, D.; Sorli, L.; Cobo, J.; Fernandez-Sampedro, M.; Del Toro, M.D.; Guio, L.; et al. The Different Microbial Etiology of Prosthetic Joint Infections according to Route of Acquisition and Time after Prosthesis Implantation, Including the Role of Multidrug-Resistant Organisms. *J. Clin. Med.* **2019**, *8*, 673. [CrossRef] [PubMed]

16. Peel, T.N.; Cole, N.C.; Dylla, B.L.; Patel, R. Matrix-assisted laser desorption ionization time of flight mass spectrometry and diagnostic testing for prosthetic joint infection in the clinical microbiology laboratory. *Diagn Microbiol. Infect. Dis.* **2014**. [CrossRef]

17. Lourtet-Hascoet, J.; Felice, M.P.; Bicart-See, A.; Bouige, A.; Giordano, G.; Bonnet, E. Species and antimicrobial susceptibility testing of coagulase-negative staphylococci in periprosthetic joint infections. *Epidemiol. Infect.* **2018**, *146*, 1771–1776. [CrossRef]

18. Flurin, L.; Greenwood-Quaintance, K.E.; Patel, R. Microbiology of polymicrobial prosthetic joint infection. *Diagn Microbiol. Infect. Dis.* **2019**. [CrossRef]

19. Cizmic, Z.; Feng, J.E.; Huang, R.; Iorio, R.; Komnos, G.; Kunutsor, S.K.; Metwaly, R.G.; Saleh, U.H.; Sheth, N.; Sloan, M. Hip and Knee Section, Prevention, Host Related: Proceedings of International Consensus on Orthopedic Infections. *J. Arthroplast.* **2019**, *34*, S255–S270. [CrossRef]

20. Abouljoud, M.M.; Alvand, A.; Boscainos, P.; Chen, A.F.; Garcia, G.A.; Gehrke, T.; Granger, J.; Kheir, M.; Kinov, P.; Malo, M.; et al. Hip and Knee Section, Prevention, Operating Room Environment: Proceedings of International Consensus on Orthopedic Infections. *J. Arthroplast.* **2019**, *34*, S293–S300. [CrossRef]

21. Aboltins, C.A.; Berdal, J.E.; Casas, F.; Corona, P.S.; Cuellar, D.; Ferrari, M.C.; Hendershot, E.; Huang, W.; Kuo, F.C.; Malkani, A.; et al. Hip and Knee Section, Prevention, Antimicrobials (Systemic): Proceedings of International Consensus on Orthopedic Infections. *J. Arthroplast.* **2019**, *34*, S279–S288. [CrossRef] [PubMed]

22. Hellmark, B.; Söderquist, B.; Unemo, M.; Nilsdotter-Augustinsson, Å. Comparison of *Staphylococcus epidermidis* isolated from prosthetic joint infections and commensal isolates in regard to antibiotic susceptibility, agr type, biofilm production, and epidemiology. *Int. J. Med. Microbiol.* **2013**, *303*, 32–39. [CrossRef] [PubMed]

23. Both, A.; Huang, J.; Lausmann, C.; Wisselberg, S.; Buettner, H.; Zahar, A.; Alawi, M.; Luoto, L.; Kendoff, D.; Rohde, H. Inter- and Intra-clonal Diversity in *S. epidermidis* from Prosthetic Joint Infections. In Proceedings of the 29th ECCMID, Amsterdam, The Netherlands, 13–16 April 2019.

24. Kyung Wee, S.; Leboreiro, C.; Aydin, A.; Warren, S.; Mchugh, T.D.; Sharma, H. The Molecular Epidemiology of *Staphylococcus Epidermidis* Prosthetic Joint Infection in a UK Specialist Orthopaedic Hospital. In Proceedings of the 29th ECCMID, Amsterdam, The Netherlands, 13–16 April 2019.

25. Sanchez Morillo, A.; Espinal, P.; Benito, N.; Rivera, A.; Gutierrez, C.; Pere, B.; Mirelis, B.; Coll, P.; Navarro, F. *Staphylococcus epidermidis* in prosthetic joint infections: Epidemiology and pathogenicity factors. In Proceedings of the 28th ECCMID, Madrid, Spain, 21–24 April 2018.

26. Månsson, E.; Bech Johannesen, T.; Nilsdotter-Augustinsson, Å.; Söderquist, B.; Stegger, M. Comparative genomics of *Staphylococcus epidermidis* from prosthetic joint infections and nares highlights genetic traits associated to antimicrobial resistance, not virulence. *Microb. Genom.* **2020**, 000504. [CrossRef]

27. Hischebeth, G.T.; Randau, T.M.; Ploeger, M.M.; Friedrich, M.J.; Kaup, E.; Jacobs, C.; Molitor, E.; Hoerauf, A.; Gravius, S.; Wimmer, M.D. *Staphylococcus aureus* versus *Staphylococcus epidermidis* in periprosthetic joint infection—Outcome analysis of methicillin-resistant versus methicillin-susceptible strains. *Diagn Microbiol. Infect. Dis.* **2019**, *93*, 125–130. [CrossRef] [PubMed]

28. Post, V.; Harris, L.G.; Morgenstern, M.; Mageiros, L.; Hitchings, M.D.; Meric, G.; Pascoe, B.; Sheppard, S.K.; Richards, R.G.; Moriarty, T.F. Comparative Genomics Study of *Staphylococcus epidermidis* Isolates from Orthopedic-Device-Related Infections Correlated with Patient Outcome. *J. Clin. Microbiol.* **2017**, *55*, 3089–3103. [CrossRef]

29. Oh, J.; Byrd, A.L.; Deming, C.; Conlan, S.; Program, N.C.S.; Kong, H.H.; Segre, J.A. Biogeography and individuality shape function in the human skin metagenome. *Nature* **2014**, *514*, 59–64. [CrossRef]

30. Sanzen, L.; Walder, M. Antibiotic resistance of coagulase-negative staphylococci in an orthopaedic department. *J. Hosp. Infect.* **1988**, *12*, 103–108. [CrossRef]

31. Thore, M.; Kühn, I.; Löfdahl, S.; Burman, L.G. Drug-resistant coagulase-negative skin staphylococci: Evaluation of four marker systems and epidemiology in an orthopaedic ward. *Epidemiol. Infect.* **1990**, *105*, 95–105. [CrossRef]

32. Stefánsdóttir, A.; Johansson, Å.; Lidgren, L.; Wagner, P.; W-Dahl, A. Bacterial colonization and resistance patterns in 133 patients undergoing a primary hip- or knee replacement in Southern Sweden. *Acta Orthop.* **2013**, *84*, 87–91. [CrossRef]

33. James, P.J.; Butcher, I.A.; Gardner, E.R.; Hamblen, D.L. Methicillin-resistant *Staphylococcus epidermidis* in infection of hip arthroplasties. *J. Bone Jt. Surg. Br.* **1994**, *76*, 725–727. [CrossRef]

34. Stegger, M.; Andersen, P.S.; Kearns, A.; Pichon, B.; Holmes, M.A.; Edwards, G.; Laurent, F.; Teale, C.; Skov, R.; Larsen, A.R. Rapid detection, differentiation and typing of methicillin-resistant *Staphylococcus aureus* harbouring either *mecA* or the new *mecA* homologue *mecA*(LGA251). *Clin. Microbiol. Infect.* **2012**, *18*, 395–400. [CrossRef] [PubMed]

35. Sahl, J.W.; Lemmer, D.; Travis, J.; Schupp, J.M.; Gillece, J.D.; Aziz, M.; Driebe, E.M.; Drees, K.P.; Hicks, N.D.; Williamson, C.H.D.; et al. NASP: An accurate, rapid method for the identification of SNPs in WGS datasets that supports flexible input and output formats. *Microb. Genom.* **2016**, *2*, e000074. [CrossRef] [PubMed]

36. Li, H.; Durbin, R. Fast and accurate short read alignment with Burrows-Wheeler transform. *Bioinformatics* **2009**, *25*, 1754–1760. [CrossRef] [PubMed]

37. McKenna, A.; Hanna, M.; Banks, E.; Sivachenko, A.; Cibulskis, K.; Kernytsky, A.; Garimella, K.; Altshuler, D.; Gabriel, S.; Daly, M.; et al. The Genome Analysis Toolkit: A MapReduce framework for analyzing next-generation DNA sequencing data. *Genome Res.* **2010**, *20*, 1297–1303. [CrossRef] [PubMed]

38. Stamatakis, A. RAxML version 8: A tool for phylogenetic analysis and post-analysis of large phylogenies. *Bioinformatics* **2014**, *30*, 1312–1313. [CrossRef]

39. Roach, D.J.; Burton, J.N.; Lee, C.; Stackhouse, B.; Butler-Wu, S.M.; Cookson, B.T.; Shendure, J.; Salipante, S.J. A Year of Infection in the Intensive Care Unit: Prospective Whole Genome Sequencing of Bacterial Clinical Isolates Reveals Cryptic Transmissions and Novel Microbiota. *PLoS Genet.* **2015**, *11*, e1005413. [CrossRef]

40. Salipante, S.J.; SenGupta, D.J.; Cummings, L.A.; Land, T.A.; Hoogestraat, D.R.; Cookson, B.T. Application of whole-genome sequencing for bacterial strain typing in molecular epidemiology. *J. Clin. Microbiol.* **2015**, *53*, 1072–1079. [CrossRef]

41. Bankevich, A.; Nurk, S.; Antipov, D.; Gurevich, A.A.; Dvorkin, M.; Kulikov, A.S.; Lesin, V.M.; Nikolenko, S.I.; Pham, S.; Prjibelski, A.D.; et al. SPAdes: A new genome assembly algorithm and its applications to single-cell sequencing. *J. Comput. Biol.* **2012**, *19*, 455–477. [CrossRef]

42. Seemann, T. Prokka: Rapid prokaryotic genome annotation. *Bioinformatics* **2014**, *30*, 2068–2069. [CrossRef]

43. Zankari, E.; Hasman, H.; Cosentino, S.; Vestergaard, M.; Rasmussen, S.; Lund, O.; Aarestrup, F.M.; Larsen, M.V. Identification of acquired antimicrobial resistance genes. *J. Antimicrob. Chemother.* **2012**, *67*, 2640–2644. [CrossRef]

44. Castanheira, M.; Watters, A.A.; Mendes, R.E.; Farrell, D.J.; Jones, R.N. Occurrence and molecular characterization of fusidic acid resistance mechanisms among *Staphylococcus* spp. from European countries (2008). *J. Antimicrob. Chemother.* **2010**, *65*, 1353–1358. [CrossRef] [PubMed]

45. Zolfo, M.; Tett, A.; Jousson, O.; Donati, C.; Segata, N. MetaMLST: Multi-locus strain-level bacterial typing from metagenomic samples. *Nucleic Acids Res.* **2017**, *45*, e7. [CrossRef] [PubMed]

46. Hellmark, B.; Unemo, M.; Nilsdotter-Augustinsson, å.; Söderquist, B. Antibiotic susceptibility among *Staphylococcus epidermidis* isolated from prosthetic joint infections with special focus on rifampicin and variability of the *rpoB* gene. *Clin. Microbiol. Infect.* **2009**, *15*, 238–244. [CrossRef] [PubMed]

47. Yamada, M.; Yoshida, J.; Hatou, S.; Yoshida, T.; Minagawa, Y. Mutations in the quinolone resistance determining region in *Staphylococcus epidermidis* recovered from conjunctiva and their association with susceptibility to various fluoroquinolones. *Br. J. Ophthalmol.* **2008**, *92*, 848–851. [CrossRef] [PubMed]

48. Trong, H.N.; Prunier, A.L.; Leclercq, R. Hypermutable and fluoroquinolone-resistant clinical isolates of *Staphylococcus aureus*. *Antimicrob. Agents Chemother.* **2005**, *49*, 2098–2101. [CrossRef] [PubMed]

49. Dale, G.E.; Broger, C.; D'Arcy, A.; Hartman, P.G.; DeHoogt, R.; Jolidon, S.; Kompis, I.; Labhardt, A.M.; Langen, H.; Locher, H.; et al. A single amino acid substitution in *Staphylococcus aureus* dihydrofolate reductase determines trimethoprim resistance. *J. Mol. Biol.* **1997**, *266*, 23–30. [CrossRef]

50. Jolley, K.A.; Bray, J.E.; Maiden, M.C.J. Open-access bacterial population genomics: BIGSdb software, the PubMLST.org website and their applications. *Wellcome Open Res.* **2018**, *3*, 124. [CrossRef]

51. Bradley, P.; Gordon, N.C.; Walker, T.M.; Dunn, L.; Heys, S.; Huang, B.; Earle, S.; Pankhurst, L.J.; Anson, L.; de Cesare, M.; et al. Rapid antibiotic-resistance predictions from genome sequence data for *Staphylococcus aureus* and *Mycobacterium tuberculosis*. *Nat. Commun.* **2015**, *6*, 10063. [CrossRef]

52. Kock, R.; Friedrich, A.; On Behalf of The Original Author Group, C. Systematic literature analysis and review of targeted preventive measures to limit healthcare-associated infections by meticillin-resistant *Staphylococcus aureus*. *Euro Surveill* **2014**, *19*, 20860. [CrossRef]

53. Mohanty, S.S.; Kay, P.R. Infection in total joint replacements. Why we screen MRSA when MRSE is the problem? *J. Bone Jt. Surg. Br.* **2004**, *86*, 266–268. [CrossRef]

54. European Centre for Disease Prevention and Control. *Multidrug-Resistant Staphylococcus Epidermidis—8 November 2018*; ECDC: Stockholm, Sweden, 2018.

55. Muhlhofer, H.M.L.; Deiss, L.; Mayer-Kuckuk, P.; Pohlig, F.; Harrasser, N.; Lenze, U.; Gollwitzer, H.; Suren, C.; Prodinger, P.; R, V.O.N.E.-R.; et al. Increased Resistance of Skin Flora to Antimicrobial Prophylaxis in Patients Undergoing Hip Revision Arthroplasty. *In Vivo* **2017**, *31*, 673–676. [CrossRef]

56. Atkins, G.J.; Alberdi, M.T.; Beswick, A.; Blaha, J.D.; Bingham, J.; Cashman, J.; Chen, A.F.; Cooper, A.M.; Cotacio, G.L.; Fraguas, T.; et al. General Assembly, Prevention, Surgical Site Preparation: Proceedings of International Consensus on Orthopedic Infections. *J. Arthroplast.* **2019**, *34*, S85–S92. [CrossRef] [PubMed]

57. Addetia, A.; Greninger, A.L.; Adler, A.; Yuan, S.; Makhsous, N.; Qin, X.; Zerr, D.M. A novel *qacA* allele results in an elevated chlorhexidine gluconate minimum inhibitory concentration in cutaneous *Staphylococcus epidermidis* isolates. *Antimicrob Agents Chemother.* **2019**, *6*, e02607-18. [CrossRef]

58. Kärrholm, J.; Mohaddes, M.; Odin, D.; Vinblad, J.; Rogmark, C.; Rolfson, O. *Swedish Hip Arthroplasty Register Annual Report 2017*; Swedish Hip Arhroplasty Registry: Göteborg, Sweden, 2018.

59. Engesaeter, L.B.; Espehaug, B.; Lie, S.A.; Furnes, O.; Havelin, L.I. Does cement increase the risk of infection in primary total hip arthroplasty? Revision rates in 56,275 cemented and uncemented primary THAs followed for 0–16 years in the Norwegian Arthroplasty Register. *Acta Orthop.* **2006**, *77*, 351–358. [CrossRef] [PubMed]

60. Otto, M. *Staphylococcus epidermidis*—The 'accidental' pathogen. *Nat. Rev. Microbiol.* **2009**, *7*, 555–567. [CrossRef]
61. Tornero, E.; Garcia-Ramiro, S.; Martinez-Pastor, J.C.; Bori, G.; Bosch, J.; Morata, L.; Sala, M.; Basora, M.; Mensa, J.; Soriano, A. Prophylaxis with teicoplanin and cefuroxime reduces the rate of prosthetic joint infection after primary arthroplasty. *Antimicrob. Agents Chemother.* **2015**, *59*, 831–837. [CrossRef]

 *microorganisms*

Communication

Molecular Characterization of *Staphylococcus aureus* Isolated from Chronic Infected Wounds in Rural Ghana

Manuel Wolters [1], Hagen Frickmann [2,3], Martin Christner [1], Anna Both [1], Holger Rohde [1], Kwabena Oppong [4], Charity Wiafe Akenten [4], Jürgen May [5,6] and Denise Dekker [6,7,*]

[1] Institute of Medical Microbiology, Virology and Hygiene, Universitiy Medical Center Hamburg-Eppendorf (UKE), 20251 Hamburg, Germany; m.wolters@uke.de (M.W.); mchristner@uke.de (M.C.); a.both@uke.de (A.B.); Rohde@uke.de (H.R.)

[2] Department of Microbiology and Hospital Hygiene, Bundeswehr Hospital Hamburg, 20359 Hamburg, Germany; frickmann@bnitm.de

[3] Institute for Medical Microbiology, Virology and Hygiene, University Medicine Rostock, 18057 Rostock, Germany

[4] Kumasi Centre for Collaborative Research in Tropical Medicine (KCCR), Kumasi, Ghana; oppong@kccr.de (K.O.); danquah01@yahoo.co.uk (C.W.A.)

[5] Tropical Medicine II, Universitiy Medical Center Hamburg-Eppendorf (UKE), 20251 Hamburg, Germany; j.may@uke.de

[6] Department of Infectious Disease Epidemiology, Bernhard Nocht Institute for Tropical Medicine (BNITM), 20359 Hamburg, Germany; may@bnitm.de

[7] German Centre for Infection Research (DZIF), Hamburg-Lübeck-Borstel-Riems, 38124 Braunschweig, Germany

* Correspondence: dekker@bnitm.de

Received: 2 December 2020; Accepted: 17 December 2020; Published: 21 December 2020

Abstract: Background: Globally, *Staphylococcus aureus* is an important bacterial pathogen causing a wide range of community and hospital acquired infections. In Ghana, resistance of *S. aureus* to locally available antibiotics is increasing but the molecular basis of resistance and the population structure of *S. aureus* in particular in chronic wounds are poorly described. However, this information is essential to understand the underlying mechanisms of resistance and spread of resistant clones. We therefore subjected 28 *S. aureus* isolates from chronic infected wounds in a rural area of Ghana to whole genome sequencing. Results: Overall, resistance of *S. aureus* to locally available antibiotics was high and 29% were Methicillin resistant *Staphylococcus aureus* (MRSA). The most abundant sequence type was ST88 (29%, 8/28) followed by ST152 (18%, 5/28). All ST88 carried the *mecA* gene, which was associated with this sequence type only. Chloramphenicol resistance gene *fexB* was exclusively associated with the methicillin-resistant ST88 strains. Panton Valentine leukocidin (PVL) carriage was associated with ST121 and ST152. Other detected mechanisms of resistance included *dfrG*, conferring resistance to trimethoprim. Conclusions: This study provides valuable information for understanding the population structure and resistance mechanisms of *S. aureus* isolated from chronic wound infections in rural Ghana.

Keywords: rural Ghana; molecular epidemiology; chronic wounds; *Staphylococcus aureus*

1. Introduction

Staphylococcus aureus is an important bacterial pathogen in all parts of the world, causing both community and hospital acquired infections. In particular methicillin-resistant *S. aureus* (MRSA) has evolved as a global health threat due to its resistance to beta lactam and other classes of antibiotics [1]. In the last 20 years the prevalence of MRSA appears to be increasing in many African countries as suggested by data from the first decade of the present century [2]. More recent reviews indicate ongoing epidemiological relevance of this resistance type in Africa [3], with increased reporting of outbreak-association in Western African Ghana [4]. In Ghana, the abundance of MRSA in carriage studies or clinical samples demonstrated large geographical differences [5–7]. Moreover, resistance of *S. aureus* to a variety of other locally available oral antibiotics such as tetracyclines, trimethoprim/sulfamethoxazole and penicillins is frequently observed in Ghana [8,9]. However, the underlying mechanisms of resistance in this region are not well understood and to our knowledge has not been described for *S. aureus* from chronic wounds.

Effective surveillance of antimicrobial resistance in bacteria including *S. aureus* is essential for estimating the burden of resistance and molecular strain typing provides important information for understanding the spread of resistant clones. However, in Africa both surveillance and strain typing information are scarce due to the limited diagnostic microbiology infrastructure generally available in large parts of the continent. Molecular typing including *spa*-typing, multi-locus sequencing and also whole genome sequence typing has been applied in only a few studies in *S. aureus* in humans and livestock in Ghana [3,10–14], identifying strains of multiple clonal clusters [14]. In particular, the MRSA clone sequence type (ST)88-IV (2B) is not only abundant in Ghana, but also in other African counties: Angola, Cameroon, Gabon, Madagascar, Nigeria, as well as São Tomé e Príncipe [15]. In Ghana, reported rates range between 24.2–83.3% of all MRSA isolates [12].

In a previous study *S. aureus* was isolated from 14.0% (*n* = 28) of samples from patients with chronic wounds in Ghana [5]. In that study, a high frequency of methicillin-resistance (29%) was noted. Moreover, resistance to other commonly used antibiotics like penicillins, tetracyclines and trimethoprim/sulfamethoxazole was frequently observed. For Ghanaian *S. aureus* isolates from wounds, data on prevalent clones, resistance mechanisms and pathogenicity-associated genetic determinants is limited. To fill this gap in the global epidemiological picture, we have subjected the 28 *S. aureus* isolates from our previous study to whole genome sequencing (WGS), aiming at analyzing the underlying molecular basis of antimicrobial resistance and the population structure of this strain collection.

2. Materials and Methods

2.1. Sample Collection, Microbiology and Antibiotic Susceptibility Testing

S. aureus was isolated from female and male patients ≥15 years with an infected wound at the Outpatient Department of the Agogo Presbyterian Hospital, in the Asante Akim North District of Ghana from January to November 2016. Sample collection and microbiological investigations were reported previously [5]. Antibiotic susceptibility was tested by the disk diffusion method and interpreted following the European Committee on Antimicrobial Susceptibility Testing (EUCAST) guidelines v.10.0 (http://www.eucast.org).

2.2. Whole Genome Sequencing and Data Analysis

Whole genome sequencing of the isolates was performed using the Illumina NextSeq platform. WGS data were analyzed using the Nullarbor pipeline (vers. 2.0.20181010; Seemann T, available at: https://github.com/tseemann/nullarbor). Reads were assembled with spades [16] (vers. 3.13.1) and annotated with Prokka [17] (vers. 1.13.3). The resistance and virulence gene profiles were determined with ABRicate (https://github.com/tseemann/abricate) (vers 0.9.9) employing NCBI AMR (7th October 2020; 5283 sequences), Resfinder [18] (3077 sequences) and VFDB [19] (2597 sequences) databases. SCCmec types were determined with the SCCmecFinder web tool (https://cge.cbs.dtu.dk/services/SCCmedFinder/10.1128/mSphere.00612-17).

The MLST sequence types were extracted from the WGS data using the MLST tool (vers. 2.16.1). Sequencing reads have been deposited in NCBI's small reads archive (BioProject: PRJNA670821).

2.3. Ethical Considerations

The Committee on Human Research, Publications and Ethics, School of Medical Science, Kwame Nkrumah University of Science and Technology in Kumasi, Ghana, approved this study (approval number CHRPE/AP/078/16) on 14th December, 2015.

3. Results

The identified MLST sequence types and selected associated resistance and virulence genes are summarized in Table 1; the complete MLST, virulence factor and resistance gene datasets (VFDB, Resfinder and NCBI AMR), as well as AST results are available in the supplemental material (Table S1) The most abundant sequence type was ST88 (8/28). All ST88 isolates were *mecA* positive, SCC*mec* type IV(2B), *agr* type 3 and negative for the Pantone–Valentine Leukocidin toxin (PVL) genes *lukS-PV/lukF-PV*. Seven of eight ST88 isolates carried the *fexB* gene, which confers resistance to chloramphenicol [20]. The five isolates of the second most abundant clone ST152 were all *mecA* negative and carried the PVL genes *lukS-PV/lukF-PV*.

Table 1. Sequence types and associated virulence and resistance genes.

Sequence Type	n	*lukS-PV/lukF-PV*	*mecA*	*fexB*
ST88	8	0	8	7
ST152	5	5	0	0
ST15	3	0	0	0
ST1	2	0	0	0
ST5	2	0	0	0
ST45	2	0	0	0
ST2434	2	0	0	0
ST72	1	0	0	0
ST121	1	1	0	0
ST3248	1	0	0	0
ST3249	1	0	0	0
Totals	28	6	8	7

Highest rates of phenotypic antimicrobial resistance were detected for penicillin (100%, 28/28), tetracycline (57%, 16/28) and trimethoprim/sulfamethoxazole (39%, 11/28) (Table 2). All isolates were susceptible to linezolid, rifampicin, fosfomycin, tigecycline, ciprofloxacin, levofloxacin and daptomycin. WGS resistance gene profiling identified corresponding acquired resistance genes in 97% (65/67) of all phenotypically detected antimicrobial resistances (Table 2). In all but one penicillin-resistant isolate penicillinase-encoding *blaZ* was detected. The single penicillin resistant, *blaZ* negative strain carried *mecA*, reasonably explaining the observed betalactam-resistant phenotype. All oxacillin-resistant isolates were *mecA* positive and belonged to ST88. MecC was not detected in the strain collection. Two main mechanisms of resistance to tetracyclines have been described in *S. aureus*: active efflux, resulting from plasmid-located *tetK* and *tetL* genes and ribosomal protection mediated by *tetM* or *tetO* genes. In our collection tetracycline resistant isolates carried either tetK (7/16) alone or *tetL* (8/16) in combination with *tetM* (8/16). The *tetL* and *tetM* genes were exclusively detected in the ST88 MRSA isolates, while tetK was found in various clonal backgrounds. Beside mutation of the chromosomal dihydrofolate reductase (DHFR) gene, three acquired dihydrofolate reductase gene variants are known to confer resistance to trimethoprim in *S. aureus* of human origin: *dfrA*, *dfrG* and *dfrK*. In our collection all trimethoprim/sulfamethoxazole resistant isolates carried *dfrG*. Of the two erythromycin resistant isolates one carried *msrA* and one *ermC*. As expected, the latter isolate was also resistant to clindamycin. Gentamicin resistance is most commonly conferred by aminoglycoside-modifying enzymes. However, in the single case of a gentamicin-resistant isolate no corresponding resistance gene was found.

Table 2. Phenotypic antibiotic resistance and associated genetic resistance markers.

Antibiotic	Phenotypic AST Resistant n (%)	WGS Resistance Gene or Mutation	Positive n (%)
Penicillin	28 (100)	*blaZ*	27 (96)
		blaZ neg, *mecA* pos	1 (4)
Tetracycline	16 (57)	*tetK*	7 (44)
		tetL	8 (50)
		tetM	8 (50)
Trimethoprim/ Sulfamethoxazole	11 (39)	*dfrG*	11 (100)
Oxacillin	8 (29)	*mecA*	8 (100)
Erythromycin	2 (7)	*msrA*	1 (50)
		ermC	1 (50)
Clindamycin	1 (4)	*ermC*	1 (100)
Gentamicin	1 (4)	not detected	none

4. Discussion

In this study we describe the molecular epidemiology of *S. aureus* isolated from chronic infected wounds in outpatients in a rural area of Ghana.

Overall, the frequency of MRSA (29%) was high, comparable to other clinical studies conducted in Ghana [6,21,22]. Nevertheless, the frequencies of MRSA seen in patients in Ghana seem subject to geographical variations [5–7]. Moreover, the previously reported high rates of resistance to orally available antibiotics including penicillin, tetracycline and cotrimoxazole in *S. aureus* [9] were confirmed by our study results. The particularly high frequencies of penicillin resistance might be attributed to the fact that penicillin-based antibiotics are amongst the most frequently prescribed drugs in Ghana and available over the counter without prescription [23]. High rates of resistance inevitably reduce effective antibiotic treatment in areas where resources are scarce. This favors the use of cleaning and disinfecting procedures for the management of wound infections whenever clinically possible.

All isolated MRSA belonged to ST88, described as the dominant clone in various African countries, including Ghana [15]. In addition, all but one MRSA strain also carried the *fexB* gene, conferring resistance to chloramphenicol, which was not found in any of the other sequence types. Previously, *fexB* has only been described in *S. aureus* strains from Ghanaian patients with Buruli ulcer [20], also a chronic wound. As previously described in 2007 [24], chloramphenicol has been extensively prescribed in Africa, although a low risk of 0.002% for chloramphenicol-induced aplastic anemia had been described in Nigeria as early as in 1993 with an associated recommendation for strict risk-benefit-assessments prior to its prescription [25]. To the authors best knowledge, little has changed in the meantime and the substance is still in broad use in Sub-Saharan Africa, as it is readily available and shows excellent penetration even in difficult to reach compartments including bradytroph tissue like bone [26]. It is therefore possible that that frequent application of chloramphenicol in patients that did not respond to beta-lactam antibiotics, due to *mecA*-carriage of their *S. aureus* strains, might have facilitated the selection of *fexB* carrying bacteria. Due to the lack of reliable clinical data on previous antibiotic treatment this hypothesis could not be confirmed. Other mechanisms of resistance detected included *dfrG*, conferring resistance to trimethoprim, frequently found in strains isolated from Ghana [27,28]. Earlier this was regarded as an infrequent cause of trimethoprim resistance in *S. aureus* isolated from patients but is now widespread in Africa and common in *S. aureus* from ill travelers returning to Europe [27].

Panton-Valentine Leukocidin (PVL), which has been proposed as an epidemiological marker for severe skin infections [29], was encoded in strains of the ST121 and ST152 clonal lineages. The ST152 clonal lineage, in particular, is both known to be associated with PVL expression [30] and wide distribution in Ghana [10,31].

5. Conclusions

This study provides insight into the molecular epidemiology of *S. aureus* sequence types found in chronic infected wounds in a rural area of Ghana.

However, the number of samples used was quite small, and they were taken from outpatients in one hospital, so they may not be representative of the community or the wider area. Moreover, we did not have reliable information about prior use of antibiotics in these patients.

Nevertheless, this study stipulates valuable information for understanding the spread of resistant clones found in patients visiting the study hospital, which is important for effective surveillance of antibiotic resistant *S. aureus* and vital for estimating the burden of resistance.

Supplementary Materials: The following are available online at http://www.mdpi.com/2076-2607/8/12/2052/s1, Table S1: Complete MLST, *agr* and *SCCmec* typing, virulence factor and resistance gene datasets (VFDB, Resfinder and NCBI AMR) and phenotypic AST results. VFDB, Resfinder and NCBI AMR: The numbers represent the percent identity in the alignment between the best matching resistance or virulence gene in ResFinder or VFDB and the corresponding sequence in the input genome. A perfect alignment is 100% and must also cover the entire length of the resistance gene in the database.

Author Contributions: M.W., D.D. and J.M. designed and coordinated this study. M.W., M.C., A.B. and H.R. performed data analyses analysis. H.F., D.D. and M.W. wrote the first draft of this manuscript. K.O. conducted and supervised fieldwork. C.W.A. conducted and supervised lab work. M.C. and H.R. supported writing and editing the manuscript. All authors have read and agreed to the published version of the manuscript.

Funding: This study was funded by institutional funds of the Bernhard Nocht Institute for Tropical medicine (BNITM).

Acknowledgments: We are grateful to all patients, who participated in this study and to the personnel at the Agogo Presbyterian Hospital. Without their efforts, this research study would not have been possible.

Conflicts of Interest: The authors declare no conflict of interest.

References

1. Andersson, H.; Lindholm, C.; Fossum, B. MRSA—Global threat and personal disaster: Patients' experiences. *Int. Nurs. Rev.* **2011**, *58*, 47–53. [CrossRef] [PubMed]

2. Falagas, M.E.; Karageorgopoulos, D.E.; Leptidis, J.; Korbila, I.P. MRSA in Africa: Filling the global map of antimicrobial resistance. *PLoS ONE* **2013**, *8*, e68024. [CrossRef] [PubMed]

3. Osei Sekyere, J.; Mensah, E. Molecular epidemiology and mechanisms of antibiotic resistance in *Enterococcus* spp., *Staphylococcus* spp., and *Streptococcus* spp. in Africa: A systematic review from a One Health perspective. *Ann. N. Y. Acad. Sci.* **2020**, *1465*, 29–58. [CrossRef] [PubMed]

4. Donkor, E.S.; Dayie, N.T.K.D.; Tette, E.M.A. Methicillin-Resistant *Staphylococcus aureus* in Ghana: Past, Present, and Future. *Microb. Drug. Resist.* **2019**, *25*, 717–724. [CrossRef] [PubMed]

5. Krumkamp, R.; Oppong, K.; Hogan, B.; Strauss, R.; Frickmann, H.; Wiafe-Akenten, C.; Boahen, K.G.; Rickerts, V.; McCormick Smith, I.; Groß, U.; et al. Spectrum of antibiotic resistant bacteria and fungi isolated from chronically infected wounds in a rural district hospital in Ghana. *PLoS ONE* **2020**, *15*, e0237263. [CrossRef]

6. Amissah, N.A.; van Dam, L.; Ablordey, A.; Ampomah, O.W.; Prah, I.; Tetteh, C.S.; van der Werf, T.S.; Friedrich, A.W.; Rossen, J.W.; van Dijl, J.M.; et al. Epidemiology of *Staphylococcus aureus* in a burn unit of a tertiary care center in Ghana. *PLoS ONE* **2017**, *12*, e0181072. [CrossRef]

7. Janssen, H.; Janssen, I.; Cooper, P.; Kainyah, C.; Pellio, T.; Quintel, M.; Monnheimer, M.; Groß, U.; Schulze, M.H. Antimicrobial-Resistant Bacteria in Infected Wounds, Ghana, 2014. *Emerg Infect. Dis.* **2018**, *24*, 916–919. [CrossRef]

8. Dekker, D.; Wolters, M.; Mertens, E.; Boahen, K.G.; Krumkamp, R.; Eibach, D.; Schwarz, N.G.; Adu-Sarkodie, Y.; Rohde, H.; Christner, M.; et al. Antibiotic resistance and clonal diversity of invasive *Staphylococcus aureus* in the rural Ashanti Region, Ghana. *BMC Infect. Dis.* **2016**, *16*, 720. [CrossRef]

9. Newman, M.J.; Frimpong, E.; Donkor, E.S.; Opintan, J.A.; Asamoah-Adu, A. Resistance to antimicrobial drugs in Ghana. *Infect. Drug Resist.* **2011**, *4*, 215–220.

10. Egyir, B.; Hadjirin, N.F.; Gupta, S.; Owusu, F.; Agbodzi, B.; Adogla-Bessa, T.; Addo, K.K.; Stegger, M.; Larsen, A.R.; Holmes, M.A. Whole-genome sequence profiling of antibiotic-resistant *Staphylococcus aureus* isolates from livestock and farm attendants in Ghana. *J. Glob. Antimicrob. Resist.* **2020**, *22*, 527–532. [CrossRef]

11. Donkor, E.S.; Jamrozy, D.; Mills, R.O.; Dankwah, T.; Amoo, P.K.; Egyir, B.; Badoe, E.V.; Twasam, J.; Bentley, S.D. A genomic infection control study for Staphylococcus aureus in two Ghanaian hospitals. *Infect. Drug Resist.* **2018**, *11*, 1757–1765. [CrossRef] [PubMed]

12. Kpeli, G.; Buultjens, A.H.; Giulieri, S.; Owusu-Mireku, E.; Aboagye, S.Y.; Baines, S.L.; Seemann, T.; Bulach, D.; Gonçalves da Silva, A.; Monk, I.R.; et al. Genomic analysis of ST88 community-acquired methicillin resistant *Staphylococcus aureus* in Ghana. *Peer J.* **2017**, *5*, e3047. [CrossRef] [PubMed]

13. Egyir, B.; Guardabassi, L.; Monecke, S.; Addo, K.K.; Newman, M.J.; Larsen, A.R. Methicillin-resistant *Staphylococcus aureus* strains from Ghana include USA300. *J. Glob. Antimicrob. Resist.* **2015**, *3*, 26–30. [CrossRef] [PubMed]

14. Egyir, B.; Guardabassi, L.; Sørum, M.; Nielsen, S.S.; Kolekang, A.; Frimpong, E.; Addo, K.K.; Newman, M.J.; Larsen, A.R. Molecular epidemiology and antimicrobial susceptibility of clinical *Staphylococcus aureus* from healthcare institutions in Ghana. *PLoS ONE* **2014**, *9*, e89716. [CrossRef] [PubMed]

15. Abdulgader, S.M.; Shittu, A.O.; Nicol, M.P.; Kaba, M. Molecular epidemiology of Methicillin-resistant *Staphylococcus aureus* in Africa: A systematic review. *Front. Microbiol.* **2015**, *6*, 34. [CrossRef]

16. Bankevich, A.; Nurk, S.; Antipov, D.; Gurevich, A.; Dvorkin, M.; Kulikov, A.S.; Lesin, V.M.; Nikolenko, S.I.; Pham, S.; Prjibelski, A.D.; et al. SPAdes: A new genome assembly algorithm and its applications to single-cell sequencing. *J. Comput Biol.* **2012**, *19*, 455–477. [CrossRef]

17. Seeman, T. Prokka: Rapid prokaryotic genome annotation. *Bioinformatics* **2014**, *30*, 2068–2069. [CrossRef]

18. Zankari, E.; Hasman, H.; Cosentino, S.; Vestergaard, M.; Rasmussen, S.; Lund, O.; Aarestrup, F.M.; Larsen, M.V. Identification of acquired antimicrobial resistance genes. *J. Antimicrob. Chemother.* **2012**, *67*, 2640–2644. [CrossRef]

19. Chen, L.; Zheng, D.; Liu, B.; Yang, J.; Jin, Q. VFDB 2016: Hierarchical and refined dataset for big data analysis—10 years on. *Nucleic Acids Res.* **2016**, *44*, 694–697. [CrossRef]

20. Amisssah, N.A.; Chlebowicz, A.A.; Ablordey, A.; Sabat, A.J.; Tetteh, C.S.; Prah, I.; van der Wrf, T.S.; Fruedrich, A.W.; van Djil, J.M.; Rossen, J.W.; et al. Molecular characterization of *Staphylococcus aureus* isolates transmitted between patients with Buruli ulcer. *PLoS ONE Trop Dis.* **2015**, *9*, e0004049.

21. Opintan, J.A.; Newman, M.J.; Arhin, R.E.; Donkor, E.S.; Gyansa-Lutterodt, M.; Mills-Pappoe, W. Laboratory-based nationwide surveillance of antimicrobial resistance in Ghana. *Infect. Drug Resist.* **2015**, *8*, 379–389. [CrossRef] [PubMed]

22. Amissah, N.A.; Buultjens, A.H.; Ablordey, A.; van Dam, L.; Opoku-Ware, A.; Baines, S.L.; Bulach, D.; Tetteh, C.S.; Prah, I.; van der Werf, T.S.; et al. Methicillin Resistant *Staphylococcus aureus* Transmission in a Ghanaian Burn Unit: The Importance of Active Surveillance in Resource-Limited Settings. *Front. Microbiol.* **2017**, *8*, 1906. [CrossRef] [PubMed]

23. Tagoe, D.N.A.; Attah, C.O. A study of antibiotic use and abuse in Ghana: A case study of the Cape Coast Metropolis. *Internet J. Health* **2010**, *11*. [CrossRef]

24. Falagas, M.E.; Kopterides, P. Old antibiotics for infections in critically ill patients. *Curr. Opin. Crit. Care* **2007**, *13*, 592–597. [CrossRef] [PubMed]

25. Durosinmi, M.A.; Ajayi, A.A. A prospective study of chloramphenicol induced aplastic anaemia in Nigerians. *Trop. Geogr. Med.* **1993**, *45*, 159–161.

26. Summersgill, J.T.; Schupp, L.G.; Raff, M.J. Comparative penetration of metronidazole, clindamycin, chloramphenicol, cefoxitin, ticarcillin, and moxalactam into bone. *Antimicrob. Agents Chemother.* **1982**, *21*, 601–603. [CrossRef]

27. Nurjadi, D.; Olalekan, A.O.; Layer, F.; Shittu, A.O.; Alabi, A.; Ghebremedhin, B.; Schaumburg, F.; Hofmann-Eifler, J.; Van Genderen, P.J.; Caumes, E.; et al. Emergence of trimethoprim resistance gene *dfrG* in *Staphylococcus aureus* causing human infection and colonization in sub-Saharan Africa and its import to Europe. *J. Antimicrob. Chemother.* **2014**, *69*, 2361–2368. [CrossRef]

28. Nurjadi, D.; Schäfer, J.; Friedrich-Jänicke, B.; Mueller, A.; Neumayr, A.; Calvo-Cano, A.; Goorhuis, A.; Molhoek, N.; Lagler, H.; Kantele, A.; et al. Predominance of *dfrG* as determinant of trimethoprim resistance in imported *Staphylococcus aureus*. *Clin. Microbiol Infect.* **2015**, *21*, 1095.e5–1095.e9. [CrossRef]

29. Tong, A.; Tong, S.Y.; Zhang, Y.; Lamlertthon, S.; Sharma-Kuinkel, B.K.; Rude, T.; Ahn, S.H.; Ruffin, F.; Llorens, L.; Tamarana, G.; et al. Panton-Valentine leukocidin is not the primary determinant of outcome for *Staphylococcus aureus* skin infections: Evaluation from the CANVAS studies. *PLoS ONE* **2012**, *7*, e37212. [CrossRef]

30. Ruimy, R.; Maiga, A.; Armand-Lefevre, L.; Maiga, I.; Diallo, A.; Koumaré, A.K.; Ouattara, K.; Soumaré, S.; Gaillard, K.; Lucet, J.C.; et al. The carriage population of *Staphylococcus aureus* from Mali is composed of a combination of pandemic clones and the divergent Panton-Valentine leukocidin-positive genotype ST152. *J. Bacteriol.* **2008**, *190*, 3962–3968. [CrossRef]

31. Eibach, D.; Nagel, M.; Hogan, B.; Azuure, C.; Krumkamp, R.; Dekker, D.; Gajdiss, M.; Brunke, M.; Sarpong, N.; Owusu-Dabo, E.; et al. Nasal Carriage of *Staphylococcus aureus* among Children in the Ashanti Region of Ghana. *PLoS ONE* **2017**, *12*, e0170320. [CrossRef] [PubMed]

Publisher's Note: MDPI stays neutral with regard to jurisdictional claims in published maps and institutional affiliations.

microorganisms

MDPI

Article

A Murine Skin Infection Model Capable of Differentiating the Dermatopathology of Community-Associated MRSA Strain USA300 from Other MRSA Strains

Jack Zhang [1,2], John Conly [1,2,3,4,5], JoAnn McClure [1,4], Kaiyu Wu [1,2], Björn Petri [2,5], Duane Barber [1], Sameer Elsayed [2,6], Glen Armstrong [2] and Kunyan Zhang [1,2,3,4,5,*]

1   Department of Pathology & Laboratory Medicine, University of Calgary, Calgary, AB T2N4N1, Canada; jack20110915@gmail.com (J.Z.); john.conly@albertahealthservices.ca (J.C.); jammcclu@ucalgary.ca (J.M.); kwu@ucalgary.ca (K.W.); dfbarber1@gmail.com (D.B.)
2   Department of Microbiology, Immunology and Infectious Diseases, University of Calgary, Calgary, AB T2N4N1, Canada; bpetri@ucalgary.ca (B.P.); sameer.elsayed@lhsc.on.ca (S.E.); armstrog@ucalgary.ca (G.A.)
3   Department of Medicine, University of Calgary, Calgary, AB T2N4N1, Canada
4   Centre for Antimicrobial Resistance, Alberta Health Services, Alberta Precision Laboratories, University of Calgary, Calgary, AB T2N4N1, Canada
5   The Calvin, Phoebe and Joan Snyder Institute for Chronic Diseases, University of Calgary, Calgary, AB T2N4N1, Canada
6   Department of Medicine, University of Western Ontario, London, ON N6A5C1, Canada
*   Correspondence: kzhang@ucalgary.ca; Tel.: +1-403-210-8484

Citation: Zhang, J.; Conly, J.; McClure, J.; Wu, K.; Petri, B.; Barber, D.; Elsayed, S.; Armstrong, G.; Zhang, K. A Murine Skin Infection Model Capable of Differentiating the Dermatopathology of Community-Associated MRSA Strain USA300 from Other MRSA Strains. *Microorganisms* **2021**, *9*, 287. https://doi.org/10.3390/microorganisms9020287

Academic Editor: Rajan P. Adhikari

Received: 22 December 2020
Accepted: 26 January 2021
Published: 30 January 2021

Publisher's Note: MDPI stays neutral with regard to jurisdictional claims in published maps and institutional affiliations.

Abstract: USA300 is a predominant and highly virulent community-associated methicillin-resistant *Staphylococcus aureus* (CA-MRSA) strain that is a leading cause of skin and soft tissue infections. We established a murine intradermal infection model capable of demonstrating dermatopathological differences between USA300 and other MRSA strains. In this model, USA300 induced dermonecrosis, uniformly presenting as extensive open lesions with a histologically documented profound inflammatory cell infiltrate extending below the subcutis. In contrast, USA400 and a colonizing control strain M92 caused only localized non-ulcerated skin infections associated with a mild focal inflammatory infiltrate. It was also determined that the dermonecrosis induced by USA300 was associated with significantly increased neutrophil recruitment, inhibition of an antibacterial response, and increased production of cytokines/chemokines associated with disease severity. These results suggest that induction of severe skin lesions by USA300 is related to over-activation of neutrophils, inhibition of host antibacterial responses, and selective alteration of host cytokine/chemokine profiles.

Keywords: methicillin-resistant *Staphylococcus aureus* (MRSA); community-associated MRSA (CA-MRSA); CA-MRSA strain USA300; murine skin infection model; dermatopathology; dermonecrosis; neutrophil; host antibacterial response; cytokine; chemokine

1. Introduction

Methicillin-resistant *Staphylococcus aureus* (MRSA) is a leading cause of infection worldwide. It was traditionally considered a nosocomial pathogen in healthcare facilities (known as hospital-associated MRSA (HA-MRSA)). However, community-associated MRSA (CA-MRSA) emerged in the 1990s, and quickly increased and then replaced HA-MRSA in the United States and Canada. Prior to 2000, the predominant CA-MRSA strain in North America was pulsotype USA400 (ST1-MRSA-IV). It was quickly replaced by pulsotype USA300 (ST8-MRSA-IV), which is the strain responsible for the increase in CA-MRSA infections in North America [1,2], as well as globally [3–5]. USA300 causes primarily acute bacterial skin and skin structure infections (ABSSSIs) [6], with an estimated 44.6% of ABSSSIs in North America caused by *S. aureus* [7], and the majority of those caused by USA300 [8].

Microorganisms **2021**, *9*, 287. https://doi.org/10.3390/microorganisms9020287   https://www.mdpi.com/journal/microorganisms

Additionally, infections induced by USA300 are believed to be more invasive and more severe when compared with those induced by other MRSA strains [9–14].

Several skin and soft tissue animal infection models have been developed to study CA-MRSA virulence, such as the murine surgical wound infection model [15], mouse foreign body infection model [16], mouse muscle infection model with microspheres as an *S. aureus* carrier [17], mouse subcutaneous (skin or ear) infection model [18,19], and mouse burn-injured infection model [20]. However, most of these animal models were designed to test the role of individual *S. aureus* virulence factors, or the effect of antimicrobials, rather than for comparing virulence between strains. In fact, there are limited studies comparing the pathogenic propensity of CA-MRSA strains responsible for serious skin and soft tissue infections, with no mouse infection model capable of differentiating USA300 virulence from that of other CA-MRSA strains. Building on previously described models, we established a mouse skin infection model capable of differentiating the virulence of USA300 from other CA-MRSA strains (USA400) as well as from a colonization strain (M92), which simulates human skin and soft tissue infection characteristics. Furthermore, we compared the immune response induced by the various strains to explore the potential mechanisms contributing to USA300 hypervirulence.

2. Materials and Methods

2.1. MRSA Strains and Their Phenotypic and Genotypic Characterization

USA300-C2406 was isolated from a patient with a lethal case of necrotizing pneumonia during our local CA-MRSA strain USA300 outbreak in Calgary in 2004 [21]. Control strain USA400-CMRSA7 was provided by the National Microbiology Laboratory, Health Canada, Winnipeg, Canada [22]. Control strain M92 was found as a nasal colonizer of staff in our local hospitals, but was never associated with invasive infection. Phenotypic and genotypic characterization of the isolates was done as previously described [23]. Whole genome sequencing was performed for the isolates using Pacific Biosciences (PacBio, Menlo Park, CA, USA) RSII sequencing technology, at the McGill University Genome Quebec Innovation Center. Genome assembly was accomplished using the hierarchical genome assembly process (HGAP v. 2.3.0.140936.p5), with the complete genome of C2406 (GenBank No.: PRJNA345240; CP019590.1), CMRSA7 (GenBank No.: PRJNA362898), and M92 (GenBank No.: PRJNA319679; CP015447.1) submitted to GenBank and annotated using the prokaryotic genome annotation pipeline [24,25]. Virulence gene profiles for the isolates, including 17 toxin genes, 12 adhesin genes, and 4 exoenzyme genes, were identified and analyzed through the strain whole genome sequences.

2.2. Murine Skin Infection Model

Single colonies of each bacterial strain were grown overnight in brain–heart infusion (BHI) broth at 37 °C, followed by subculture in 50 mL BHI at 37 °C until the optical density at 600 nm (OD600) was 0.7. Bacteria were centrifuged and the pellet washed twice with saline, then resuspended in 0.5 volume of saline (roughly 2×10^8 CFU/mL). Animal infection experiments were performed at the Animal Resource Centre at the University of Calgary, in accordance with institutional and national guidelines of the Canadian Council on Animal Care, under protocol numbers M06074, M09115, and AC13-0076 (approved by the Animal Care Committee, University of Calgary). The experiments were repeated 3 times. Female BALB/c mice (Charles River Laboratories, Inc., Wilmington, MA, USA), aged 6 to 8 weeks, were shaved in the intrascapular region with electrical clippers prior to injection. Fifteen mice were assigned to each of the 4 groups and injected intradermally with 1×10^7 CFU/50 µL of the appropriate strain or mock-infected control (saline control), in the center of the shaved area. Mice were carefully monitored for skin (wound) infection and signs of distress, and euthanized on days 4, 7, and 15~17. The skin lesion areas (mm^2) were estimated by multiplying the width (mm) and length (mm) of the lesion. For each mouse, the whole spleen was aseptically harvested and homogenized in 1 ml sterile saline. Quantitative cultures [serial dilutions and spread on Tryptic soy agar (TSA)plates] were performed to determine bacterial load, as splenic bacterial load was used as an indicator of systemic MRSA dissemination. Full thickness biopsies of the core cutaneous lesions were performed, fixed in 10% neutral buffered formalin, and subjected to detailed histopathological examination.

2.3. Histology

The lesional skin tissue was cut into tissue blocks by the Airway Inflammation Research Group (AIRG) Histology Services at the University of Calgary. Tissue sections 4 μm thick were affixed to microscope slides and deparaffinized. The sections were stained with hematoxylin and eosin (H&E), Gram, and chloracetate esterase staining for routine histology, to identify bacteria, and to highlight neutrophils, respectively. For the H&E staining (NovaUltra™ H&E Stain Kit, Rockville, MD, USA), fixed slides were deparaffinized by 2 changes of xylene for 5 min, and re-hydrated in 2 changes of 100% alcohol for 5 min each, then 95% and 70% alcohol for 1 min each. The slides were stained with Mayer's hematoxylin solution for 2 min, washed with running tap water for 2 min, and rinsed in 95% alcohol. The slides were further counterstained with eosin solution for 45 s, and rinsed in 95% alcohol. The stained slides were dehydrated through 2 changes of 100% alcohol, 5 min each, and cleared with 2 changes of xylene, 5 min each. Finally, the slides were mounted with xylene-based mounting medium. For the Gram staining (BD Gram Stain Kits, Franklin Lakes, NJ, USA), fixed slides were deparaffinized and re-hydrated as described above. The slides were stained with Gram crystal violet for 1 min, and flooded with Gram iodine for 1 min. The slides were decolorized by Gram decolorizer, then counterstained with Gram safranin for 1 min. For the chloracetate esterase staining, deparaffinized sections were stained with the freshly prepared staining solution (5.0 mg naphthol AS-D chloracetate, 5 mL N-N dimethylformamide (NNDMF), 6 drops 4.0% sodium nitrite, 6 drops 4% new fuchsin in 47.5 mL phosphate buffer) for 25 min, followed by counterstaining with hematoxylin nuclear stain. The stained slides were dehydrated and cleared as described above.

2.4. Myeloperoxidase (MPO) Assay

The tissue myeloperoxidase activity assay was performed as an index of neutrophil recruitment, as previously described [26]. The test was repeated twice, with at least 5 mice from each group and each time point included from each experiment. Briefly, the skin was cut and weighed prior to homogenization in a 0.5% hexadecyltrimethylammonium bromide phosphate-buffered (pH 6.0) solution using a polytron PT1300D homogenizer (Kinematica, Lucerne, Switzerland). The homogenates were centrifuged at 14,000 rpm for 5 min at 4 °C in a microcentrifuge and five aliquots of each supernatant were transferred into 96-well plates, followed by the addition of a 3, 3′-dimethoxybenzidine and 1% hydrogen peroxide solution. Standard dilutions of pure myeloperoxidase were also tested for their activity to construct a standard curve (OD as a function of units of enzyme activity). Optical density readings at 450 nm were taken at 1 min (which corresponds to the linear portion of the enzymatic reaction) using a Spectra Max Plus plate reader using SOFTmax Pro v. 3.0 software (Molecular Devices Corp., Sunnyvale, CA, USA). Myeloperoxidase activity was expressed as units of enzyme per gram of tissue.

2.5. Spinning Disk Confocal Microscopy

An amount of 10^7/CFU of bacteria were injected intradermally into the lateral abdominal skin, to avoid forming lesions on the mid-line. Examination of the microcirculation of the lateral abdominal skin was prepared for microscopy in order to determine the recruitment of immune cells in lesions, as described [27]. Flank skin microvasculature was visualized using a spinning disk confocal microscope, using an Olympus BX51 upright microscope with an x20/0.95 XLUM Plan Fl water immersion objective. The microscope was equipped with a confocal light path (WaveFx; Quorum, ON, Canada) based on a modified Yokogawa CSU-10 head (Yokogawa Electric Corporation, Tokyo, Japan). Anti-

CD4-FITC (L3T4; BD Biosciences; 3 µg/mouse) and anti-mouse Gr-1 efluo660 (RB6-8C5; eBioscience, CA, USA; 2 µg/mouse) and anti-mouse CD31 coupled to Alexa647 (390; BD Biosciences conjugated via an Invitrogen protein labeling kit: 10 µg/mouse) were injected intravenously (IV) into BALB/c mice to image CD4+ T lymphocytes, neutrophils, and endothelial cells, respectively. Both 491 and 640 nm laser excitation wavelengths (Cobalt, Stockholm, Sweden) were used in rapid succession and visualized with the appropriate long-pass filters (520 +/− 35 nm and 640 +/− 40 nm, respectively, Semrock, IL, USA). Typical exposure time for both excitation wavelengths was 900 ms. A 512 × 512 pixel back-thinned electron-multiplying charge-coupled device camera (C9100-13, Hamamatsu, Japan) was used for fluorescence detection, with Volocity Acquisition software (Improvision, MA, USA) used to drive the confocal microscope. CD4 T lymphocyte and neutrophil rolling and adhesion were simultaneously assessed in 5–10 random fields of view in the postcapillary venules of the skin. The adherent cells were defined as cells which remained in the same location for 30 s in 100 µm of the venule, while the emigrated cells were defined as the cells which stayed outside of the vasculature in the field of view (FOV).

2.6. In Vivo Neutrophil Depletion

To determine the role of neutrophils in USA300-C2406 infection, a depletion process was employed using the antibody RB6-8C5 (BioXCell, West Lebanon, NH, USA), which mainly depletes neutrophils, at different time points in mice with a USA300-C2406. There were at least 3 mice for each group and each time point. As noted above, mice were injected with 10^7 CFU of USA300-C2406 intradermally in the intrascapular region on day 0. Mice in the early depletion groups received an intraperitoneal (i.p.) injection of 200 µg RB6-8C5 24 h before infection, followed by injections every 48~72 h. Mice in the late depletion groups received an i.p. injection of 200 µg RB6-8C5 24 h after infection, followed by injections every 48~72 h. Mice in the control group were injected i.p. with 200 µg of rat IgG2 isotype control monoclonal antibody (BioXCell), on the same schedule with the depletion antibody RB6-8C5. All mice were carefully monitored (including obtaining weights), then euthanized on days 4, 7, and 14 post-infection. Lung, liver, and spleen were harvested and homogenized in 1 mL sterile saline, then quantitative cultures (serial dilutions and spread on TSA plates) were performed to determine bacterial load, as described above. The skin samples were processed into tissue sections and affixed to microscope slides for Gram staining, as described above.

2.7. Mouse Antibacterial Response PCR Array

To assess whether the bacterial strains could inhibit host immune responses, the transcriptional levels of 84 host genes related to the antibacterial response were determined in local skin using a Qiagen mouse antibacterial response PCR array (Qiagen Inc., Germantown, MD, USA). Included were genes involved in Toll-like receptor (TLR) signaling, NOD-like receptor (NLR) signaling, apoptosis, inflammatory response, cytokines/chemokines, and antimicrobial peptides. Samples from at least 3 mice in each group on day 4 were assessed. RNA from lesion samples was isolated using the RNeasy Plus Mini Kit (Qiagen Inc.). Ten milligrams of each sample were homogenized in RLT buffer. The contaminating DNA was removed with gDNA eliminator spin columns. RNA was purified by washing with RW1 buffer, and collected with the RNeasy spin column. RNA was further purified and reverse transcribed to cDNA using the RT2 First Strand Kit (Qiagen Inc.). Eight microliters of raw RNA were mixed with 2 µL buffer GE and incubated for 5 min at 42 °C to eliminate contaminating DNA. Four microliters of buffer BC3, 1 µL control P2, 2 µL RE3 reverse transcriptase mix, and 3 µL RNase-free water were added and the reaction was incubated at 42 °C for 15 min. The reaction was stopped by incubation at 95 °C for 5 min. The expression of 84 host genes in the mouse antibacterial response PCR array (Qiagen Inc.) was assessed by qRT-PCR using RT2 SYBR Green qPCR Mastermix (Qiagen Inc.) on a CFX96 Real-Time Detection System (Bio-Rad, Hercules, CA, USA). Thermal cycle conditions were performed as described for the RT2 Profiler PCR Array (Qiagen Inc.). One

hundred and two microliters of the cDNA synthesis reaction were mixed with 1350 µL 2×
RT2 SYBR green mastermix and 1248 µL RNase-free water. Twenty-five microliters were
then aliquoted into each well of the RT2 Profiler PCR Array. The PCR plates were heated to
95 °C for 15 min, followed by 40 cycles of denaturation–extension (95 °C for 15 s and 60 °C
for 1 min). Because the transcription of housekeeping genes during wound forming and
healing varies [28], expression of the 84 genes was normalized to the mean of the 5 most
stably expressing housekeeping genes (*actb, b2m, gapdh, gusb*, and *hsp90ab1*). The PCR array
data were further uploaded to the Qiagen RT2 Profiler PCR Array Data Analysis Center
(http://www.qiagen.com/geneglobe. Samples). Ct = 35 was set as the cut-off Ct value,
while 5 housekeeping genes were selected as normalization factors. Relative target gene
expression was calculated according to the ΔΔCt method [29], in which the fold difference
in expression was $2^{-\Delta\Delta Ct}$.

2.8. Cytokine and Chemokine Assay

The production of various cytokines and chemokines in local skin samples was determined using a Luminex 32-plex by Eve Technologies (University of Calgary, AB, Canada).
The 32-plex consisted of eotaxin, G-CSF, GM-CSF, IFN-gamma, IL-1alpha, IL-1beta, IL-2,
IL-3, IL-4, IL-5, IL-6, IL-7, IL-9, IL-10, IL-12 (p40), IL-12 (p70), IL-13, IL-15, IL-17, IP-10,
KC, LIF, LIX, MCP-1, M-CSF, MIG, MIP-1alpha, MIP-1beta, MIP-2, RANTES, TNF-alpha,
and VEGF. Samples were from 5 mice in each group at each time point. Tissue lesions
were homogenized in 20 mM Tris HCl (pH 7.5), 0.5% Tween 20, 150 mM NaCl, and
Sigma protease inhibitor 1:100 buffer. The supernatant was collected and cytokines and
chemokines were separated into several groups for analysis, as follows: (1) protective
cytokines and chemokines whose depletion results in severe infection in a mouse skin
infection model [30,31]; (2) cytokines and chemokines related to *S. aureus* disease severity, but which lack depletion experiments to confirm their role in a mouse skin infection
model [32–38]; (3) growth factors which are important for wound healing;(4) cytokines and
chemokines related to the Th1 and Th2 response; and (5) cytokines and chemokines whose
roles are unclear in MRSA infection. The latter was further divided into 2 sub-groups; one
sub-group generally believed to be related to tissue injury, infection, and allergic disease
and the other believed to be related to immune cell development, virus infection, and aging.

2.9. Statistical Analysis

All analyses were performed using SPSS 20.0 and Prism v. 5. Differences in CFU,
diameter of lesion, and MPO activity were evaluated using one way ANOVA. The Qiagen
RT2 Profiler PCR Array Data Analysis Center was used to analyze the Qiagen PCR Array.
p values of $p < 0.05$ were considered statistically significant.

3. Results

3.1. Genotypic and Phenotypic Characterization and Virulence Gene Profiles of MRSA Strains

The genotype, antibiotic resistance (Figure 1A), and virulence factor profiles
(Figure 1B) of USA300-C2406 were compared to those of the control strains. Colonization strain M92 belonged to ST239-MRSA-III, carried a non-typeable *spa* gene, and was
Panton–Valentine leucocidin (PVL) negative and *agr* type I. USA400-CMRSA7 belonged to
pulsotype USA400 (ST1-MRSA-IVa), was PVL positive, belonged to *spa* type t128, and *agr*
type III. USA300-C2406 belonged to pulsotype USA300 (ST8-MRSA-IVa), was PVL positive,
belonged to spa type t008, and *agr* type I. Phenotypic tests indicated that USA400-CMRSA7
was sensitive to erythromycin and ciprofloxacin, unlike USA300-C2406, which was resistant to both. Likewise, M92 was resistant to clindamycin, gentamicin, and tetracycline,
while USA300-C2406 was sensitive to these agents. Virulence factors were also assessed
and compared between the strains. They were similar, with the differences being that
USA300-C2406 carried the *chp* gene and CMRSA7 carried the *sea* and *seh* genes.

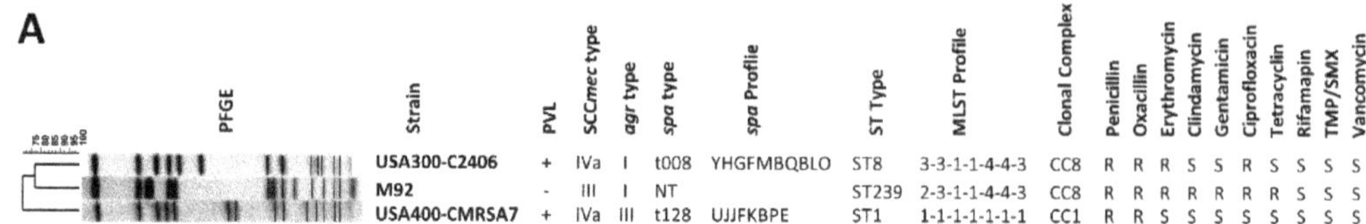

A

PFGE	Strain	PVL	SCC*mec* type	*agr* type	*spa* type	*spa* Profile	ST Type	MLST Profile	Clonal Complex	Penicillin	Oxacillin	Erythromycin	Clindamycin	Gentamicin	Ciprofloxacin	Tetracyclin	Rifampin	TMP/SMX	Vancomycin
(gel)	USA300-C2406	+	IVa	I	t008	YHGFMBQBLO	ST8	3-3-1-1-4-4-3	CC8	R	R	R	S	S	R	S	S	S	S
(gel)	M92	-	III	I	NT		ST239	2-3-1-1-4-4-3	CC8	R	R	R	R	R	R	R	S	S	S
(gel)	USA400-CMRSA7	+	IVa	III	t128	UJJFKBPE	ST1	1-1-1-1-1-1-1	CC1	R	R	S	S	S	S	S	S	S	S

B

PFGE Type	Isolate	*sea*	*sec*	*sed*	*see*	*seg*	*seh*	*sei*	*sej*	*sek*	*seq*	*tst*	*chp*	*scn*	*hla*	*hlb*	*hld*	*hlg*	*clfA*	*fnbA*	*fnbB*	*cnaA*	*cnaB*	*sdrC*	*sdrD*	*sdrE*	*bbp*	*ebpS*	*map/eap*	*ica*	*coa*	*V8*	*hysA*	*sak*
USA300	C2406	-	-	-	-	-	-	-	-	+	+	-	+	+	+	-	+	+	+	+	+	-	-	+	+	+	-	+	+	+	+	+	+	+
USA400	CMRSA7	+	-	-	-	-	+	-	-	+	+	-	-	+	+	-	+	+	+	+	+	-	-	+	+	+	-	+	+	+	+	+	+	+
Colonization Strain	M92	-	-	-	-	-	-	-	-	+	+	-	-	+	+	-	+	+	+	+	+	-	-	+	+	+	-	+	+	+	+	+	+	+

Figure 1. Methicillin-resistant *Staphylococcus aureus* (MRSA) strain genotypic and phenotypic characteristics, and virulence factor profiles. (**A**) Pulsed Field Gel Elecctrophoresis (PFGE) profiles for the MRSA strains, along with genotypic and phenotypic typing results. PVL, Panton–Valentine leucocidin (+, positive; -, negative); SCC*mec*, staphylococcal cassette chromosome *mec*; *agr*, accessory gene regulator; *spa*, staphylococcal protein A (non-typable, NT); MLST, multilocus sequence type; CC, clonal complex; S, susceptible; R, resistant. (**B**) Virulence gene profiles show that the isolates differ by only 3 genes (*sea, seh, chp*). *sea*/c/d/e/g/h/i/j/k/q, staphylococcal enterotoxin A/C/D/E/G/H/I/J/K/Q; *tst*, toxic shock syndrome toxin; *chp*, chemotaxis inhibitory protein; *scn*, staphylococcal complement inhibitory protein; *hla/b/d/g*, α/β/δ/γ-hemolysin; *clfA*, clumping factor; *fnbA/B*, fibronectin adhesive molecule A/B; *cnaA/B*, collagen adhesive molecule A/B; *sdrC/D/E*, putative adhesin; *bbp*, bone sialoprotein adhesin; *ebpS*, elastin adhesin; *map*, major histocompatibility complex class II analog protein; *ica*, polysaccharide intercellular adhesin. *coa*, coagulase; *V8*, serine protease; *hysA*, hyaluronidase; *sak*, staphylokinase; +, positive; -, negative.

3.2. USA300 Induced Extensive Open Lesions (Dermonecrosis) while M92 and USA400 Caused Localized Infection

Our mouse skin infection model differentiated USA300-C2406 hypervirulence from the control strains in that it demonstrated more severe skin lesions than M92 and USA400-CMRSA7 (Figure 2). Intradermal inoculation in either M92 or USA400-CMRSA7 resulted in confined dermal abscesses with average areas of 15.5 mm^2 and 16.36 mm^2, respectively. The sizes of these abscesses remained relatively unchanged from day 1 to day 7. Healing began after day 7, with full recovery on day 14~17. In contrast, USA300-C2406 caused abscesses with an average area of 51.40 mm^2 ($p < 0.01$ compared with M92 and USA400-CMRSA7) on day 1. This was followed by the development of an ulcerated open wound with underlying necrosis that reached its maximum size on day 5. The size of the lesion did not change significantly from day 4 to day 7. The wound started to heal after day 7, with recovery on day 14~17. Cultures of the spleen demonstrated that there was no systemic infection in this model.

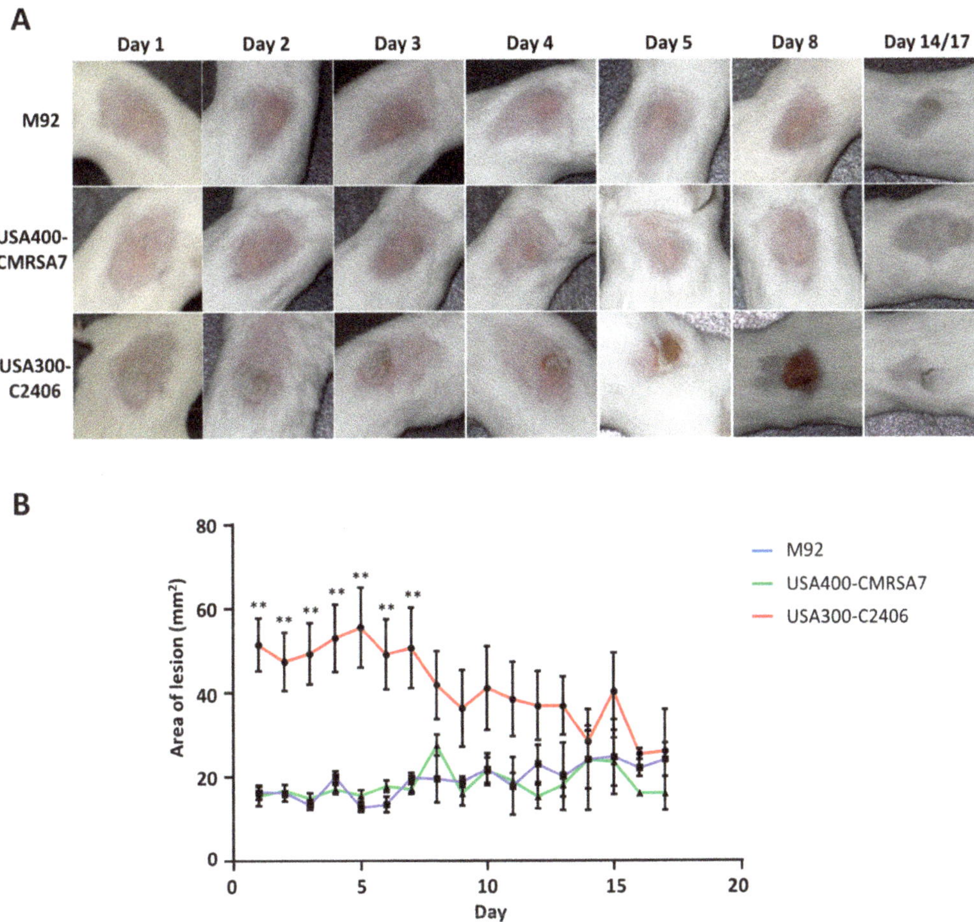

Figure 2. Disease progression in acute intradermal infection model showing localized infections caused by M92 and USA400-CMRSA7, and necrosis caused by USA300-C2406. (**A**) Representative photos for select time points (day 1, 2, 3, 4, 5, 8, 14, or 17) for each of the MRSA strains (M92, USA400-CMRSA7, USA300-C2406) showed that M92 and USA400 caused localized infections, while USA300 caused ulceration and necrosis (open wound). (**B**) The lesion diameter for each of the 3 strains. The average lesion area (mm^2) for M92 (blue), USA400-CMRSA7 (green), and USA300-C2406 (red) at each time point was expressed as mean + SEM. The experiment was repeated twice, with 5 mice in each time point and each group. Not significant between USA400-CMRSA7 and M92. Statistical comparisons between USA300-C2406 and M92 or USA400-CMRSA7 were indicated with asterisk: ** $p < 0.01$.

3.3. Profound Inflammatory Cell Infiltration by USA300

The histological changes induced by various MRSA strains were assessed on day 4 since this was the point when the skin lesions reached their maximum diameter/area. The severe tissue damage induced by USA300-C2406 was associated with more disseminated inflammation when compared with that caused by M92 and USA400-CMRSA7. As shown in the Figure 3, M92-infected skin was associated with mild changes in the epidermis and dermis (panel A). The inflammatory cell infiltration extended into the adipose tissue, superficial muscle, and fascia (D and G). USA400-CMRSA7 infection formed a well-circumscribed area of skin necrosis (B), which was associated with a mild inflammatory cell infiltration in the surrounding adipose tissue, superficial muscle, and fascia (E and

H). In contrast, USA300-C2406 formed extensive lesions with ulceration (C) and dense neutrophil-rich infiltrates that extended into the dermis, adipose tissue, fascia, and skeletal muscle (F and I). The necrotic debris was located at the edge and bottom (fascial level) of the ulcer and the disseminated inflammatory cells infiltrated not only the skin but also the skeletal muscle beneath the fascial plane.

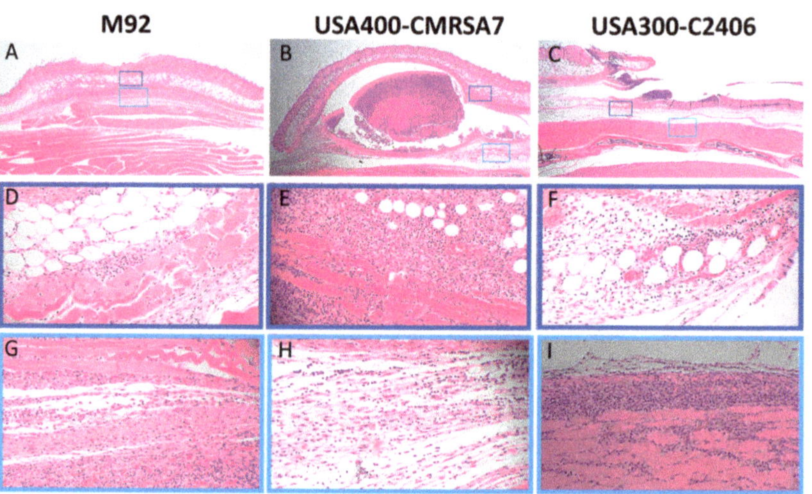

Figure 3. Histopathology (hematoxylin and eosin (H&E) stain) of the core lesions, induced on day 4, showing more severe inflammatory changes secondary to infection with USA300-C2406. The overall lesions (10×) induced by M92, USA400-CMRSA7, and USA300-C2406 are presented in panels (**A–C**), respectively. The 40X amplifications of select regions are presented in panels (**D–I**). M92 induced minor inflammatory cell infiltration in adipose layer, muscular layer, and fascia (**D,G**). USA400-CMRSA7 induced abscess formation in the fascia, surrounded by inflammatory cell infiltration (**E,H**). In contrast, USA300-C2406 induced ulceration, more tissue necrosis, and a denser, more deep-seated inflammatory cell infiltrate that extended into the skeletal muscle (**F,I**).

3.4. Primary Infiltrating Inflammatory Cell Type Was the Neutrophil

An esterase stain was done on samples collected on day 4 (Figure 4). This confirmed the histological findings of a neutrophil-rich inflammatory infiltrate in the case of USA300-C2406 that extended deeper than was the case with M92 and USA400-CMRSA7. In the former, a dense neutrophilic infiltrate was located not only in the skin but also in the deep skeletal muscle.

3.5. USA300 Induced Prolonged Periods of Excessive Neutrophil Infiltration

An MPO assay was employed to quantify neutrophil infiltration in local skin samples (Figure 5). All three MRSA strains induced higher levels of MPO activity relative to the mock-infected control. On day 4, M92, USA400-CMRSA7, and USA300-C2406 induced 386.15 U/g, 276.75 U/g, and 244.85 U/g, respectively ($p > 0.05$). On day 7, however, the MPO activity of M92 and USA400-CMRSA7 was 150.88 U/g and 85.97 U/g, respectively, while the MPO activity of USA300-C2406 reached 415.61 U/g ($p = 0.007$ and 0.001 compared with M92 and CMRSA7, respectively). On day 17, the MPO level of M92, USA400-CMRSA7, and USA300-C2406 decreased to 3.87 U/g, 5.00 U/g, and 16.96 U/g, respectively, with no significant differences noted ($p > 0.05$).

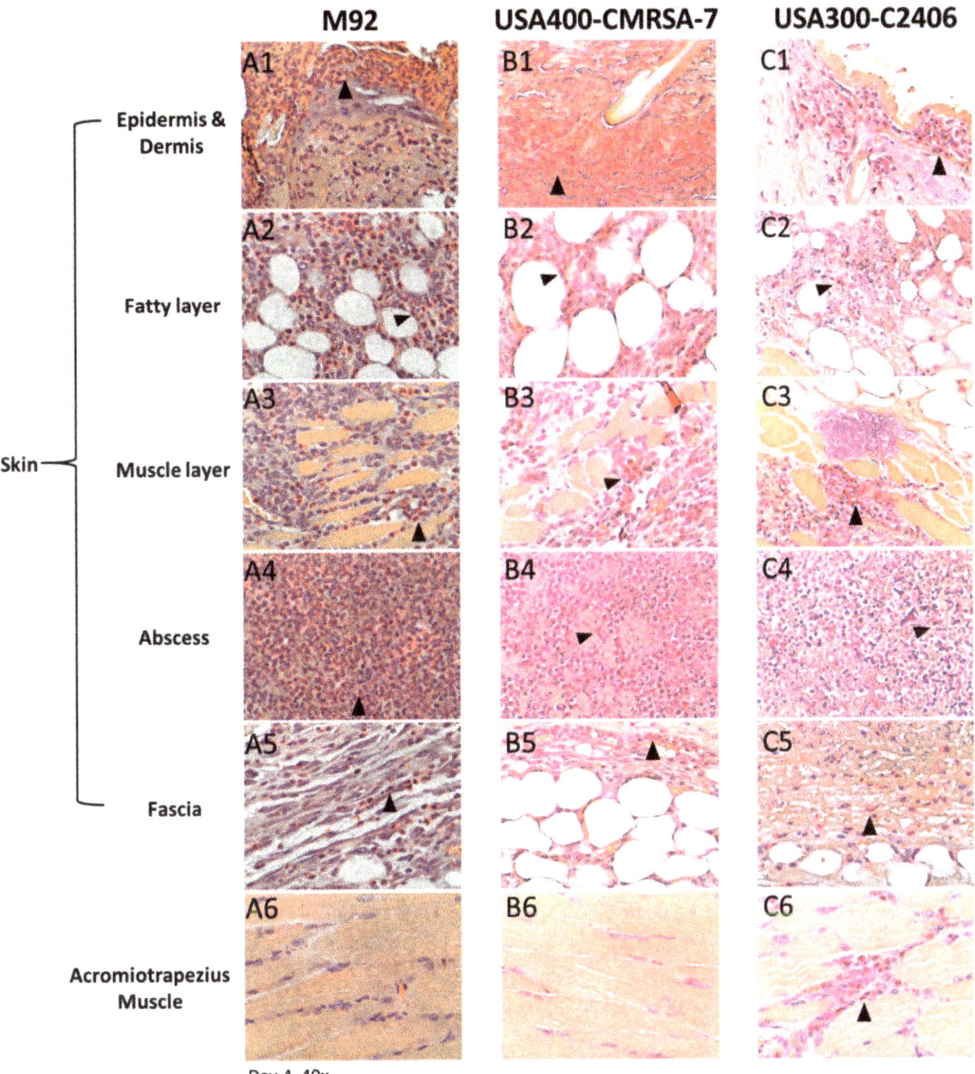

Figure 4. Histopathology (esterase staining) of skin lesions from day 4. Representative photos (40×) from different layers of lesions induced by M92 (A1–A6), USA400-CMRSA7 (B1–B6), and USA300-C2406 (C1–C6). Positive cells (primarily neutrophils) were located at the center of the abscess (A4, B4, and C4) and edges of lesions (A1–A3, B1–B3, and C1–C3). Positive cells are labeled with arrows. The skin lesion induced by USA300-C2406 presented a denser and deeper neutrophil infiltrate than did that of USA400-CMRSA7 and M92.

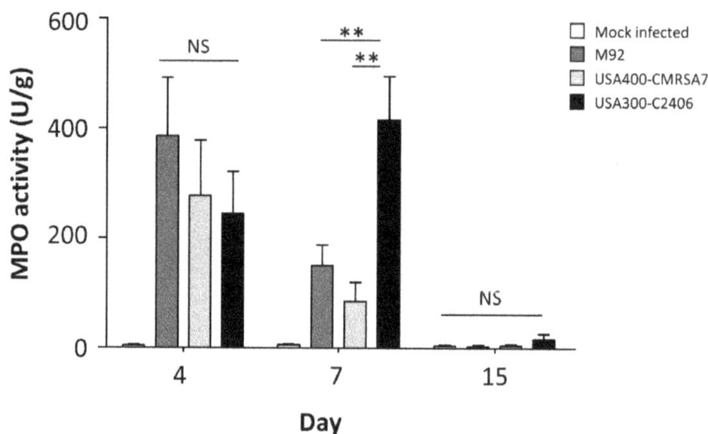

Figure 5. USA300 infection leads to significantly more myeloperoxidase (MPO) activity in skin samples on day 7. The myeloperoxidase activity (U/g) in local skin samples from mock-infected control (white bars), M92- (dark gray bars), USA400-CMRSA7- (light gray bars), and USA300-C2406-infected (black bars) mice at different time points (days 4, 7, and 15) is shown as mean + SEM. The experiment was repeated twice, with 5 mice in each time point and each group. NS, not significant; ** $p < 0.01$.

3.6. Significantly Greater Neutrophil (But Not CD4 T Cell) Adhesion and Emigration Triggered by USA300

Neutrophil adherence and emigration were assessed using spinning disk confocal microscopy (Figure 6). Infected mice had elevated numbers of adherent and emigrated neutrophils as compared to mock-infected controls, with greater neutrophil adhesion and emigration induced by USA300-C2406 in comparison to what was seen in the other two strains. On day 1, M92 and USA400-CMRSA7 induced 3.00 cells/100 μm and 8.00 cells/100 μm, respectively, while USA300-C2406 induced 11.60 cells/100 μm ($p = 0.055$ and 0.000 compared with USA400-CMRSA7 and M92, respectively). On day 7, the number of neutrophils induced by all three strains was reduced; M92 and USA400-CMRSA7 induced 0.60 cells/100 μm and 3.60 cells/100 μm, respectively, and USA300-C2406 induced 7.40 cells/100 μm, which was significantly greater than M92 and USA400-CMRSA7 ($p = 0.000$ and 0.001, respectively). On day 14, there was no significant difference among the three strains. In terms of neutrophil emigration, on day 1, M92 and USA400-CMRSA7 induced 8.40 cells/field and 24.56 cells/field, respectively, while USA300-C2406 induced 50.60 cells/field ($p = 0.000$ compared with both M92 and CMRSA7). On day 7, M92 and USA400-CMRSA7 induced 2.00 cells/field and 10.60 cells/field, respectively, while USA300-C2406 induced 20.20 cells/field ($p = 0.000$ compared with both M92 and USA400-CMRSA7). On day 14, there was no significance among the three strains. No significant trend was observed for CD4+ T cell adhesion and emigration among the three strains.

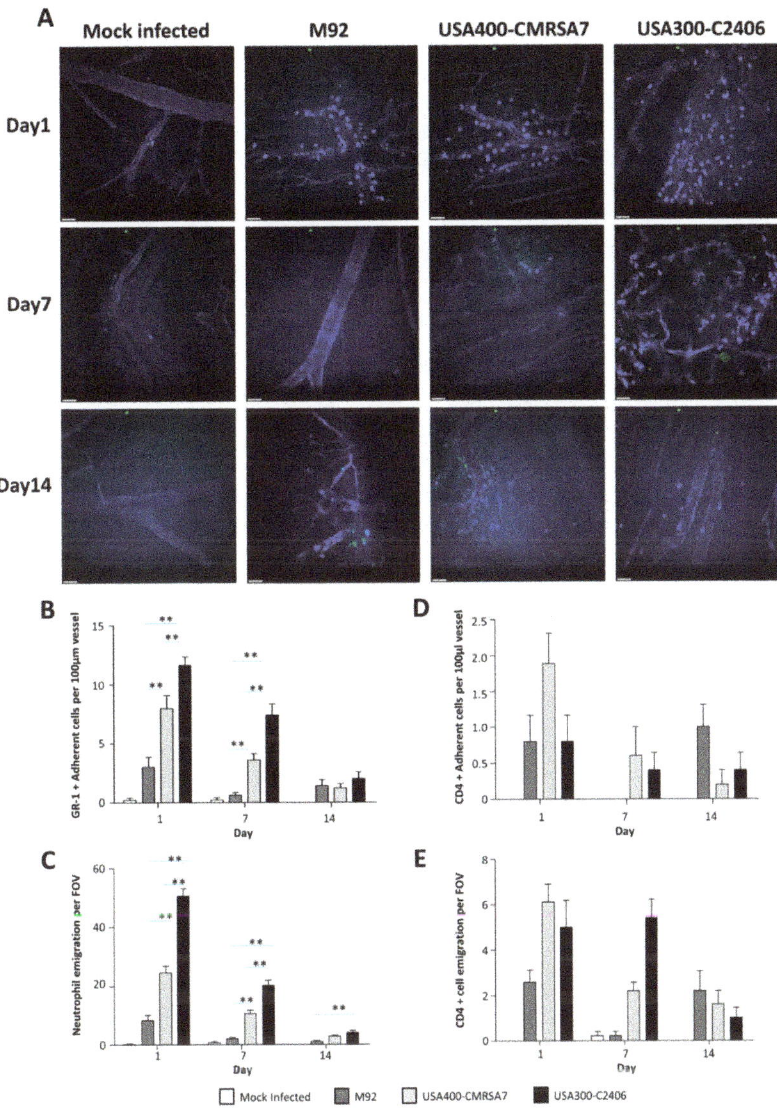

Figure 6. Confocal microscopy revealed that USA300 induced greater neutrophil (but not CD4 cell) adherence and emigration. (**A**) Representative examples for visualizing GR-1+ cells, CD4 + cells, and endothelial cells in mouse flank skin with spinning disc confocal microscopy on days 1, 7, and 14. Gr-1+ efluo660 neutrophils are seen as bright blue ovals whereas endothelial cells stained with CD31-A647 can be seen as thin blue lines outlining the vessel walls. CD4+ cells are depicted in FITC (green). (**B–E**) Bar graphs showing quantification (mean + SEM) of adherent and emigrated neutrophils (**B** and **C**) and CD4+ T cells (**D** and **E**) on days 1, 7, and 14 for the mock-infected control (white bars), M92 (dark gray bars), USA400-CMRSA7 (light gray bars), and USA300-C2406 (black bars). USA300-C2406 induced significantly greater neutrophil adherence and emigration when compared with USA400-CMRSA7 and M92 on day 1 and 7 ($p < 0.01$). Experiments with 5 mice in each group and each time point. ** $p < 0.01$. Scale bar represents 40 μm.

3.7. Co-localization of Higher Bacterial Load with Neutrophil Infiltration in USA300 Infection

To determine if neutrophils were targeting bacteria, tissue slides (from day 4 lesions) were stained with Gram stain to determine bacterial localization (Figure 7). Deeper bacterial dissemination in USA300-C2406-infected tissue was noted, with bacteria found throughout the skin tissue and invading the skeletal muscle. In contrast, bacteria were predominantly located in the abscess of lesions caused by M92 and USA400-CMRSA7. There was a direct correlation between the presence of bacteria and the presence of neutrophils.

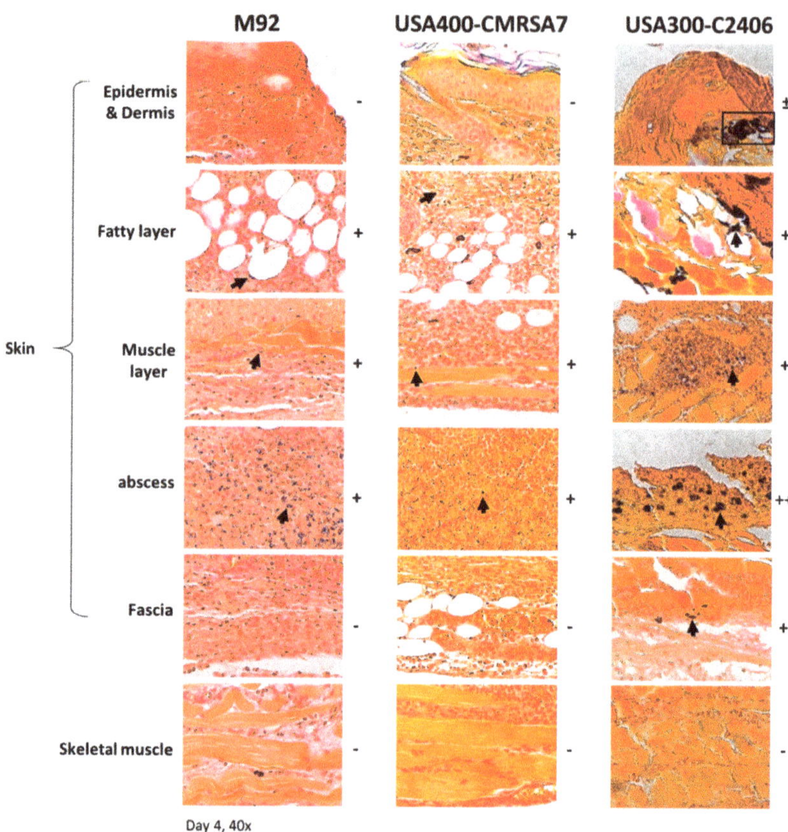

Figure 7. Histopathology (Gram staining) of skin lesions shows that USA300 infections have a higher bacterial load. Representative images (40× magnification) from various layers of skin and subcuticular tissue in the lesion from M92, USA400-CMRSA7, and USA300-C2406 infections on day 4 are presented. USA300-C2406 infections have a higher bacterial load as compared to USA400-CMRSA7 and M92 infections. Positive spots (bacteria) are labeled with a black box or arrow. The relative quantity of positive spots is indicated with + (positive), ++, and – (negative for bacteria).

3.8. Neutrophil Depletion Resulted in a More Severe Infection

We depleted the neutrophils in mice infected with USA300-C2406 to examine the role they played in disease progression. We observed that early neutrophil depletion (24 h before infection) resulted in a significantly more severe infection as compared to the isotype control (Figure 8). With the isotype control group, the wound formed on day 1 and remained constant with an average area of 42.53 mm^2 until day 10, at which time it gradually decreased and was healed around day 17. Similarly, with late neutrophil depletion (24 h after infection), the average wound area on day 1 was 65.67 mm^2, which was not statistically different from the isotype control group ($p = 0.216$). The wound maintained an average area of 50 mm^2 until day 8, at which time it decreased, similar to the isotype control. When neutrophils were depleted prior to infection, however, the mice showed an enlarged wound with an average area of 164.37 mm^2 on day 1 ($p = 0.001$ and $= 0.005$, compared with isotype control and late depletion group, respectively), which remained constant until day 11, then gradually reduced in size and healed by day 17 (Figure 8 B).

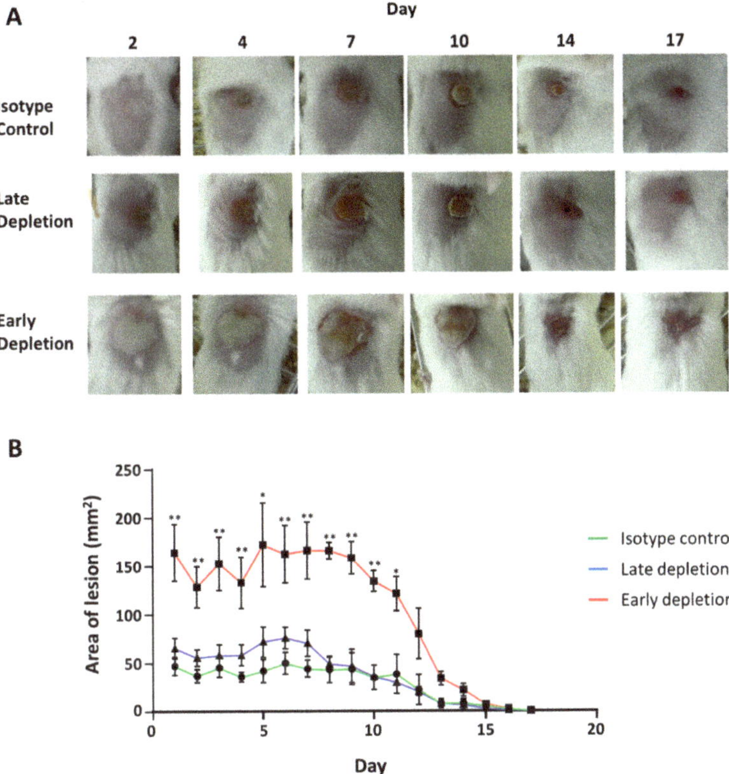

Figure 8. Early neutrophil depletion results in more severe lesions in mice infected with USA300. (**A**) Representative photos from days 2, 4, 7, 10, 14, and 17 showing the lesion in isotype control, late depletion, and early depletion groups infected with USA300-C2406. (**B**) Lesion area, expressed as mean area of skin lesion (mm^2) + SEM, for the isotype control (green), late depletion (blue), and early depletion (red) groups. Experiments with 9 mice in each group. * $p < 0.05$; ** $p < 0.01$.

3.9. Neutrophil Depletion Resulted in Bacterial Dissemination and Invasive Infection

To investigate the cause of increased tissue damage following neutrophil depletion, we used Gram staining to identify bacteria in local skin (Figure 9A), and organ culture to identify bacterial load in the internal organs (Figure 9B). Gram stains indicated that there were more bacteria present in the tissue in both early and late neutrophil depletion groups as compared to the isotype control group, and that the bacteria invaded deeply into skeletal muscle in depletion groups, but were limited to the skin in the isotype control group. Organ cultures demonstrated that the bacterial load in the lung, liver, and spleen was less than 1 CFU/mg in the control group, whereas the bacterial load from the early and late depletion groups ranged from 5–100 CFU/mg. Weight change during infection was used as a marker for infection severity (Figure 9C), and showed that the late depletion group had significantly more weight loss from days 3 to 4, while the early depletion group had significantly more weight loss from days 1 to 8.

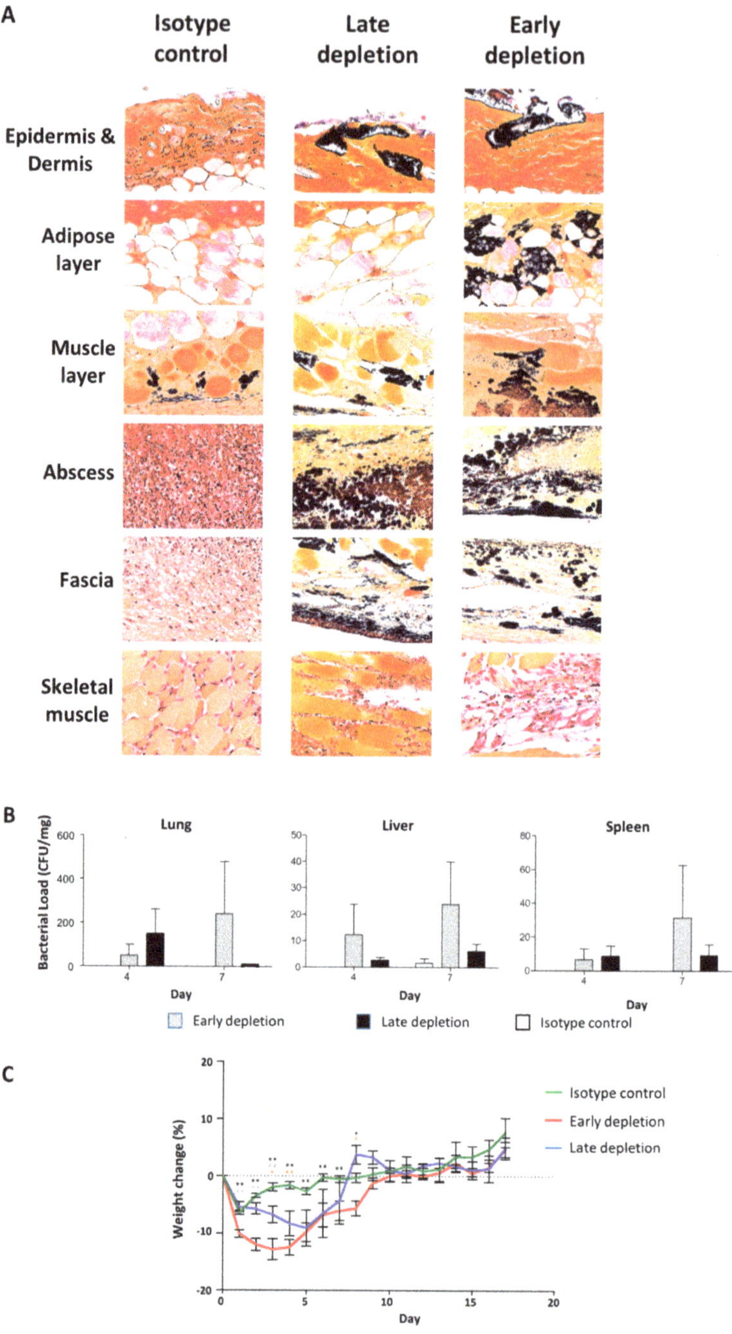

Figure 9. Neutrophil depletion resulted in bacterial dissemination and weight loss. (**A**) Representative images (40×) from the skin and subcuticular tissue following early neutrophil depletion, late neutrophil depletion, and isotype control group on day 4 after infection with USA300-C2406. Gram

staining indicates that there were high bacterial loads when neutrophils were depleted. (**B**) USA300-C2406 bacterial counts (CFU/mg) recovered from lung, liver, and spleen at day 4, 7, and 17 time points, in early neutrophil depletion (gray bars), late neutrophil depletion (black bars), and isotype control (white bars) groups. (**C**) Percentage of weight change following infection with USA300-C2406 in the isotype (green), early neutrophil depletion (red), and late neutrophil depletion groups (blue), showing that severe weight loss was associated with neutrophil depletion. Weight changes were expressed as mean + SEM. Experiments with 9 mice in each group. Statistical comparisons between mock infection and early depletion (black asterisk), mock infection and late depletion (orange asterisk), early and late depletion (grey asterisk); *$p < 0.05$; **$p < 0.01$.

3.10. Unique Pattern of Molecular Transcription Results from Infection with USA300

Because USA300 infections were associated with extensive neutrophil infiltration and increased bacterial load, we hypothesized that USA300 might employ a mechanism designed to modify or inhibit the antibacterial response in the mice. Using an antibacterial response PCR array, we examined the expression of factors such as Toll-like receptor signaling, NOD-like receptor (NLR) signaling, bacterial pattern recognition receptors (PRRs), signaling downstream of antibacterial responses, apoptosis, inflammatory responses, cytokines/chemokines, and antimicrobial peptides in local skin samples. Our results demonstrated that MRSA USA300 infection could activate the transcription of most of the components involved in the antibacterial responses. However, when compared with USA400-CMRSA7 and M92, USA300-C2406 induced lower degrees of upregulation.

For components related to TLR signaling, USA300-C2406 induced less transcription of *akt1*, *irak3*, and *tlr9* (2.03–4.49-fold) when compared with both USA400-CMRSA7 and M92, less transcription of *tlr6* and *tollip* ($p < 0.05$) (2.69–2.74-fold) when compared with USA400-CMRSA7, and less transcription of *fadd* (2.07-fold) when compared with M92. Only the transcription of LPS binding protein (*lbp*) was higher in USA300-C2406 infections than in USA400-CMRSA7 and M92 infections ($p < 0.05$) (3.31-fold and 2.26-fold, respectively).

For components related to NLR signaling, USA300-C2406 induced less transcription of *card6*, *naip1*, *nlrp1a*, and *nod2* (2.0–6.41-fold) when compared with USA400-CMRSA7, and less transcription of *casp1*, *pycard*, and *ripk2* ($p < 0.01$) (2.06–2.92-fold) when compared with M92.

For components related to signaling downstream of antibacterial responses, USA300-C2406 induced less transcription of *map2k3* ($p < 0.05$), *mapk3* ($p < 0.05$), and *nfkbia* (2.08–2.13-fold) when compared with USA400-CMRSA7.

For components related to apoptosis, USA300-C2406 induced less transcription of *akt1*, *ifnb1*, *il12b* (2.07–5.83-fold) when compared with both USA400-CMRSA7 and M92. Furthermore, USA300-C2406 induced less transcription of *card6*, *il12a*, and *nfkbia* (2.01–3.2-fold) when compared with USA400-CMRSA7 and less transcription of *casp1*, *fadd*, *il6*, *pycard*, and *ripk2* (2.05–2.92-fold) when compared with M92.

For components related to inflammatory response, USA300-C2406 induced less transcription of *akt1* and *tlr9* (2.07–4.49-fold) when compared with USA400-CMRSA7 and M92, less transcription of *tlr6* and *tollip* (2.69–2.74-fold) when compared with USA400-CMRSA7, and less transcription of *ccl5*, *cxcl1*, *il6*, and *ripk2* when compared with M92. Only the transcription of *cxcl1* and *lbp* was higher (2.23–3.31-fold) in USA300-C2406 infections than in USA400-CMRSA7 and M92 infections.

For cytokines and chemokines, USA300-C2406 induced less transcription of *ifna9*, *ifnb1*, *il12b*, and *il18* (2.07–5.83-fold) when compared with USA400-CMRSA7 and M92, and less transcription of *ccl5*, *cxcl3*, and *il6* (2.05–4.39-fold) when compared with M92. Only the transcription of *cxcl1* (2.23–2.55-fold) was higher in USA300-C2406 infections compared to USA400-CMRSA7 and M92 infections.

For antimicrobial peptides, USA300-C2406 induced less transcription of *bpi* (2.07–4.49-fold) when compared with USA400-CMRSA7 and M92, and less transcription of *camp*, *ctsg*, *ltf*, and *prtn3* when compared with USA400-CMRSA7. Transcription of *slpi* was higher

(2.52–3.69-fold) in USA300-C2406 infections when compared with USA400-CMRSA7 and M92 infections.

There was, however, no statistically significant difference in the transcription of most of these aforementioned genes, with the exception of *il-18*. With *il-18*, the transcription in USA400-CMRSA7 and M92 infected mice was close to that of the mock-infected control, while USA300-C2406 induced a 2.69-fold downregulation ($p = 0.0211$). The level of transcription in USA300-C2406-infected mice was downregulated 3.22- and 3.39-fold ($p = 0.011$ and 0.030) when compared with USA400-CMRSA7 and M92, respectively.

The detailed list of transcriptional levels for genes related to each response is summarized in Supplementary Table S1.

3.11. Unique Cytokine and Chemokine Profiles in USA300-Infected Lesions

After determining that USA300-C2406 infections were associated with unique antibacterial responses, we further compared the local cytokine and chemokine profiles (protein profile) with a Luminex assay. On day 4, most of the cytokines and chemokines (IFNy, IL-17, IL-1b, G-CSF, GM-CSF, M-CSF, KC, TNFa, MIP2, RANTES, IL-4, IL-13, IL-6, IL-12p70, IL-12p40, VEGF, MIP1a, MIP1b, IL-5, LIF, MCP1, MIG, IL-3, eotaxin, and IP10) were increased (1.50–1584.13-fold increase) in mice infected with any of the MRSA strains as compared to the mock-infected control. The details are listed in Table 1.

For cytokines/chemokines reported to be associated with protection (IL-17, IL-1a, and IL-1b), there was no difference between USA300-C2406 and M92 or USA400-CMRSA7 ($p = 0.8809$–0.9996 and $p = 0.344$–0.769, respectively). For cytokines/chemokines associated with disease severity, USA300-C2406 significantly increased production of G-CSF, M-CSF, KC, MIP2, IFNy, GM-CSF, TNFa, IL-4, and IL-6 when compared to M92 and USA400-CMRSA7 (1.99–40.0-fold, $p = 0.000$–0.027, and 3.52–20.17-fold, $p = 0.000$–0.123, respectively), but there was no difference in the production of IL-10, IL-12p70, IL-12p40, and IL-15. For VEGF, USA300-C2406 induced 2.70- and 12.23-fold increases compared to M92 and USA400-CMRSA7, respectively ($p = 0.013$ and <0.001, respectively). Cytokines and chemokines involved in tissue injury, infection, and allergic disease, such as IL-5, LIF, LIX, MCP-1, MIP-1a, and MIP-1b, were increased in the USA300-C2406-infected mice. USA300-C2406 induced 3.26–15.99-fold and 7.67–1825.60-fold increases compared with M92 and USA400-CMRSA7, respectively ($p = 0.000$–0.093, and 0.000–0.02, respectively). Cytokines and chemokines associated with immune cell development, virus infection, and aging, including IP-10, eotaxin, and IL-3, showed no difference among these strains ($p = 0.343$–0.918 and $p = 0.089$–1.000 for USA300-C2406 compared with M92 and USA400-CMRSA7). There were no differences in the other Th1 and Th2 cytokines (IL-2, IL-9, and IL-13) between infected groups and the mock-infected control group. On day 7, most cytokines and chemokines were reduced compared to day 4, but USA300-C2406 still induced more G-CSF, IL-6, and LIF than M92 and USA400-CMRSA7 (4.86–8.53-fold, $p = 0.000$–0.046, and 2.62–6.37-fold, $p = 0.000$–0.185, respectively) (Supplementary Table S2). On day 15, the overall cytokines and chemokines were reduced, and there was no difference between the strains (Supplementary Table S3).

Table 1. Cytokine and chemokine profiles for mice on day 4 post-infection with M92, USA400, and USA300 MRSA strains.

| Role | Cytokine/Chemokine | Mock Infected conc. | M92 conc. | USA400-CMRSA7 conc. | USA300-C2406 conc. | USA300-C2406 | | |
						Vs. Mock Infected	Vs. M92	Vs. USA400-CMRSA7
Protection	IL-17A	0.0185 (0.0017–0.0416)	1.072 † (0.0768–2.2690)	0.1850 (0.0454–0.3744)	0.8985 (0.4565–1.4555)			
	IL-1a	37.9831 (25.0558–45.3151)	29.6027 (10.8897–54.1207)	20.2297 (6.1071–33.1123)	30.2987 (20.9430–39.5406)			
	IL-1b	0.5015 (0.2672–0.8363)	9.356 † (2.06583–18.1579)	1.3075 (0.1432–2.9846)	7.3872 (3.9964–11.6633)			
Severity	IL-4	0.0129 (0.0097–0.0183)	0.0273 (0.0122–0.0553)	0.0215 (0.0119–0.0339)	0.0918 (0.0680–0.1042)	**	**	**
	IL-6	0.0270 (0.0000–0.049)	1.418 (0.1269–4.8495)	1.1344 (0.1020–2.7832)	16.3018 (5.1485–37.0610)	*	*	*
	IFNγ	0.0000 (0.0000–0.0000)	0.0666 (0.0000–0.2207)	0.0150 (0.0000–0.0748)	0.1793 (0.0664–0.3274)	**		**
	TNFa	0.0684 (0.000–0.1614)	0.7811 (0.2561–2.0464)	0.2644 (0.2373–0.5366)	1.6897 (0.9215–2.6344)	**		**
	GM-CSF	0.1779 (0.0000–0.5163)	0.6738 (0.0000–1.2190)	0.5376 (0.0000–0.9983)	1.8187 (0.6250–2.9474)	**	*	**
	G-CSF	0.1111 (0.0000–0.4109)	81.6 † (19.2325–115.1702)	8.7264 (3.864253–22.4210)	175.9966 (102.2350–224.3600)	**	*	**
	M-CSF	0.0775 (0.0492–0.1206)	0.3669 (0.2111–0.6281)	0.4481 (0.1236–0.8643)	14.6826 (2.4079–40.4880)	*	*	*
	KC	2.3430 (1.7789–3.2617)	19.8096 (2.0290–47.4260)	5.7795 (2.2214–12.5455)	70.1897 (37.6055–135.1504)	**	*	**
	MIP-2	1.0161 (0.3052–1.7494)	223.4 † (154.0767–271.0803)	230.60† (219.8402–241.4560)	357.2018 (246.9371–480.9945)	**	*	
	IL-12(p70)	0.0702 (0.0169–0.1461)	0.1498 (0.0987–0.2221)	0.1746 (0.0593–0.1578)	0.3607 (0.0000–0.6981)	*		
	IL-12(p40)	0.0086 (0.0000–0.0305)	0.0160 (0.0000–0.0392)	0.01974 (0.0000–0.0765)	0.0736 (0.0000–0.2041)			
	Rantes	0.0190 (0.0000–0.0335)	0.4183 † (0.2961–0.6175)	0.4276 † (0.2293–0.7087)	0.3085 (0.1040–0.4368)	*		
	IL-15	1.8561 (0.7573–3.7306)	0.5533 † (0.0000–1.5679)	0.7991 (0.0496–1.1612)	0.9902 (0.5022–1.3107)			
	IL–10	0.1976 (0.0873–0.3272)	0.2875 (0.1620–0.3606)	0.2817 (0.1698–0.4221)	0.3899 (0.1375–0.5556)			

149

Table 1. *Cont.*

Role	Cytokine/ Chemokine	Mock Infected conc.	M92 conc.	USA400-CMRSA7 conc.	USA300-C2406			
					USA300- C2406 conc.	Vs. Mock Infected	Vs. M92	Vs. USA400- CMRSA7
Growth factor	VEGF	0.0621 (0.0425–0.0883)	1.1501 (0.2439–2.8221)	0.2536 (0.0873–0.6286)	3.1011 (1.6067–4.7437)	**	*	**
Remaining Th1 and Th2 cytokines	IL-2	0.5800 (0.3685–0.7409)	0.3426 (0.1942–0.5813)	0.4792 (0.3380–0.7139)	0.4275 (0.2350–0.6595)			
	IL-9	3.3111 (2.3307–4.6925)	2.5649 (1.7161–4.4603)	2.8168 (1.7129–4.9587)	2.6974 (1.6700–3.9570)			
	IL-13	0.0000 (0.0000–0.0000)	0.0794 (0.0000–0.2386)	0.0327 (0.0000–0.1635)	0.1067 (0.0000–0.2262)			
Other cytokines (tissue injury, infection, allergic diseases)	IL-5	0.0033 (0.0000–0.0131)	0.0142 (0.0000–0.0195)	0.0173 (0.0139–0.0203)	0.2271 (0.1089–0.5159)	**	**	**
	LIF	0.0241 (0.0072–0.0560)	0.2109 (0.0948–0.3703)	0.0890 (0.0356–0.1318)	3.9190 (2.0239–7.280)	**	**	**
	LIX	0.0086 (0.0000–0.0431)	0.0000 (0.0000–0.0000)	0.000 (0.0000–0.0000)	1.8256 (0.0690–3.9411)	*	*	*
	MCP-1	0.3565 (0.2727–0.5333)	3.2274 (1.2843–4.9421)	1.5886 (0.9464–1.8880)	12.1896 (6.0745–21.0912)	**	**	**
	MIP-1a	0.6512 (0.4869–0.8522)	5.2930 (0.6879–12.8468)	2.9084 (0.5998–6.7366)	37.4366 (12.7200–85.4010)	**	*	*
	MIP-1b	0.0000 (0.0000–0.0000)	2.058 (0.4973–5.3016)	1.3585 (0.6601–3.5958)	13.7665 (6.1825–25.9812)	**	**	**
	MIG	3.3655 (1.4177–5.2275)	8.3413 (2.6754–17.4490)	11.3525 (3.9570–34.2732)	5.4001 (3.3130–8.5667)			
Other cytokines (immune cell development, virus infection, aging)	IP-10	0.0963 (0.0529–0.1781)	0.8319 (0.0987–2.0276)	0.4538 (0.15725–0.7421)	1.7451 (0.7332–3.9248)	*		
	Eotaxin	0.6111 (0.2139–1.2606)	3.5192 † (0.9022–5.4013)	1.9032 (1.1310–2.8953)	4.3125 (1.0695–7.3215)	**		
	IL-3	0.0015 (0.0000–0.0075)	0.0681 (0.0183–0.1806)	0.1669 (0.0475–0.3370)	0.1577 (0.0395–0.2731)	*		
	IL-7	0.1743 (0.0000–0.7500)	0.1810 (0.0817–0.2839)	0.0696 (0.0057–0.2083)	0.2900 (0.1815–0.3531)	*		

Note: Conc., mean concentration (pg/mg) and range (minimum–maximum); †, $p < 0.05$ when compared to mock-infected group; ** $p < 0.01$; * $p < 0.05$ when USA300-C2406 compared with mock-infected group and the other 2 MRSA strains.

4. Discussion

There are several mouse models that have been used to study *S. aureus* infections [14–19]. Most of the models, however, were either designed to study the mechanisms of *S. aureus* virulence by comparing mutants to their corresponding wild type strains, or to study more invasive infections similar to what is seen in hospital-acquired diseases. *S. aureus* USA300 is the most common pathogen responsible for skin and soft tissue infections (SSTIs), which differ from wound or surgical site infections in that they generally begin as small lesions with areas of necrosis [36,39,40]. In this study, building on previously described models, we established a mouse skin infection model that mimics the clinical presentations seen with CA-MRSA SSTIs, and which is able to differentiate between two CA-MRSA strains, USA400 and USA300, as well as the colonization control strain, M92. With this mouse intradermal infection model, M92 and USA400-CMRSA7 caused localized infections, while USA300-C2406 caused ulceration with necrosis. The model was also able to show that the severe infections induced by USA300 were associated with a specific pattern of immune response, as well as an increased bacterial burden.

USA300 induced ulcer formation on day 1, reaching a maximum size and severity on day 4, remaining constant until day 7, and then healing by day 14–17. The USA300 infection was marked and lasted longer with more pronounced inflammation and a higher bacterial burden than the other MRSA infections, suggesting that USA300 might employ a mechanism to modify/inhibit the host's immune response. The role of neutrophils in MRSA infection is, however, complex. We showed that neutrophil depletion resulted in more severe tissue damage and invasive infections, possibly from loss of the protective role of neutrophils, resulting in uncontrolled bacterial infections. Neutrophils play an important role in the response to *S. aureus* SSTIs [41–43], however, extensive infiltration and activation of neutrophils at the site of infection is believed to cause chronic inflammation, impaired injury repair, and loss of organ function [44].

Transcriptional PCR array results further supported the notion that USA300 is capable of modulating the mouse immune system and neutrophil response. Il-18, which has been shown to restore neutrophil phagocytosis during severe skin damage [20], and Il-12, which is related to natural killer NK cell and macrophage defensive function in mice [45], each showed decreased transcriptional levels in the USA300-infected mice. USA300 also induced higher transcription of *slpi*, which was shown to be involved with the inhibition of neutrophil extracellular trap (NET) formation [46]. In addition, several factors involved with Toll-like and NOD-like receptor signaling showed decreased transcriptional levels in USA300-infected mice. Recognition of *S. aureus* by Toll-like and NOD-like receptors is required for the activation of protective host responses, such as the production of inflammatory cytokines and chemokines, and is important in the defense against *S. aureus*-induced SSTIs [47,48]. It was demonstrated that some staphylococcal virulence factors, such as staphylococcal TIR domain protein (TirS) that can inhibit TLR signaling via molecular mimicry [49], and staphylococcal superantigen-like protein 3 (SSL3) that can negatively interfere with TLR2 recognition and heterodimer formation [50,51], can inhibit and block TLR2-mediated inflammatory responses and impair the cytokine production, as well as neutrophil and macrophage activity, and consequently inhibit polymorphonuclear leukocyte (PMN) infiltration [52]. Taken together, these results indicate that USA300 might possess a mechanism to inhibit the defensive immune responses, especially involving neutrophils. However, the PCR array assayed the overall transcription of factors in whole skin samples, which includes various immune cells, keratinocytes, and other structural cells. This may not represent the transcriptional level for specific cell types and may explain why there was no difference in *mpo* transcription among the various strains and mock-infected control.

Cytokine and chemokine profiles in local lesions were consistent with our hypothesis that USA300 is modulating the mouse immune response. IL-17A, IL-1α, and IL-1β are believed to be protective during mouse MRSA skin infections [30,31]. The cytokines not only play a role in neutrophil recruitment, but can also promote the production of antimicrobial peptides (such as β-defensins and cathelicidins) [53]. In this mouse skin infection model, USA300 induced the same level of production for these cytokines on day 4 when compared with the other strains, yet more neutrophils were recruited, and there was a lower level of antimicrobial peptide production in USA300-infected mice. USA300 appears to be suppressing certain factors involved with the protective response. However, infection with USA300 did trigger increased production of cytokines and chemokines related to severity, including IFNy, GM-CSF, TNFa, IL-4, and IL-6. These are believed to be related to severe infections [35,36]. Other cytokines and chemokines such as IL-5 [54], LIF [55], LIX, MCP-1 [56], MIP-1a, and MIP-1b [57], which are believed to be involved in tissue damage, infection, and allergic diseases, also showed increased production in USA300 infections. Interestingly, the expression of MIG (also related to tissue damage, infection, or allergic diseases), was the only member of this group whose expression in USA300 infections was less than that in infections with the other MRSA strains. This phenomenon deserves further study. Macrophages expressing MIG (also known as CXCL9) are believed to protect against abscess formation [58] and promote tissue repair. Our USA300 resulted in lower production of MIG on day 7 and 14 when compared with other strains, which could indicate that USA300 is inhibiting MIG to increase tissue damage. We also recognize that the level of infiltration was very different between the different strains, as shown in this study. This could complicate the transcriptomic analysis as the composition of the material used for analysis was different with more or less immune cells. For example, a lack of a statistically significant difference in IL-17a, IL1a, and IL-1b should be viewed in light of the fact that there was more immune cell infiltration in the lesions for USA300 than the other strains. Further study by using tissue type-specific control genes is needed to address this issue.

5. Conclusions

Our results demonstrated that USA300 infections induce an intense neutrophilic response, inhibits host antibacterial responses, selectively inhibits the production of protective cytokines, and selectively activates the production of pro-inflammatory cytokines and chemokines. This unique pattern of immune responses could cause increased tissue damage at the site of the wound and increased bacterial persistence, which would facilitate the spread of USA300. This may provide some insight into why USA300 is more dominant and more virulent than other MRSA strains.

Supplementary Materials: The following are available online at https://www.mdpi.com/2076-260 7/9/2/287/s1, Table S1: List of transcription levels for host genes related to the antibacterial response, for mice infected with USA300, USA400, and M92 MRSA strains (day 4); Table S2: Cytokine and chemokine profiles for mice on day 7 post-infection with M92, USA400, and USA300 MRSA strains; Table S3: Cytokine and chemokine profiles for mice on day 15 post-infection with M92, USA400, and USA300 MRSA strains.

Author Contributions: Conceived the idea and designed the experiments: K.Z. Performed the experiments and analyzed data: J.Z., J.M., K.W., B.P., D.B., S.E. Supervised the study: G.A., J.C., K.Z. Structured and drafted the manuscript: J.Z., K.Z. Reviewed and edited the manuscript: J.M., J.C., K.Z. Commented on and approved the final manuscript: all authors. All authors have read and agreed to the published version of the manuscript.

Funding: This research was supported in part by the Alberta Sepsis Network by Alberta Innovations Health Solutions (Project # RT716258) and in part by an operation fund from the Centre for Antimicrobial Resistance (CAR), Alberta Health Services.

Institutional Review Board Statement: The animal infection experiments in this study were performed at the Animal Resource Centre at the University of Calgary, in accordance with institutional and national guidelines of the Canadian Council on Animal Care, under protocol numbers M06074, M09115, and AC13-0076, approved by the Animal Care Committee at the University of Calgary.

Data Availability Statement: The complete genome sequence data have been deposited at Gen-Bank under the accession numbers: USA300-C2406 (PRJNA345240; CP019590.1), USA400-CMRSA7 (PRJNA362898), and M92 (PRJNA319679; CP015447.1).

Acknowledgments: Our experiments were supported by the Mouse Phenomics Resource Laboratory, funded by the Snyder Institute for Chronic Diseases. Intravital imaging was performed at and supported by the Live Cell Imaging Laboratory, Snyder Institute for Chronic Diseases.

Conflicts of Interest: The authors declare no conflict of interest.

References

1. David, M.Z.; Cadilla, A.; Boyle-Vavra, S.; Daum, R.S. Replacement of HA-MRSA by CA-MRSA infections at an academic medical center in the midwestern United States, 2004–2005 to 2008. *PLoS ONE* **2014**, *9*, e92760. [CrossRef] [PubMed]
2. Huang, H.; Flynn, N.M.; King, J.H.; Monchaud, C.; Morita, M.; Cohen, S.H. Comparisons of community-associated methicillin-resistant *Staphylococcus aureus* (MRSA) and hospital-associated MSRA infections in Sacramento, California. *J. Clin. Microbiol.* **2006**, *44*, 2423–2427. [CrossRef] [PubMed]
3. Baud, O.; Giron, S.; Aumeran, C.; Mouly, D.; Bardon, G.; Besson, M.; Delmas, J.; Coignard, B.; Tristan, A.; Vandenesch, F.; et al. First outbreak of community-acquired MRSA USA300 in France: Failure to suppress prolonged MRSA carriage despite decontamination procedures. *Eur. J. Clin. Microbiol. Infect. Dis.* **2014**, *33*, 1757–1762. [CrossRef]
4. Lee, H.; Kim, E.S.; Choi, C.; Seo, H.; Shin, M.; Bok, J.H.; Cho, J.E.; Kim, C.J.; Shin, J.W.; Kim, T.S.; et al. Outbreak among healthy newborns due to a new variant of USA300-related meticillin-resistant *Staphylococcus aureus*. *J. Hosp. Infect.* **2014**, *87*, 145–151. [CrossRef] [PubMed]
5. Vindel, A.; Trincado, P.; Cuevas, O.; Ballesteros, C.; Bouza, E.; Cercenado, E. Molecular epidemiology of community-associated methicillin-resistant *Staphylococcus aureus* in Spain: 2004–12. *J. Antimicrob. Chemother.* **2014**, *69*, 2913–2919. [CrossRef]
6. Suaya, J.A.; Mera, R.M.; Cassidy, A.; O'Hara, P.; Amrine-Madsen, H.; Burstin, S.; Miller, L.G. Incidence and cost of hospitalizations associated with *Staphylococcus aureus* skin and soft tissue infections in the United States from 2001 through 2009. *BMC Infect. Dis.* **2014**, *14*, 296. [CrossRef]
7. Thom, H.; Thompson, J.C.; Scott, D.A.; Halfpenny, N.; Sulham, K.; Corey, G.R. Comparative efficacy of antibiotics for the treatment of acute bacterial skin and skin structure infections (ABSSSI): A systematic review and network meta-analysis. *Curr. Med. Res. Opin.* **2015**, *31*, 1539–1551. [CrossRef]
8. Moran, G.J.; Krishnadasan, A.; Gorwitz, R.J.; Fosheim, G.E.; McDougal, L.K.; Carey, R.B.; Talan, D.A. Methicillin-*resistant S. aureus* infections among patients in the emergency department. *N. Engl. J. Med.* **2006**, *355*, 666–674. [CrossRef]
9. Carrillo-Marquez, M.A.; Hulten, K.G.; Hammerman, W.; Mason, E.O.; Kaplan, S.L. USA300 is the predominant genotype causing *Staphylococcus aureus* septic arthritis in children. *Pediatr. Infect. Dis. J.* **2009**, *28*, 1076–1080. [CrossRef]
10. Kempker, R.R.; Farley, M.M.; Ladson, J.L.; Satola, S.; Ray, S.M. Association of methicillin-resistant *Staphylococcus aureus* (MRSA) USA300 genotype with mortality in MRSA bacteremia. *J. Infect.* **2010**, *61*, 372–381. [CrossRef]
11. Kreisel, K.M.; Stine, O.C.; Johnson, J.K.; Perencevich, E.N.; Shardell, M.D.; Lesse, A.J.; Gordin, F.M.; Climo, M.W.; Roghmann, M.C. USA300 methicillin-resistant *Staphylococcus aureus* bacteremia and the risk of severe sepsis: Is USA300 methicillin-resistant *Staphylococcus aureus* associated with more severe infections? *Diagn. Microbiol. Infect. Dis.* **2011**, *70*, 285–290. [CrossRef] [PubMed]
12. Seybold, U.; Kourbatova, E.V.; Johnson, J.G.; Halvosa, S.J.; Wang, Y.F.; King, M.D.; Ray, S.M.; Blumberg, H.M. Emergence of community-associated methicillin-resistant *Staphylococcus aureus* USA300 genotype as a major cause of health care-associated blood stream infections. *Clin. Infect. Dis.* **2006**, *42*, 647–656. [CrossRef] [PubMed]
13. Sherwood, J.; Park, M.; Robben, P.; Whitman, T.; Ellis, M.W. USA300 methicillin-resistant *Staphylococcus aureus* emerging as a cause of bloodstream infections at military medical centers. *Infect. Control Hosp. Epidemiol.* **2013**, *34*, 393–399. [CrossRef] [PubMed]
14. Kobayashi, T.; Nakaminami, H.; Ohtani, H.; Yamada, K.; Nasu, Y.; Takadama, S.; Noguchi, N.; Fujii, T.; Matsumoto, T. An outbreak of severe infectious diseases caused by methicillin-resistant *Staphylococcus aureus* USA300 clone among hospitalized patients and nursing staff in a tertiary care university hospital. *J. Infect. Chemother.* **2020**, *26*, 76–81. [CrossRef]
15. McRipley, R.J.; Whitney, R.R. Characterization and quantitation of experimental surgical-wound infections used to evaluate topical antibacterial agents. *Antimicrob. Agents Chemother.* **1976**, *10*, 38–44. [CrossRef]
16. Espersen, F.; Frimodt-Moller, N.; Corneliussen, L.; Riber, U.; Rosdahl, V.T.; Skinhoj, P. Effect of treatment with methicillin and gentamicin in a new experimental mouse model of foreign body infection. *Antimicrob. Agents Chemother.* **1994**, *38*, 2047–2053. [CrossRef]
17. Fallon, M.T.; Shafer, W.; Jacob, E. Use of cefazolin microspheres to treat localized methicillin-resistant *Staphylococcus aureus* infections in rats. *J. Surg. Res.* **1999**, *86*, 97–102. [CrossRef]
18. Abtin, A.; Jain, R.; Mitchell, A.J.; Roediger, B.; Brzoska, A.J.; Tikoo, S.; Cheng, Q.; Ng, L.G.; Cavanagh, L.L.; von Andrian, U.H.; et al. Perivascular macrophages mediate neutrophil recruitment during bacterial skin infection. *Nat. Immunol.* **2014**, *15*, 45–53. [CrossRef]
19. Tseng, C.W.; Sanchez-Martinez, M.; Arruda, A.; Liu, G.Y. Subcutaneous infection of methicillin resistant *Staphylococcus aureus* (MRSA). *J. Vis. Exp.* **2011**, *48*, 2528. [CrossRef]

20. Kinoshita, M.; Miyazaki, H.; Ono, S.; Inatsu, A.; Nakashima, H.; Tsujimoto, H.; Shinomiya, N.; Saitoh, D.; Seki, S. Enhancement of neutrophil function by interleukin-18 therapy protects burn-injured mice from methicillin-resistant *Staphylococcus aureus*. *Infect. Immun.* **2011**, *79*, 2670–2680. [CrossRef]
21. Gilbert, M.; MacDonald, J.; Gregson, D.; Siushansian, J.; Zhang, K.; Elsayed, S.; Laupland, K.; Louie, T.; Hope, K.; Mulvey, M.; et al. Outbreak in Alberta of community-acquired (USA300) methicillin-resistant *Staphylococcus aureus* in people with a history of drug use, homelessness or incarceration. *CMAJ* **2006**, *175*, 149–154. [CrossRef] [PubMed]
22. Christianson, S.; Golding, G.R.; Campbell, J.; Mulvey, M.R. Comparative genomics of Canadian epidemic lineages of methicillin-resistant *Staphylococcus aureus*. *J. Clin. Microbiol.* **2007**, *45*, 1904–1911. [CrossRef] [PubMed]
23. Wu, K.; Conly, J.; McClure, J.A.; Elsayed, S.; Louie, T.; Zhang, K. *Caenorhabditis elegans* as a host model for community-associated methicillin-resistant *Staphylococcus aureus*. *Clin. Microbiol. Infect.* **2010**, *16*, 245–254. [CrossRef] [PubMed]
24. McClure, J.A.; Zhang, K. Complete Genome Sequence of a Community-Associated Methicillin-Resistant *Staphylococcus aureus* Hypervirulent Strain, USA300-C2406, Isolated from a Patient with a Lethal Case of Necrotizing Pneumonia. *Genome Announc.* **2017**, *5*, e00461-17. [CrossRef]
25. McClure, J.A.; Zhang, K. Complete Genome Sequence of the Methicillin-Resistant *Staphylococcus aureus* Colonizing Strain M92. *Genome Announc.* **2017**, *5*, e00478-17. [CrossRef]
26. Houle, S.; Papez, M.D.; Ferazzini, M.; Hollenberg, M.D.; Vergnolle, N. Neutrophils and the kallikrein-kinin system in proteinase-activated receptor 4-mediated inflammation in rodents. *Br. J. Pharmacol.* **2005**, *146*, 670–678. [CrossRef]
27. Hickey, M.J.; Kanwar, S.; McCafferty, D.M.; Granger, D.N.; Eppihimer, M.J.; Kubes, P. Varying roles of E-selectin and P-selectin in different microvascular beds in response to antigen. *J. Immunol.* **1999**, *162*, 1137–1143.
28. Turabelidze, A.; Guo, S.; DiPietro, L.A. Importance of housekeeping gene selection for accurate reverse transcription-quantitative polymerase chain reaction in a wound healing model. *Wound Repair Regen.* **2010**, *18*, 460–466. [CrossRef]
29. Livak, K.J.; Schmittgen, T.D. Analysis of relative gene expression data using real-time quantitative PCR and the 2(-Delta Delta C(T)) Method. *Methods* **2001**, *25*, 402–408. [CrossRef]
30. Cho, J.S.; Guo, Y.; Ramos, R.I.; Hebroni, F.; Plaisier, S.B.; Xuan, C.; Granick, J.L.; Matsushima, H.; Takashima, A.; Iwakura, Y.; et al. Neutrophil-derived IL-1beta is sufficient for abscess formation in immunity against *Staphylococcus aureus* in mice. *PLoS Pathog.* **2012**, *8*, e1003047. [CrossRef]
31. Cho, J.S.; Pietras, E.M.; Garcia, N.C.; Ramos, R.I.; Farzam, D.M.; Monroe, H.R.; Magorien, J.E.; Blauvelt, A.; Kolls, J.K.; Cheung, A.L.; et al. IL-17 is essential for host defense against cutaneous *Staphylococcus aureus* infection in mice. *J. Clin. Investig.* **2010**, *120*, 1762–1773. [CrossRef] [PubMed]
32. Agalar, C.; Eroglu, E.; Sari, M.; Sari, A.; Daphan, C.; Agalar, F. The effect of G-CSF in an experimental MRSA graft infection in mice. *J. Investig. Surg.* **2005**, *18*, 227–231. [CrossRef] [PubMed]
33. Hume, E.B.; Cole, N.; Khan, S.; Garthwaite, L.L.; Aliwarga, Y.; Schubert, T.L.; Willcox, M.D. A *Staphylococcus aureus* mouse keratitis topical infection model: Cytokine balance in different strains of mice. *Immunol. Cell Biol.* **2005**, *83*, 294–300. [CrossRef] [PubMed]
34. Kielian, T.; Barry, B.; Hickey, W.F. CXC chemokine receptor-2 ligands are required for neutrophil-mediated host defense in experimental brain abscesses. *J. Immunol.* **2001**, *166*, 4634–4643. [CrossRef] [PubMed]
35. Kobayashi, M.; Koyama, A. Methicillin-resistant *Staphylococcus aureus*-related glomerulonephritis. *Nephrol. Dial. Transplant.* **1999**, *14* (Suppl. 1), 17–18. [CrossRef]
36. McNicholas, S.; Talento, A.F.; O'Gorman, J.; Hannan, M.M.; Lynch, M.; Greene, C.M.; Humphreys, H.; Fitzgerald-Hughes, D. Cytokine responses to *Staphylococcus aureus* bloodstream infection differ between patient cohorts that have different clinical courses of infection. *BMC Infect. Dis.* **2014**, *14*, 580. [CrossRef]
37. Tseng, C.W.; Kyme, P.; Low, J.; Rocha, M.A.; Alsabeh, R.; Miller, L.G.; Otto, M.; Arditi, M.; Diep, B.A.; Nizet, V.; et al. *Staphylococcus aureus* Panton-Valentine leukocidin contributes to inflammation and muscle tissue injury. *PLoS ONE* **2009**, *4*, e6387. [CrossRef]
38. Yoshii, T.; Magara, S.; Miyai, D.; Nishimura, H.; Kuroki, E.; Furudoi, S.; Komori, T.; Ohbayashi, C. Local levels of interleukin-1beta, -4, -6 and tumor necrosis factor alpha in an experimental model of murine osteomyelitis due to *Staphylococcus aureus*. *Cytokine* **2002**, *19*, 59–65. [CrossRef]
39. Cohen, P.R. Community-acquired methicillin-resistant *Staphylococcus aureus* skin infections: Implications for patients and practitioners. *Am. J. Clin. Dermatol.* **2007**, *8*, 259–270. [CrossRef]
40. Gorwitz, R.J. A review of community-associated methicillin-resistant *Staphylococcus aureus* skin and soft tissue infections. *Pediatr. Infect. Dis. J.* **2008**, *27*, 1–7. [CrossRef]
41. Kim, M.H.; Granick, J.L.; Kwok, C.; Walker, N.J.; Borjesson, D.L.; Curry, F.R.; Miller, L.S.; Simon, S.I. Neutrophil survival and c-kit(+)-progenitor proliferation in *Staphylococcus aureus*-infected skin wounds promote resolution. *Blood* **2011**, *117*, 3343–3352. [CrossRef] [PubMed]
42. Molne, L.; Verdrengh, M.; Tarkowski, A. Role of neutrophil leukocytes in cutaneous infection caused by *Staphylococcus aureus*. *Infect. Immun.* **2000**, *68*, 6162–6167. [CrossRef] [PubMed]
43. Rigby, K.M.; DeLeo, F.R. Neutrophils in innate host defense against *Staphylococcus aureus* infections. *Semin. Immunopathol.* **2012**, *34*, 237–259. [CrossRef]
44. de Oliveira, S.; Rosowski, E.E.; Huttenlocher, A. Neutrophil migration in infection and wound repair: Going forward in reverse. *Nat. Rev. Immunol.* **2016**, *16*, 378–391. [CrossRef] [PubMed]

45. Nguyen, Q.T.; Furuya, Y.; Roberts, S.; Metzger, D.W. Role of Interleukin-12 in Protection against Pulmonary Infection with Methicillin-Resistant *Staphylococcus aureus*. *Antimicrob. Agents Chemother.* **2015**, *59*, 6308–6316. [CrossRef]
46. Zabieglo, K.; Majewski, P.; Majchrzak-Gorecka, M.; Wlodarczyk, A.; Grygier, B.; Zegar, A.; Kapinska-Mrowiecka, M.; Naskalska, A.; Pyrc, K.; Dubin, A.; et al. The inhibitory effect of secretory leukocyte protease inhibitor (SLPI) on formation of neutrophil extracellular traps. *J. Leukoc. Biol.* **2015**, *98*, 99–106. [CrossRef]
47. Hruz, P.; Zinkernagel, A.S.; Jenikova, G.; Botwin, G.J.; Hugot, J.P.; Karin, M.; Nizet, V.; Eckmann, L. NOD2 contributes to cutaneous defense against *Staphylococcus aureus* through alpha-toxin-dependent innate immune activation. *Proc. Natl. Acad. Sci. USA* **2009**, *106*, 12873–12878. [CrossRef]
48. Takeuchi, O.; Hoshino, K.; Kawai, T.; Sanjo, H.; Takada, H.; Ogawa, T.; Takeda, K.; Akira, S. Differential roles of TLR2 and TLR4 in recognition of gram-negative and gram-positive bacterial cell wall components. *Immunity* **1999**, *11*, 443–451. [CrossRef]
49. Patot, S.; Imbert, P.R.; Baude, J.; Martins Simoes, P.; Campergue, J.B.; Louche, A.; Nijland, R.; Bes, M.; Tristan, A.; Laurent, F.; et al. The TIR homologue lies near resistance genes in *Staphylococcus aureus*, coupling modulation of virulence and antimicrobial susceptibility. *PLoS Pathog.* **2017**, *13*, e1006092. [CrossRef]
50. Bridenbaugh, L.D. The 1990 John J. Bonica lecture. Acute pain therapy–whose responsibility? *Reg. Anesth.* **1990**, *15*, 219–222.
51. Koymans, K.J.; Feitsma, L.J.; Brondijk, T.H.; Aerts, P.C.; Lukkien, E.; Lossl, P.; van Kessel, K.P.; de Haas, C.J.; van Strijp, J.A.; Huizinga, E.G. Structural basis for inhibition of TLR2 by staphylococcal superantigen-like protein 3 (SSL3). *Proc. Natl. Acad. Sci. USA* **2015**, *112*, 11018–11023. [CrossRef] [PubMed]
52. Yokoyama, R.; Itoh, S.; Kamoshida, G.; Takii, T.; Fujii, S.; Tsuji, T.; Onozaki, K. Staphylococcal superantigen-like protein 3 binds to the Toll-like receptor 2 extracellular domain and inhibits cytokine production induced by *Staphylococcus aureus*, cell wall component, or lipopeptides in murine macrophages. *Infect. Immun.* **2012**, *80*, 2816–2825. [CrossRef] [PubMed]
53. Miller, L.S.; Cho, J.S. Immunity against *Staphylococcus aureus* cutaneous infections. *Nat. Rev. Immunol.* **2011**, *11*, 505–518. [CrossRef] [PubMed]
54. Greenfeder, S.; Umland, S.P.; Cuss, F.M.; Chapman, R.W.; Egan, R.W. Th2 cytokines and asthma. The role of interleukin-5 in allergic eosinophilic disease. *Respir. Res.* **2001**, *2*, 71–79. [CrossRef] [PubMed]
55. Kodama, H.; Fukuda, K.; Pan, J.; Makino, S.; Baba, A.; Hori, S.; Ogawa, S. Leukemia inhibitory factor, a potent cardiac hypertrophic cytokine, activates the JAK/STAT pathway in rat cardiomyocytes. *Circ. Res.* **1997**, *81*, 656–663. [CrossRef] [PubMed]
56. Deshmane, S.L.; Kremlev, S.; Amini, S.; Sawaya, B.E. Monocyte chemoattractant protein-1 (MCP-1): An overview. *J. Interferon Cytokine Res.* **2009**, *29*, 313–326. [CrossRef] [PubMed]
57. Jennes, W.; Vereecken, C.; Fransen, K.; de Roo, A.; Kestens, L. Disturbed secretory capacity for macrophage inflammatory protein (MIP)-1 alpha and MIP-1 beta in progressive HIV infection. *AIDS Res. Hum. Retrovir.* **2004**, *20*, 1087–1091. [CrossRef]
58. Asai, A.; Tsuda, Y.; Kobayashi, M.; Hanafusa, T.; Herndon, D.N.; Suzuki, F. Pathogenic role of macrophages in intradermal infection of methicillin-resistant *Staphylococcus aureus* in thermally injured mice. *Infect. Immun.* **2010**, *78*, 4311–4319. [CrossRef]

 *microorganisms*

Article

Impact of *Staphylococcus aureus* Small Colony Variants on Human Lung Epithelial Cells with Subsequent Influenza Virus Infection

Janine J. Wilden [1], Eike R. Hrincius [1], Silke Niemann [2], Yvonne Boergeling [1], Bettina Löffler [3,4], Stephan Ludwig [1,5] and Christina Ehrhardt [6,*]

1   Institute of Virology Muenster (IVM), Westfaelische Wilhelms-University Muenster,
    48149 Muenster, Germany; wildenja@ukmuenster.de (J.J.W.); hrincius@uni-muenster.de (E.R.H.);
    borgelin@uni-muenster.de (Y.B.); ludwigs@uni-muenster.de (S.L.)
2   Institute of Medical Microbiology, Westfaelische Wilhelms-University Muenster, 48149 Muenster, Germany;
    Silke.Niemann@uni-muenster.de
3   Institute of Medical Microbiology, Jena University Hospital, 07747 Jena, Germany;
    bettina.loeffler@med.uni-jena.de
4   Cluster of Excellence EXC 2051 "Balance of the Microverse", FSU Jena, 07743 Jena, Germany
5   Cluster of Excellence EXC 1003 "Cells in Motion", WWU Muenster, 48149 Muenster, Germany
6   Section of Experimental Virology, Institute of Medical Microbiology, Jena University Hospital,
    07745 Jena, Germany
*   Correspondence: christina.ehrhardt@med.uni-jena.de; Tel.: +49-(0)3641-9395700

Received: 13 November 2020; Accepted: 11 December 2020; Published: 15 December 2020

Abstract: Human beings are exposed to microorganisms every day. Among those, diverse commensals and potential pathogens including *Staphylococcus aureus* (*S. aureus*) compose a significant part of the respiratory tract microbiota. Remarkably, bacterial colonization is supposed to affect the outcome of viral respiratory tract infections, including those caused by influenza viruses (IV). Since 30% of the world's population is already colonized with *S. aureus* that can develop metabolically inactive dormant phenotypes and seasonal IV circulate every year, super-infections are likely to occur. Although IV and *S. aureus* super-infections are widely described in the literature, the interactions of these pathogens with each other and the host cell are only scarcely understood. Especially, the effect of quasi-dormant bacterial subpopulations on IV infections is barely investigated. In the present study, we aimed to investigate the impact of *S. aureus* small colony variants on the cell intrinsic immune response during a subsequent IV infection in vitro. In fact, we observed a significant impact on the regulation of pro-inflammatory factors, contributing to a synergistic effect on cell intrinsic innate immune response and induction of harmful cell death. Interestingly, the cytopathic effect, which was observed in presence of both pathogens, was not due to an increased pathogen load.

Keywords: *Staphylococcus aureus*; small colony variants; influenza virus; super-infection; pro-inflammatory response

---

1. Introduction

The respiratory tract is a major portal for microorganisms, through which virus infections can cause non-symptomatic, mild, and self-limiting but also severe diseases, sometimes with fatal outcomes [1]. A growing body of evidence shows that the human respiratory tract contains a highly adapted microbiota including commensal and opportunistic pathogens. Among those, *Staphylococcus aureus* (*S. aureus*) is of special importance, forming quasi-dormant subpopulations characterized by increased fitness compared to other phenotypes [2]. Colonization of *S. aureus* could either be persistent or non-persistent, whereby nasal colonization appears to be the most prominent localization [3]. *S. aureus* as a community-acquired pathogen is already colonized on approximately 30% of the human population, some without causing any symptoms [4]. During long-term colonization or infection, *S. aureus* can change phenotypes to so-called small colony variants (SCVs), which adapt in their metabolic and phenotypic characteristics, allowing them to evade the host's immune system. SCVs can be localized intracellularly and are characterized by a slow growth rate, non-pigmentation, less hemolytic activity, and decreased antibiotic susceptibility [5–7] but often enhanced surface presentation of adhesion molecules [8]. SCVs are often misdiagnosed [9]. Due to their slow growth, they often get overgrown by other bacteria, and an initially effective antibiotic treatment results in the development of resistances accompanied by chronic and relapsing infections [5,6,8,10,11]. The clinical relevance of colonizing SCVs gets obvious in patients with chronic respiratory diseases, such as chronic obstructive pulmonary disease (COPD) or cystic fibrosis (CF) [5]. Patients who are colonized with bacteria are more likely to suffer from recurring infections [12], as the phenotype can revert to the pathogenic phenotype.

Besides, simultaneous occurrence of different pathogens can induce or even exacerbate a pathological effect in the lung. Super-infections with influenza viruses (IV) and with the community-acquired *S. aureus* are known to be harmful and lead to increased inflammatory lung damage [13]. Due to their quick adaptation and genomic changes, both pathogens can evade the host's immune response, causing the tedious development of effective medications. Concerning super-infections, most studies describe infections with a primary viral infection that paves the path for a secondary bacterial infection [14–17]. However, there is evidence that primary bacterial colonization also occurs prior to viral infections [18].

However, the influence of colonizing *S. aureus* SCVs on subsequent IV infection is largely unexplored. Thus, in the present study, we aimed to investigate the effect of the bacterial strain *S. aureus* 3878$_{SCV}$ on cell intrinsic immune responses to a subsequent IV infection, in vitro. Here, we observed that the response of anti-viral gene expression was barely changed. However, pro-inflammatory genes were highly upregulated upon super-infection, resulting in an induction of necrotic cell death. Thus, we were able to show that colonizing SCVs could enhance severity of subsequent viral infection.

2. Materials and Methods

2.1. Cell Lines, Virus Strains, and Bacteria Strain

All cell lines were cultivated at 37 °C and 5% CO_2 under sterile conditions. Human lung epithelial cells A549 (American Type Culture Collection (ATCC), Wesel, Germany) were cultivated in Dulbeccos's modified eagle medium (DMEM; Sigma-Aldrich, St. Louis, MO, USA) and Madin-Darby canine kidney cells II (MDCKII) in minimum essential medium eagle (MEM; Sigma-Aldrich, St. Louis, MO, USA), supplemented with 10% fetal bovine serum (FBS; Biochrom, Berlin, Germany).

The human IV strains A/Puerto Rico/8/34 (H1N1, PR8-M) and A/Panama/2007/99 (H3N2, Panama) were taken from the virus stock of the Institute of Virology Muenster, 48149 Muenster, Germany, subcultured and passaged on MDCKII cells.

The persisting bacterial strain *S. aureus* 3878$_{SCV}$, wildtype phenotype strain *S. aureus* 3878$_{WT}$, and the human lung isolate of another SCV phenotype strain *S. aureus* 814$_{SCV}$ (provided by Karsten Becker, Institute of Medical Microbiology, Muenster, Germany) were stored at −80 °C in a 30% glycerol/brain-heart infusion (BHI; Merck; Darmstadt, Germany) medium. *S. aureus* 3878$_{SCV}$ and *S. aureus* 3878$_{WT}$ were already characterized and described previously [10,19–21]. Before experiments, bacteria were plated on blood agar plates to take single clones, which were inoculated in BHI medium and incubated for 24 h at 37 °C and 5% CO_2. For bacterial infection, bacterial suspension was washed with phosphate buffered saline (PBS) (4000 rpm; 4 °C; 5 min) and adjusted to an optical density of $OD_{600nm} = 1$. Growth kinetics were performed to determine a colony forming unit (CFU) of 2×10^8 CFU/mL at $OD_{600nm} = 1$ for each bacterial strain used.

2.2. Super-Infection Protocol

Human lung epithelial cells were seeded in either 6-well plates (0.5×10^6) or 12-well plates (0.2×10^6) in 2 mL or 1 mL culture medium 24 h before infection. For bacterial infection, the overnight culture was set to $OD_{600nm} = 1$ to determine the multiplicity of infection (MOI). Cells were washed with PBS and infected with *S. aureus* 3878_{SCV} in invasion media ($DMEM_{INV}$: DMEM supplemented with 1% human serum albumin, 25 nmol/L HEPES) for 24 h with a MOI of 0.01. For viral infection, supernatant was aspirated, cells were washed with PBS and incubated with IV PR8-M (MOI = 0.1) or IV Panama (MOI = 0.01) in infection PBS (PBS_{INF}: PBS supplemented with 0.2% bovine serum albumin (BSA), 1 mM $MgCl_2$, 0.9 mM $CaCl_2$, 100 U/mL penicillin, 0.1 mg/mL streptomycin) for 30 min. Viral suspension was aspirated, and cells were washed with PBS and further incubated in infection media ($DMEM_{INF}$: DMEM supplemented with 0.2% bovine serum albumin (BSA), 1 mM $MgCl_2$, 0.9 mM $CaCl_2$) up to 8 hpvi, 24 hpvi, 32 hpvi, 44 hpvi, or 48 hpvi (hours post-viral infection).

2.3. Transfection Protocol

For transfection of the 3× NFκB reporter plasmid construct as described elsewhere [22] (0.1 µg/µL) A549 cells were seeded in 12-well plates as described above. Cells were transfected with 0.1 µg/µL of the indicated plasmid for 4 h with Lipofectamine® 2000 (Invitrogen, Carlsbad, CA, USA) corresponding to the manufacturer's protocol. Afterwards, cells were washed with PBS and further incubated in cell culture media up to 24 h. Afterwards transfected cells were infected up to 8 hpvi. Performance of luciferase assay was done as described elsewhere [23].

2.4. Intra- and Extracellular Bacterial Titer Measurements

Extracellular bacterial titers were determined by collecting the supernatant of infected cells including the washing with PBS. Cells were lysed via hypotonic shock with 2 mL ddH_2O according to Tuchscherr et al. [7,8] (37 °C, 30 min) to determine intracellular bacterial titers, including adherent bacteria at the cells surface. Bacterial suspensions were centrifuged (4000 rpm, 4 °C, 10 min), pellets were resuspended in 1 mL PBS, and serial dilutions (1:10) were plated on BHI agar plates and incubated for 32 h at 37 °C.

2.5. Standard Plaque Assay

Infectious virus particles in the supernatant were titrated to determine viral titers. A standard plaque assay was performed as described earlier [24].

2.6. Quantitative Real-Time PCR (qRT-PCR)

RNA isolation was performed with RNeasy Kit (Qiagen, Hilden, Germany) according to the manufacturer's instructions. Reverse transcription was performed with 2 µg of total RNA with Revert AID H Minus Reverse Transciptase (Thermo Fisher Scientific, Karlsruhe, Germany) and oligo (dT) primers according to the manufacturer's protocol. qRT-PCR was performed using a Roche LightCycler 480 and Brilliant SYBRGreen Mastermix (Agilent, Santa Clara, CA, USA) according to the manufacturer's instructions. The following primers were used: GAPDH: fwd 5′GCAAATTCCATGGCACCGT3′, rev 5′GCCCCACTTGATTTGGAGG3′; IL-6: fwd 5′AACCTGAACCTTCCAAAGATGG3′, rev 5′TCTGGCTTGTTCCTCACTAGT3′; IL-8: fwd 5′CTTGTTCCACTGTGCCTTGGTT3′, rev 5′GCTTCCACATGTCCTCACAACAT3′; TNFα: fwd 5′-ATGAGCACTGAAAGCATGATC-3′, rev 5′-GAGGGCTGATTAGAGAGAGGT-3′; IL-1β: fwd 5′-CAGCTACGAATCTCCGACCAC-3′, rev 5′-GGCAGGGAACCAGCATCTTC-3′; IFNγ: fwd 5′AAACGAGATGACTTCGAAAAGCTG3′, rev 5′TGTTTAGCTGCTGGCGACAG3′; RIG-I: fwd 5′CCTACCTACATCCTGAGCTACAT3′, rev 5′TCTAGGGCATCCAAAAAGCCA3′; IFNβ: fwd 5′TCTGGCACAACAGGTAGTAGGC3′, rev 5′GAGAAGCACAACAGGAGAGCAA3′; MxA: fwd 5′GTTTCCGAAGTGGACATCGCA3′, rev 5′GAAGGGCAACTCCTGACAGT3′; OAS1: fwd

5′GATCTCAGAAATACCCCAGCCA3′, rev 5′AGCTACCTCGGAAGCACCTT3′. Relative changes in expression levels (n-fold) were calculated according to the $2^{-\Delta\Delta Ct}$ method [25].

Bacterial RNA was isolated with the RNeasy Protect Bacteria Mini Kit (Qiagen, Hilden, Germany), and cDNA synthesis was performed using QuantiTect Reverse Transcription Kit (Qiagen, Hilden, Germany) according to the manufacturer's instructions. qRT-PCR was performed using a Roche LightCycler 480 (Basel, Switzerland) and Brilliant SYBRGreen Mastermix (Agilent, Santa Clara, CA, USA) according to the manufacturer's instructions. The primers to determine the gene expression of *gyrB*, *aroE*, *arg*, *hla*, *sarA*, and *sigB* were already described elsewhere [7].

2.7. RT^2 Profiler Array Analysis

For pathway focused gene expression analysis, we used RT^2 *Profiler PCR Arrays* (Qiagen, Hilden, Germany). RNA isolation, cDNA synthesis and procedure were performed according to the manufacturer's protocol and instructions. Analysis of data was accomplished by using the GeneGlobe Data Analysis Center recommended by Qiagen [26].

2.8. FACS Analysis

Determination of secreted proteins in the supernatant was performed with BioLegend's LEGENDplex™ (San Diego, CA, USA) according to the manufacturer's protocol. The human anti-viral and pro-inflammatory chemokine panels were used. Results were analyzed by BioLegend's cloud-based LEGENDplex™ Data Analysis Software. To analyze apoptotic or necrotic cells, infection was performed as described above until 44 hpvi. Cells were treated with tumor necrosis factor related apoptosis inducing ligand (TRAIL; Enzo Life Sciences, Farmingdale, NY, USA) (150 ng/mL) 4.5 h before harvested and used as a positive control for apoptosis. The supernatant was collected for this purpose, and cells were detached from the wells with trypsin-EDTA and recombined with the supernatant. Cell suspension was centrifuged at $1000 \times g$ at room temperature (RT) for 5 min, and cells were washed with PBS supplemented with 5% FCS. Afterwards, cells were stained with annexin V FITC (20 µL) (ImmunoTool, Friesoythe, Germany) and 1:2000 eBioscience™ Fixable Viability Dye eFluor™ 660 (Thermo Fisher Scientific, Karlsruhe, Germany) in 100 µL 1× annexin V staining buffer (10× annexin V staining buffer: 0.1 M HEPES, 1.4 M NaCl, and 25 mM $CaCl_2$ (pH 7.5)) for 30 min at RT in the dark. Further 150 µL of staining buffer were added and the supernatant was removed after centrifugation. Cells were fixed with 500 µL PBS containing 4% formaldehyde and 1.25 mM $CaCl_2$ for 20 min at RT in the dark. Cells were finally resuspended in 150 µL staining buffer and stored at 4 °C until measurement with the FACSCalibur flow cytometer (BD Biosciences, Heidelberg, Germany), followed by the analysis with FlowJo software (v.10; Flow Jo, Ashland, OR, USA). Three gates were set as the following: annexin V positive cells (early apoptotic cells) and live/dead marker positive cells (cells with a membrane rupture tending to necrosis).

2.9. Recording of Cytopathic Effect of Infected Cells

To record the CPE at different time points, cells were visualized with Canon (EOS 500D) by light microscopy (Axiovert 40C, ZEISS, Jena, Germany) with a 10× magnification.

2.10. SDS-PAGE and Western Blot Analysis

Protein expressions were determined by separating proteins in a polyacrylamide gel and subsequent transfer on nitrocellulose membranes by western blot analysis as described earlier [27]. The following antibodies were used: pMLKL [(S353) #91689 Cell Signaling, Frankfurt, Germany], PARP (#611039 BD, Heidelberg, Germany) and ERK1/2 (#4696 Cell Signaling, Frankfurt, Germany).

2.11. Lactate Dehydrogenase (LDH) Assay

The lactate dehydrogenase assay (CellBiolabs, San Diego, CA, USA) was used to measure the cell cytotoxicity and was used according to the manufacturer's instructions. Cells were infected as described previously, and 90 µL of the supernatant was mixed with 10 µL of the LDH cytotoxicity reagent in a 96-well plate. This plate was incubated at 37 °C and 5% CO_2 for 30 min, and the OD_{450nm} was measured on a Spectromax M2 Instrument (Molecular Devices, Munich, Germany). Triton X-100 used according to the manufacturer's instructions served as a positive control.

2.12. Quantification and Statistical Analysis

All data represent the means + standard deviation (SD) of three independent experiments. Statistical significances were determined by unpaired *t*-test (Figure S4A), one-way ANOVA followed by Tukey's, (Figures 4D,E, 5, 6A,B,D,E, Figure S1B,D,E, S2D,E, S3 and S4C–E) or two-way ANOVA followed by Sidak's (Figure 2, Figures S1C and S4B) or followed by Tukey's (Figure 4A–C,F–I and Figure S2A–C,F–I) multiple comparison test using GraphPad Prism software (v.7.03, GraphPad Prism, Inc., La Jolla, CA, USA).

3. Results

3.1. Primary S. aureus 3878$_{SCV}$ Infection Provokes a Cytopathic Effect in Presence of IV

Cell death mechanisms induced by *S. aureus* or IV alone are very well investigated and described [28–33]. With respect to IV and *S. aureus* super-infection, we recently were able to show a *S. aureus*-mediated switch from IV-induced apoptosis to necrosis [27]. It is known that IV infection paves the path for secondary bacterial infection, resulting in enhanced pathogen-load [15,34], cytokine expression [35,36], and cell death [27]. Since *S. aureus* often persist in humans without any harm, we aimed to investigate the effects of colonizing *S. aureus* SCVs on secondary IV super-infection.

In a first set of experiments, we focused on the cell morphology of A549 human lung epithelial cells in absence and presence of *S. aureus* 3878$_{SCV}$ and IV. For this reason, A549 human lung epithelial cells were infected with *S. aureus* 3878$_{SCV}$, which is a well described SCV patient isolate [10,37], for 24 h followed by infection with IV strain A/Puerto Rico/8/34 (PR8-M; H1N1) for the indicated points in time. The morphology of single- and super-infected cells was analyzed by light microscopy in comparison to uninfected control (mock) (Figure 1). While the cell monolayer is still intact in un-, single-, and super-infected cells up to 32 hpvi (hours post-viral infection), first changes in the cell morphology were visible 48 hpvi in single virus-infected and super-infected cells. Pictures of virus-infected cells showed a less confluent cell monolayer compared to uninfected cells, and in super-infected samples a clear cytopathic effect was observed, indicated by cell monolayer disruption and floating cells. To be able to ascribe these findings to the SCV phenotype, we additionally specified the pathological difference between *S. aureus* wildtype phenotype and SCV phenotype (*S. aureus* 3878$_{WT}$ and *S. aureus* 3878$_{SCV}$) by infecting A549 human lung epithelial cells. Cell morphology was monitored by light-microscopy (Supplementary Figure S1A) and cell viability was quantified by lactate dehydrogenase assay (LDH) assay (Supplementary Figure S1B). Both assays indicate a massive destruction of the cell monolayer 8 h post bacterial infection (hpbi) with *S. aureus* 3878$_{WT}$ in comparison to *S. aureus* 3878$_{SCV}$. Further, the determination of the expression of distinct bacterial genes, which are involved in the virulence of the pathogens, verified the reduced virulence of *S. aureus* 3878$_{SCV}$ in comparison to the *S. aureus* 3878$_{WT}$ (Supplementary Figure S1C). Based on these results, *S. aureus* 3878$_{WT}$ was not used in the following experiments. The analysis of cell viability at 32 hpvi and 48 hpvi confirmed the cell disturbance in presence of *S. aureus* 3878$_{SCV}$ and IV infection (Supplementary Figure S1D,E).

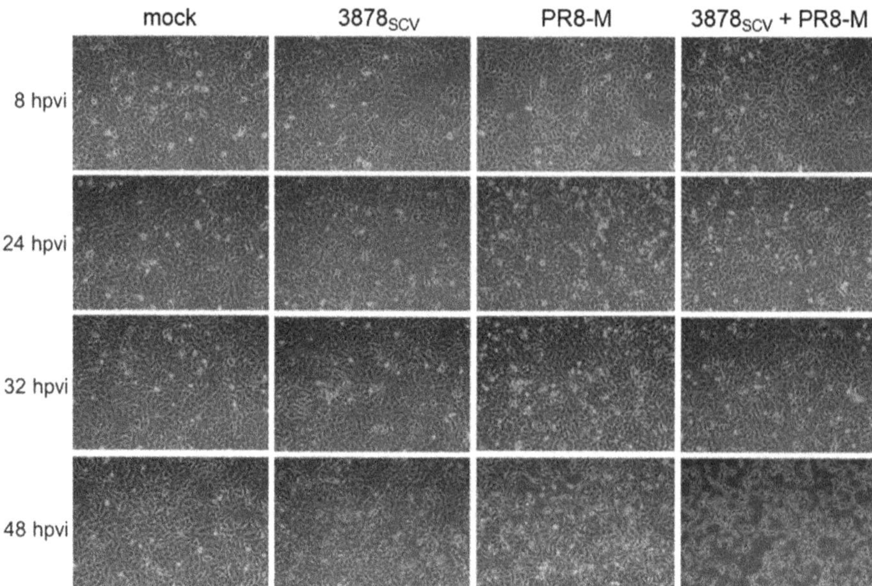

Figure 1. *S. aureus* 3878$_{SCV}$ colonization and subsequent influenza virus infection provokes a cytopathic effect. A549 human lung epithelial cells were infected with *S. aureus* 3878$_{SCV}$ (multiplicity of infection (MOI) = 0.01) for 24 h at 37 °C and 5% CO_2. Afterwards, cells were infected with influenza viruses (IV) Puerto Rico/8 (PR8)-M (MOI = 0.1) until the indicated points in time (hours post-viral infection, hpvi). Cells were visualized by light microscopy with a 10× magnification. Shown are representative images of three independent experiments (*n* = 3).

These results point to an altered cell culture environment and/or cellular signaling upon super-infection with *S. aureus* 3878$_{SCV}$ and IV, which could be triggered by increased pathogen load or cell intrinsic signaling changes in presence of both pathogens.

3.2. Primary Infection with S. aureus 3878$_{SCV}$ Followed by IV Infection Had No Impact on Bacterial or Viral Titers

First, we analyzed whether the observed cytotoxicity of co-infected A549 human lung epithelial cells with *S. aureus* 3878$_{SCV}$ and IV was due to increased pathogen load. For this, we infected A549 cells with *S. aureus* 3878$_{SCV}$ for 24 h and super-infected with two different IV strains for the indicated points in time to determine the amount of plaque forming units (PFU) or colony forming units (CFU) of viruses or bacteria, respectively (Figure 2).

In general, titers of IV and SCVs increased with time, but neither viral (Figure 2A,B) nor bacterial titers (Figure 2C–F) were significantly changed upon super-infection compared to single-infected cells, a phenomenon independent of the virus strain used [PR8-M (H1N1), A/Panama/2007/99 (Panama; H3N2)].

Thus, these data indicate that the disruption of the cell monolayer upon super-infection is not induced by increased amounts of pathogens but by a different mechanism that is altered by the presence of both pathogens.

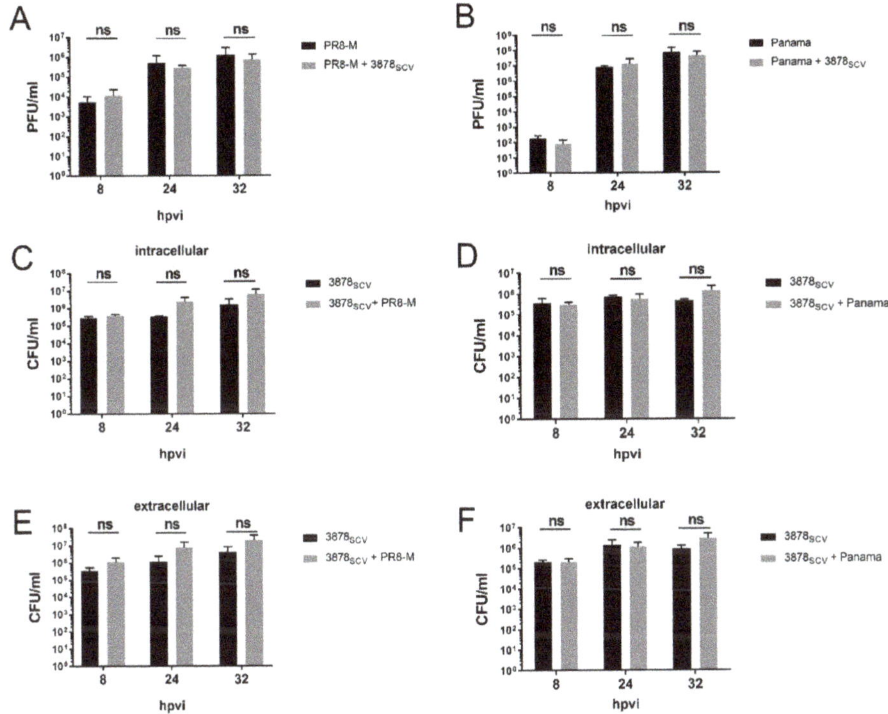

Figure 2. Pathogen load is not affected during *S. aureus* 3878$_{SCV}$ colonization and subsequent influenza virus infection. A549 human lung epithelial cells were infected with *S. aureus* 3878$_{SCV}$ (MOI = 0.01) for 24 h and/or super-infected with (**A,C,E**) IV PR8-M (H1N1; MOI = 0.1) or (**B,D,F**) IV Panama (H3N2; MOI = 0.01) for 8 hpvi, 24 hpvi, or 32 hpvi. At the indicated times post-viral infection, supernatants were collected to determine viral and extracellular bacterial titers. Afterwards, cells were lysed via hypotonic shock to analyze intracellular bacterial titers. Means + SD of three independent experiments with technical duplicates are shown (*n* = 3). Statistical significance (compared to single-IV infection (**A,B**) or single-bacteria infection (**C–F**) was analyzed by a two-way ANOVA, followed by Sidak's multiple comparison test; (hpvi = hours post-viral infection; ns = not significant).

3.3. Pro-Inflammatory Gene Expression Is Highly Upregulated after Super-Infection of S. aureus 3878$_{SCV}$ and IV

Given the observation that super-infection of *S. aureus* 3878$_{SCV}$ and IV PR8-M induced a cytopathic effect (Figure 1), which was not caused by increased pathogen load (Figure 2), we aimed to elucidate if changes of cell intrinsic signaling and inflammatory gene expression might be responsible for this phenomenon. We analyzed the gene expression of 84 different genes, involved in different signaling cascades by use of a *RT^2 profiler Array* (Qiagen, Hilden, Germany) in a single experiment to gain a first insight in the complexity of cellular signaling (Figure 3). This enables a quick analysis of expression levels of different genes that are organized by their function to be able to limit the amount of genes, altering the cell intrinsic signaling. Here, we used the anti-viral immune response panel, including pattern recognition receptors (PRRs), cytokines, and chemokines involved in pathogen recognition and immune responses. The bioinformatic analysis is based on conventional ct-values and was performed with the recommended GeneGlobe online software [26]. A clustergram was generated to visually illustrate all up- and downregulated genes that were analyzed (Figure 3A). To further interpret the results of the *RT^2 profiler Array*, we did an in silico clustering of the upregulated genes of the array that were highly upregulated (difference of an n-fold of 2) in *S. aureus* 3878$_{SCV}$ and PR8-M super-infected cells compared to single-infected cells (*APOBEC3G, CASP1, CASP10, CCL3, CCL5, CD40, CD80, CTSS, CXCL10, CXCL11, CYLD, IL1B, IL6, CXCL8, MEFV, TLR3, TNF, B2M*) (see Supplementary Table S1), with respect to their linkage to specific signaling pathways (Figure 3B) by using the Kyoto Encyclopedia of Genes and Genomes mapper (KEGG mapper). KEGG mapper is a database resource of collected information about pathways and the involved genes representing a pool of molecular interactions, reactions, and their relation to each other [38–40]. Down-regulated genes were excluded, as the gene expressions were negligble (Supplementary Table S1). Besides gene clusters connected to expected PRR pathways including TLR-(11 genes involved, out of the 18 highly upregulated genes comparing co- and single-infected cells identified (11/18), NLR- (7/18), TNFR- (6/18), RLR- (5/18), and NFκB- (5/18) signaling pathways (Figure 3B), we identified gene clusters belonging to two cell death mechanisms, necroptosis (5/18) and apoptosis (3/18). Furthermore, we identified genes involved in the IL-17 (5/18) and c-type lectin (5/18) signaling pathways. To further classify the activated genes leading to the observed cytopathic effect on human lung epithelial cells, we searched for a specific induction pattern in which super-infected cells led to upregulated genes. We, therefore, compared all upregulated genes of single-infected to super-infected samples in a Venn diagram (Figure 3C,D). We identified 11 genes that were induced in all three infection-scenarios compared to uninfected cells and 9 genes that were upregulated in super-infected cells only. We also compared the upregulated genes for super-infection with IV Panama. Here, all infection scenarios shared the induction of 12 genes, where 7 genes were exclusively induced by the super-infection of *S. aureus* 3878$_{SCV}$ and IV Panama. The upregulated genes of the Venn diagram are listed in Table S2A,B. With respect to the mRNA expression levels shown in Supplementary Table S1 and the cytopathic effect observed in super-infected cells (Figure 1), an induction of pro-inflammatory immune response can be concluded, which was further visualized by graphs, exhibiting the gene expression of the highly upregulated genes (Figure 3E).

To confirm an increased pro-inflammatory status of the human lung epithelial cells upon super-infection, we analyzed the mRNA expression of different representative pro-inflammatory cytokines and chemokines (IL-6, IL-8, TNFα, IL-1β, and IFN-γ) in detail (Figure 4A–E). Furthermore, we analyzed the mRNA expression of molecules that are involved in the induction of the type-I-IFN signaling (RIG-I, IFN-β, MxA, and OAS1) (Figure 4F–I), since it was described that IV-induced type-I-IFN signaling had an impact on bacterial infections [41].

In *S. aureus* 3878$_{SCV}$ colonized cells subsequently infected with IV PR8-M the mRNA expression of IL-6, IL-8, TNFα, and IL-1β 32 hpvi was induced if compared to uninfected cells or single-infected cells (Figure 4A–D) and IFN-γ showed the same tendency (Figure 4E). Single-infection of *S. aureus* 3878$_{SCV}$ or IV PR8-M resulted in no significant induction of the mRNA expression 8 hpvi, 24 hpvi, or 32 hpvi, except for IL-8, which was significantly induced 8 hpvi in bacteria single-infected cells (Figure 4B). Nevertheless, this induction was abolished over time. Genes, encoding key proteins involved in the recognition, and induction of type-I-IFN signaling were upregulated in IV PR8-M infected cells 8 hpvi (IFN-β by tendency) (Figure 4G) or 24 hpvi (RIG-I, MxA and OAS1) (Figure 4F,H,I). Previous colonization with *S. aureus* 3878$_{SCV}$ had no impact on IV-induced mRNA expression of factors linked to the type-I-IFN response, except for RIG-I at 32 hpvi, which was significantly decreased in super-infected cells. However, the enhanced RIG-I mRNA synthesis did not result in alterations of viral titers. Similar results were obtained upon super-infection with *S. aureus* 3878$_{SCV}$ and IV Panama, indicating a virus-independent effect (Supplementary Figure S2A–I).

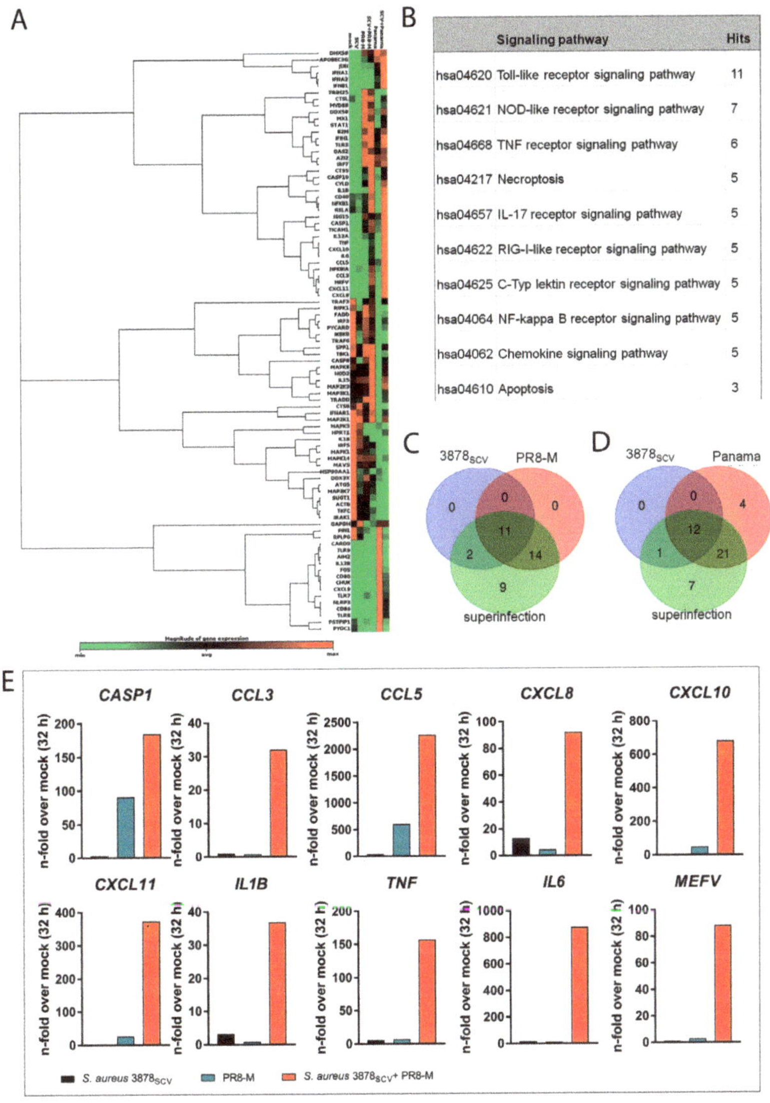

Figure 3. Gene expression analysis by *RT^2 Profiler Array*. (**A–E**) A549 lung epithelial cells were infected with *S. aureus* 3878$_{SCV}$ (MOI = 0.01) for 24 h and/or super-infected with IV PR8-M (H1N1; MOI = 0.1) or IV Panama (H3N2; MOI = 0.01) for 32 h. Subsequently, RNA was isolated and further used to perform the *RT^2 Profiler Array* (Qiagen, Hilden, Germany). Ct-values were analyzed with the recommended QIAGEN web portal [26]. (**A**) A clustergram is shown, visualizing the up- and downregulated genes of the customized 84-gene array. (**B**) Ten potential signaling pathways are listed, which can be analyzed by the *RT^2 Profiler Array* with the correlating count of genes involved. The mapping was done by using the Kyoto Encyclopedia of Genes and Genomes (KEGG) mapper [38–40]. (**C,D**) A Venn diagram of the upregulated genes in a super-infection scenario with *S. aureus* 3878$_{SCV}$ and IV PR8-M (**C**) or IV Panama (**D**) is shown. The analysis was performed by use of http://bioinformatics.psb.ugent.be/webtools/Venn/. (**E**) Gene expression of highly induced genes indicating an increased pro-inflammatory cytokine response. Values are shown as n-fold over mock (32 hpvi); (*n* = 1); (hpvi = hours post-viral infection).

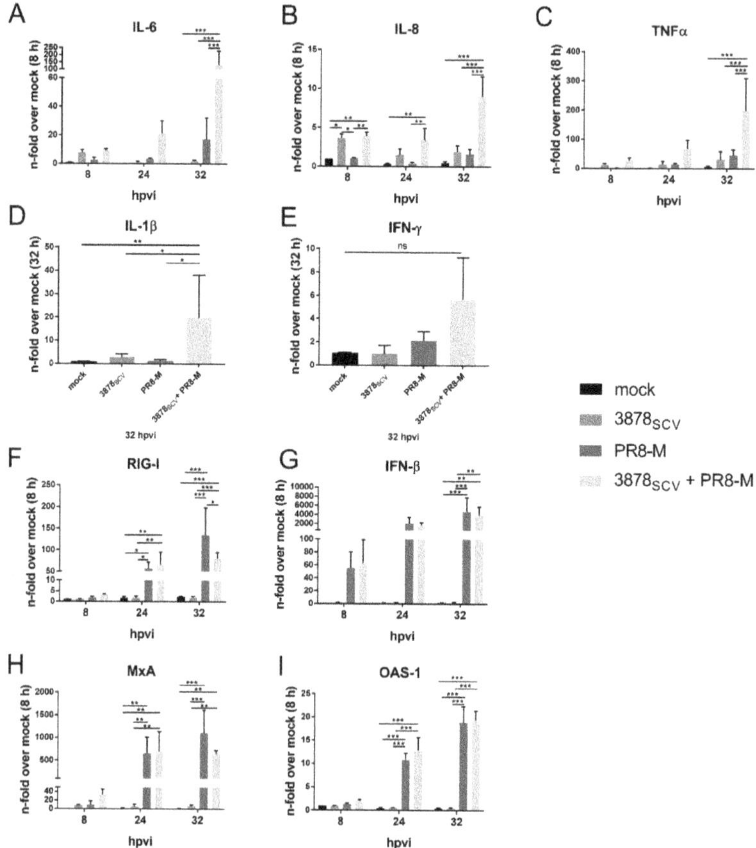

Figure 4. Pro-inflammatory cytokines and chemokines are enhanced after *S. aureus* 3878$_{SCV}$ colonization and subsequent IV PR8-M infection. (**A–I**) A549 human lung epithelial cells were infected with *S. aureus* 3878$_{SCV}$ (MOI = 0.01) for 24 h and/or super-infected with IV PR8-M (H1N1; MOI = 0.1) for 8 hpvi, 24 hpvi and/or 32 hpvi. Afterwards, RNA was isolated and mRNA levels of IL-6, IL-8, TNFα, IL-1β, IFN-γ, RIG-I, IFN-β, MxA, and OAS1 were determined by qRT-PCR. All values were correlated to the representative mock-control 8 hpvi (IL-6, IL-8, TNFα, RIG-I, IFNβ, MxA and OAS1) or 32 hpvi (IL-1β and IFN-γ). Means + SD of three independent experiments including technical duplicates are shown. Statistical significance was analyzed by a two-way (**A–C**), (**F–I**) or one-way (**D,E**) ANOVA, followed by Tukey's multiple comparison test (* $p < 0.05$, ** $p < 0.01$, *** $p < 0.001$); (hpvi = hours post-viral infection; ns = not significant).

To analyze whether the induction of mRNA synthesis of pro-inflammatory genes could also be detected on protein level and to get further insights into the cell intrinsic innate immune status of super-infected A549 cells, the protein expression of exemplary cytokines and chemokines was monitored by FACS analysis (Figure 5A–H). Remarkably, FACS analysis verified the increased pro-inflammatory response of A549 human lung epithelial cells for the secretion of representative factors. In super-infected cells, protein levels of IL-6, RANTES (CCL5), IP-10, and I-TAC were significantly induced compared to uninfected or single-infected cells with either *S. aureus* 3878$_{SCV}$ or IV PR8-M (Figure 5A,C,E,F). TNFα was also significantly upregulated upon super-infection compared to uninfected and bacteria single-infected cells but not to IV PR8-M single-infected cells. Furthermore, IV PR8-M infection provoked TNFα protein expression 32 hpvi (Figure 5B). Representative IFN protein concentrations of IFN-γ and IFNβ (Figure 5G,H) showed no alteration in the amount of secreted proteins.

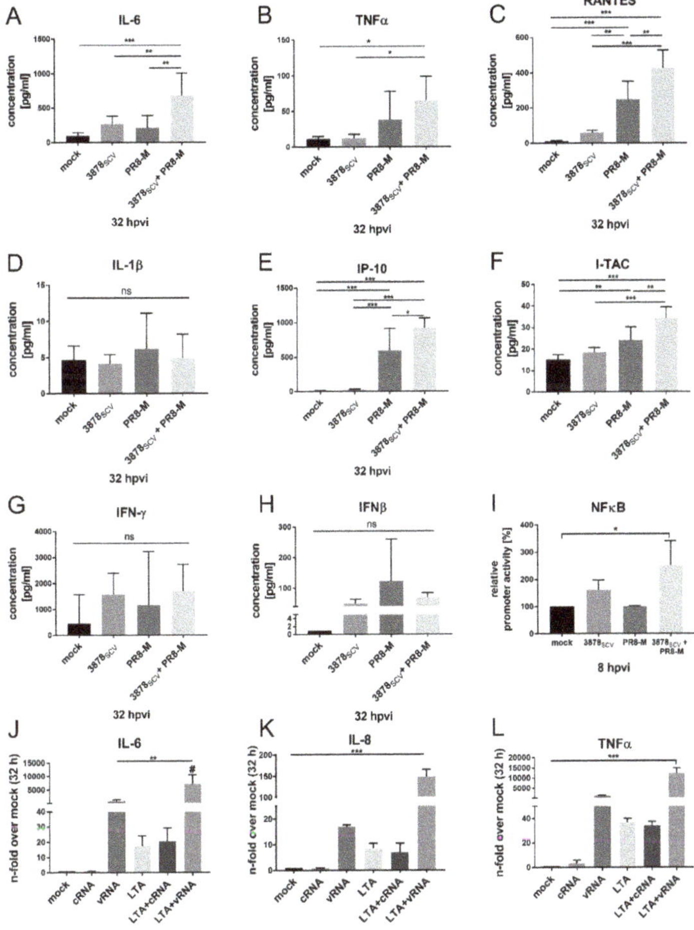

Figure 5. Secretion of the pro-inflammatory cytokines and chemokines are enhanced after *S. aureus* 3878$_{SCV}$ colonization and subsequent IV PR8-M infection regulated by TLR2- and RIG-I-mediated NFκB promoter activation. (**A–H**) A549 human lung epithelial cells were infected with *S. aureus* 3878$_{SCV}$ (MOI = 0.01) for 24 h and/or super-infected with IV PR8-M (H1N1; MOI = 0.1) for 32 h. Afterwards, supernatants were collected to measure the concentration of secreted proteins via FACS analysis. Means + SD of three independent experiments, including technical duplicates, are shown. (**I**) A549 human lung epithelial cells were transfected with 3× NFκB luciferase promoter reporter construct for 24 h prior to super-infection as described before. Afterwards cells were harvested and analyzed for luciferase activity. (**J–L**) A549 human lung epithelial cells were stimulated with LTA (100 ng/mL) for 24 h at 37 °C and 5% CO_2. Afterwards, cells were stimulated with cellular RNA (cRNA) or viral RNA (vRNA) (100 ng/mL) in the presence or absence of LTA for 4 h at 37 °C and 5% CO_2. Subsequently, RNA was isolated, and mRNA levels of IL-6, IL-8, and TNFα, were measured by qRT-PCR. All values are correlated to the respective mock-control (*n* = 3). Statistical significance was analyzed by one-way ANOVA followed by Tukey's multiple comparison test (* $p < 0.05$, ** $p < 0.01$, *** $p < 0.001$; # = *** $p < 0.001$ compared to LTA + vRNA, except for vRNA); (hpvi = hours post-viral infection; ns = not significant).

These results emphasize the enhanced cell intrinsic pro-inflammatory status of the super-infected cells. We could confirm these results by the use of IV strain Panama, verifying a viral strain independent effect (Supplementary Figure S3A–H). Based on these results and due to the fact that the induction of pro-inflammatory cytokines and chemokines is mainly driven by NFκB activation, we hypothesized an induction of the pro-inflammatory response via specific PRRs, resulting in the activation of the NFκB-signaling cascade [42]. To confirm the induction of NFκB-signaling we transfected A549 cells with an artificial NFκB promoter-dependent luciferase reporter plasmid prior to super-infection with *S. aureus* 3878$_{SCV}$ and subsequent IV PR8-M infection. An increase of NFκB activation was observed in super-infected cells compared to uninfected and IV PR8-M-infected cells, while *S. aureus* 3878$_{SCV}$ infection only resulted in an increase of NFκB activation by trend (Figure 5I). This induction pattern was also confirmed in cells super-infected with IV Panama (Supplementary Figure S3I).

The induction of pro-inflammatory responses via NFκB in epithelial cells after pathogen exposure can be triggered by both pathogens through different factors and their corresponding receptors [43–45], which were further analyzed. To exclude specific pathogen-mediated interference with cellular factors due to differences in protein expression, such as virulence factors or surface proteins, viral RNA (vRNA) and bacterial lipoteichonic acid (LTA, Invivogen, San Diego, CA, USA) were used as pathogen specific molecular stimuli. In human lung epithelial cells vRNA is mainly recognized by RIG-I, leading to a strong induction of the type-I-IFN-signaling cascade [46], while LTA is mainly recognized by TLR-2 [47]. We investigated mRNA expression of IL-6, IL-8, and TNFα after stimulating A549 cells with vRNA and LTA (Figure 5J–L). The results matched our findings obtained from super-infected cells, since significant enhancement of mRNA expression of IL-6, IL-8, and TNFα was observed in presence of both stimuli. Artificial effects caused by RNA transfection could be excluded due to equal cytokine mRNA expression induced by cellular RNA (cRNA) and cRNA + LTA stimulated cells. Stimulation with vRNA tended to induce the expression of IL-6, IL-8, and TNFα, which, however, was not significant compared to unstimulated cells.

Overall, these data suggest an induction of pro-inflammatory gene expression responses through the detection of bacterial and viral components via the pathogen-associated molecular pattern receptors (PAMP) RIG-I and TLR-2, followed by the induction of NFκB. To exclude bacterial strain-specific effects, another SCV strain (*S. aureus* 814$_{SCV}$) was used to determine pathogen loads and pro-inflammatory gene expression in IV super-infection (Supplementary Figure S4). While neither viral titers nor intra- and extracellular bacterial load were increased in presence of both pathogens, pro-inflammatory cytokine expression was enhanced, verifying the former observations.

3.4. S. aureus 3878$_{SCV}$ Provoke Enhanced Necrotic Cell Death in Presence of IV Infection

The observed disruption of the cell monolayer (Figure 1) could be induced by a variety of mechanisms. Besides the involvement of pro-inflammatory cytokines in the innate immune response, these factors are also involved in the induction of cell death mechanisms, like apoptosis and necrosis. As the results shown in Figure 3 indicate, an upregulation of pro-inflammatory cytokines, the cell death mechanisms might be triggered by TLRs or cell death receptors through PAMPs or cytokines, like TNFα, among others [48–50]. Therefore, we further investigated the induction of apoptosis and necrosis, correlating to the cell death mechanisms identified in the *RT^2 Profiler Array* analysis (Figure 3B).

As the results of the LDH assay led to the hypothesis of an induced necrotic cell death mechanism, we performed FACS analysis to determine the number of early apoptotic cells by detecting phosphatidylserine which switches to the cells' surface in early apoptotic cells and can be labeled with annexin V. Cells with a membrane rupture tending to necrosis were detected by using a viability marker comparable to 7-aminoactinomycin D and propidium iodide staining. Therefore, we performed the infection up to 44 hpvi to be able to still distinguish between early apoptosis and necrotic-like cells and stained the cells accordingly. The amount of necrotic cells significantly increased comparing un- or single-infected with super-infected cells, probably indicating necrosis (Figure 6A).

Furthermore, the amount of apoptotic cells was significantly higher in IV-infected cells compared to un-, bacteria-, or super-infected cells 44 hpvi (Figure 6B).

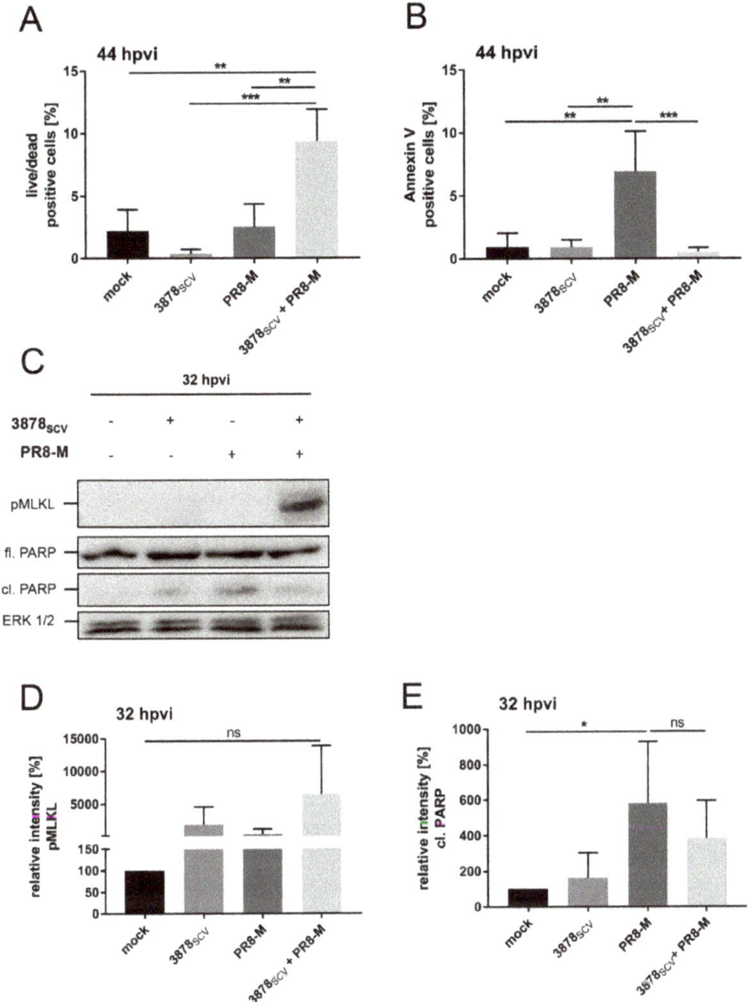

Figure 6. *S. aureus* 3878$_{SCV}$ colonization and subsequent IV infection inhibits IV-induced apoptosis but results in the induction of necrosis. (**A–E**) A549 human lung epithelial cells were infected with *S. aureus* 3878$_{SCV}$ (MOI = 0.01) for 24 h and/or super-infected with IV PR8-M (H1N1; MOI = 0.1) for 44 hpvi (**A,B**) or 32 hpvi (**C–E**). At the indicated times post-viral infection, total amount of cells was collected to perform FACS analysis to determine the relative amount of viability marker positive cells (**A**) or annexin V positive cells (**B**). Furthermore, whole cell lysates were subjected to western blot analysis (**C**). (**D,E**) Densitometrical analysis of three independent western blot experiments of cleaved pMLKL (**D**) and PARP (**E**) 32 hpvi are shown. Equal protein amounts were calculated by correlating the signal intensities to their corresponding ERK1/2 signals. Means + SD of three independent experiments are shown (*n* = 3). Statistical significance was analyzed by a one-way ANOVA, followed by Tukey's multiple comparison test (**A,B,D,E**); (* $p < 0.05$, ** $p < 0.01$, *** $p < 0.001$); (hpvi = hours post-viral infection; ns = not significant).

As cell death mechanisms like necrosis can be further defined in specific mechanisms and to compare our findings to previously described inductions of cell death mechanisms upon infection with SCV [51] or in co-infection scenarios [27], we performed western blot analysis to be able to differentiate between necroptosis and apoptosis.

Necroptosis is an inflammatory programmed form of necrosis, which was already described in a recent publication reporting its induction by *S. aureus* SCV in single-infected human primary keratinocytes [51]. Necroptosis is induced via a receptor-interacting protein (RIP) kinase-mediated activation, resulting in the phosphorylation and oligomerization of mixed lineage kinase domain like pseudokinase (MLKL) and pore formation, leading to the release of inflammatory cytokines. To distinguish necroptosis from apoptosis induction, we monitored the induction of phosphorylated MLKL and PARP cleavage, which are indications for both cell death mechanisms [52]. We infected A549 human lung epithelial cells with *S. aureus* 3878$_{SCV}$ for 24 h, followed by IV infection with PR8-M for 32 h (Figure 6C–E).

In super-infected cells, induction of pMLKL 32 hpvi was observed in comparison to uninfected, bacteria-, or virus single-infected cells, respectively. PARP cleavage was more likely to be induced in IV PR8-M-infected cells, and was slightly decreased in super-infected samples 32 hpvi (Figure 6C). However, the densitometrical analysis of three independent experiments could only confirm a trend of activated MLKL due to induced phosphorylation in super-infected cells (Figure 6D), whereas an induction of apoptosis upon IV infection could be verified (Figure 6E). Additionally, pyroptosis as another form of regulated necrotic cell death mechanism, which is activated via the induction of the inflammasome resulting in the cleavage of gasdermin D, could not be detected by cleaved gasdermin D (Supplementary Figure S5). The original blots are shown in the Supplementary Figure S6.

Thus, our results indicate a necrotic cell death induction, most likely induced by increased pro-inflammatory gene expression response after super-infection with *S. aureus* 3878$_{SCV}$, followed by secondary IV infection with PR8-M and Panama.

4. Discussion

The first occurrence of persisting bacteria or SCVs was already described about 100 years ago [53]. Even though they are known for such a long time, not many studies were undertaken to elucidate their impact on cellular responses or their impact on additional infections with other pathogens. Our aim was to investigate the interaction of *S. aureus* SCVs with a subsequent IV infection in respect to epithelial cell responses, which built the first cellular barrier for pathogens in the lung. Here, we demonstrate that invasive *S. aureus* SCVs do have an impact on the cell intrinsic response in human lung epithelial cells, as indicated by highly secreted pro-inflammatory cytokines and chemokines (Figure 5 and Supplementary Figure S3) and, furthermore, an induction of necrotic cell death of super-infected compared to single-infected cells (Figure 6). This was somehow surprising, since the majority of SCVs are not described to significantly induce cell intrinsic responses, due to decreased secretion of virulence factors [54], a feature that would match their dormant status. In particular, not much is known about the impact of SCVs on lung tissue responses and nothing so far about their impact on a secondary IV infection. In this study, we were able to show a cytopathic effect accompanied by increased pro-inflammatory cytokine and chemokine release and necrotic cell death through colonizing *S. aureus* 3878$_{SCV}$ and different IV strains, such as PR8-M and Panama.

Typically, super-infections with pathogenic *S. aureus* strains and IV led to increased pathogen loads accompanied with the induction of pro-inflammatory responses [13,35]. However, this could not be confirmed within the present study in the SCV and IV super-infection scenario. It was shown previously that super-infection with pathogenic *S. aureus* leads to the enhancement of viral titers due to the inhibition of STAT1 and STAT2 dimerization, resulting in decreased production of anti-viral factors [15]. This inhibitory effect could be excluded since mRNA expression of RIG-I, IFNβ, MxA, or OAS1 in super-infected cells compared to IV-infected cells were not altered (Figure 4 and Supplementary Figure S2). Based on these results, we could exclude an effect of the anti-viral response and the involvement of an altered pathogen load. Nevertheless, we identified a clear induction of pro-inflammatory cytokines and chemokines in human lung epithelial cells. Besides the attraction of immune cells and the induction of an anti-pathogen status of the cell, pro-inflammatory cytokines induce a stress response leading to the induction of cell death mechanisms via TLRs or death receptors [55–57].

To proof the impact of two main cell death mechanisms we performed FACS analysis to monitor early apoptotic and necrotic-like cells. Correlating to the LDH assays, we could confirm an increase in necrosis during super-infection (Figure 6A). Besides, the amount of apoptotic cells was decreased in super- compared to IV-infected cells. Further specifications of cell-death mechanisms by western blot analysis revealed the tendency for an increase of phosphorylated MLKL in super-infected cells, giving the hit of probably induced necroptosis. Concomitantly, IV-induced PARP cleavage was reduced in super-infected cells compared to IV PR8-M-infected cells by trend.

Interestingly, there are two different mechanisms described, how pathogenic *S. aureus* and *S. aureus* SCVs are able to induce necroptosis [27,51]. During the critical phase of *S. aureus* infection the virulence factor *agr* is induced [8], resulting in possible secretion of different toxins, which induces necroptosis [27,33]. In SCV-infected keratinocytes, necroptosis was driven by the activation of glycolysis [51]. *S. aureus* adopts its whole metabolism to persist within the host. The metabolic changes of *S. aureus* were already described elsewhere [58]. As the utilization of the tricarboxylic acid cycle for the host cell and the persisting bacteria is decreased, the glycolysis is stronger induced to generate adenosine triphosphate (ATP). As we observed a disruption of cell monolayer and a possible induction of phosphorylated MLKL upon super-infection, we linked our findings more to necroptosis (Figures 1 and 6). Even though *S. aureus*-induced necroptosis might be independent of TLR stimulation [59], our data indicate a synergistic effect of *S. aureus* 3878$_{SCV}$ and IV inducing cell death, which can be related to TLR2- and RIG-I-mediated pro-inflammatory response induction. In addition, our data show that the superinfection could be imitated with the stimuli LTA and vRNA. This underlines that the initial induction of the pro-inflammatory response and the subsequent cell death must be different from that of pathogenic bacterial strains that induce cell death much more quickly. In case of SCV, this indicates a lower virulence probably due to the decreased secretion of virulence factors. Nonetheless, dormant SCVs can work synergistically and affect the virus-induced immune response. As we performed pure ligand experiments, inhibitory effects of molecules of this pro-inflammatory cell intrinsic response is supposed to trigger cellular stress in the form of reactive oxygen species [60].

So far, the impact of *S. aureus* SCVs with subsequent IV infection had not been investigated. Interestingly, we could give first insights in this super-infection scenario and unravel one extraordinary role of a SCV patients' isolate *S. aureus* 3878$_{SCV}$ with subsequent IV infection. We observed an induction of pro-inflammatory cytokines and chemokines, which underlines the severity of the coincident occurrence of *S. aureus* SCVs and IV. These data point to a cross-interaction of necrotic cell death and pro-inflammatory cell intrinsic response, as the pathogens alone can induce an inflammatory response through PAMPs and secreted cell damage-associated molecular patterns (DAMPs). Upon necrotic cell death induction, further pro-inflammatory responses are induced via DAMP receptors [50], leading to an enhancement of pro-inflammatory cytokines and chemokines seen on transcriptional and translational level.

In summary, we were able to show that persistent *S. aureus* SCV and subsequent IV infection affects cell-internal immune response by inducing the release of pro-inflammatory cytokines and chemokines, resulting in cell death induction.

Supplementary Materials: The following are available online at http://www.mdpi.com/2076-2607/8/12/1998/s1, Table S1: List of ct-values analyzed by GeneGlobe online software of the *RT*^{2} *Profiler Array* plate. Table S2: (A,B) Listed gene names of the venn diagrams shown in Figure 3. Figure S1: Wildtype phenotype *S. aureus* 3878 is more virulent compared to *S. aureus* 3878$_{SCV}$, but *S. aureus* 3878$_{SCV}$ induces LDH release upon super-infection with PR8-M. Figure S2: Pro-inflammatory cytokines and chemokines are enhanced after *S. aureus* 3878$_{SCV}$ colonization and subsequent IV Panama infection. Figure S3: Secretion of the pro-inflammatory cytokines and chemokines are enhanced after *S. aureus* 3878$_{SCV}$ colonization and subsequent IV Panama infection regulated by TLR2- and

RIG-I-mediated NFκB promoter activation. Figure S4: Pathogen load and pro-inflammatory cytokines and chemokines are enhanced after super-infection with the SCV strain *S. aureus* 814$_{\text{SCV}}$. Figure S5: *S. aureus* 3878$_{\text{SCV}}$ colonization and subsequent influenza virus infection has no effect on the induction of pyroptosis. Figure S6: Original western blots of Figure 6C and S5A.

Author Contributions: Conceptualization, J.J.W., E.R.H., S.L. and C.E.; methodology, J.J.W.; software, J.J.W.; validation, J.J.W., Y.B. and E.R.H.; formal analysis, J.J.W., E.R.H., Y.B., S.L., S.N. and C.E.; investigation, J.J.W.; resources, S.L. and C.E.; data curation, S.L. and C.E.; writing—original draft preparation, J.J.W. and C.E.; writing—review and editing, J.J.W., E.R.H., S.N., Y.B., B.L., S.L. and C.E.; visualization, J.J.W.; supervision, C.E.; project administration, C.E.; funding acquisition, J.J.W., S.L. and C.E. All authors have read and agreed to the published version of the manuscript.

Funding: This work was supported by the Deutsche Forschungsgemeinschaft (SFB 1009, project B01 and B02, CRU342 P06 and under Germany´s Excellence Strategy—EXC 2051—Project-ID 390713860). We acknowledge support by the German Research Foundation and the Open Access Publication Fund of the Thueringer Universitaets- und Landesbibliothek Jena Projekt-Nr. 433052568.

Acknowledgments: We would like to thank Karsten Becker for providing us with the bacterial isolate.

Conflicts of Interest: The authors declare no conflict of interest.

Abbreviations

DAMP	damage-associated molecular pattern
hpbi	hours post bacterial infection
hpvi	hours post-viral infection
I-TAC	interferon-inducible T-cell alpha chemoattractant
IFN	interferon
IL	interleukin
IP-10	interferon gamma-induced protein 10
IV	influenza virus
LDH	Lactate dehydrogenase
MAPK	mitogen-activated protein kinase
MxA	interferon-induced GTP-binding protein MxA
NFκB	nuclear factor kappa-light-chain-enhancer of activated B-cells
NLR	NOD-like receptor
NOD	nucleotide-binding oligomerization domain
OAS1	2′-5′-oligoadenylate synthetase 1
PAMP	pathogen associated molecular pattern
PRR	pattern recognition receptor
RANTES	CC-chemokine ligand 5
RELA	nuclear factor NF-kappa-B p65 subunit
RIG-I	retinoic acid inducible gene I
RLR	RIG-I like receptor
S. aureus	*Staphylococcus aureus*
SCVs	small colony variants
TLR	toll-like receptor
TNFR	tumor necrosis factor receptor
TNFα	tumor necrosis factor alpha

References

1. Lynch, S.V. Viruses and Microbiome Alterations. *Ann. Am. Thorac. Soc.* **2014**, *11*, S57–S60. [CrossRef] [PubMed]

2. Lee, J.; Zilm, P.S.; Kidd, S.P. Novel Research Models for Staphylococcus aureus Small Colony Variants (SCV) Development: Co-pathogenesis and Growth Rate. *Front. Microbiol.* **2020**, *11*, 1–8. [CrossRef] [PubMed]

3. van Belkum, A.; Verkaik, N.J.; de Vogel, C.P.; Boelens, H.A.; Verveer, J.; Nouwen, J.L.; Verbrugh, H.A.; Wertheim, H.F.L. Reclassification of Staphylococcus aureus Nasal Carriage Types. *J. Infect. Dis.* **2009**, *199*, 1820–1826. [CrossRef] [PubMed]

4. Jenul, C.; Horswill, A.R. Regulation of Staphylococcus aureus Virulence. In *Gram-Positive Pathogens*; ASM Press: Washington, DC, USA, 2019; Volume 6, pp. 669–686.

5. Kahl, B.C.; Becker, K.; Löffler, B. Clinical Significance and Pathogenesis of Staphylococcal Small Colony Variants in Persistent Infections. *Clin. Microbiol. Rev.* **2016**, *29*, 401–427. [CrossRef] [PubMed]

6. Tuchscherr, L.; Löffler, B. Staphylococcus aureus dynamically adapts global regulators and virulence factor expression in the course from acute to chronic infection. *Curr. Genet.* **2016**, *62*, 15–17. [CrossRef] [PubMed]

7. Tuchscherr, L.; Bischoff, M.; Lattar, S.M.; Noto Llana, M.; Pförtner, H.; Niemann, S.; Geraci, J.; Van de Vyver, H.; Fraunholz, M.J.; Cheung, A.L.; et al. Sigma Factor SigB Is Crucial to Mediate Staphylococcus aureus Adaptation during Chronic Infections. *PLoS Pathog.* **2015**, *11*, 1–26. [CrossRef] [PubMed]

8. Tuchscherr, L.; Medina, E.; Hussain, M.; Völker, W.; Heitmann, V.; Niemann, S.; Holzinger, D.; Roth, J.; Proctor, R.A.; Becker, K.; et al. Staphylococcus aureus phenotype switching: An effective bacterial strategy to escape host immune response and establish a chronic infection. *EMBO Mol. Med.* **2011**, *3*, 129–141. [CrossRef]

9. Kahl, B.C.; Belling, G.; Reichelt, R.; Herrmann, M.; Proctor, R.A.; Peters, G. Thymidine-dependent small-colony variants of Staphylococcus aureus exhibit gross morphological and ultrastructural changes consistent with impaired cell separation. *J. Clin. Microbiol.* **2003**, *41*, 410–413. [CrossRef]

10. Kriegeskorte, A.; König, S.; Sander, G.; Pirkl, A.; Mahabir, E.; Proctor, R.A.; von Eiff, C.; Peters, G.; Becker, K. Small colony variants of Staphylococcus aureus reveal distinct protein profiles. *Proteomics* **2011**, *11*, 2476–2490. [CrossRef]

11. Proctor, R.A.; Balwit, J.M.; Vesga, O. Variant subpopulations of Staphylococcus aureus as cause of persistent and recurrent infections. *Infect. Agents Dis.* **1994**, *3*, 302–312.

12. Papi, A.; Bellettato, C.M.; Braccioni, F.; Romagnoli, M.; Casolari, P.; Caramori, G.; Fabbri, L.M.; Johnston, S.L. Infections and Airway Inflammation in Chronic Obstructive Pulmonary Disease Severe Exacerbations. *Am. J. Respir. Crit. Care Med.* **2006**, *173*, 1114–1121. [CrossRef] [PubMed]

13. Iverson, A.R.; Boyd, K.L.; McAuley, J.L.; Plano, L.R.; Hart, M.E.; McCullers, J.A. Influenza Virus Primes Mice for Pneumonia From Staphylococcus aureus. *J. Infect. Dis.* **2011**, *203*, 880–888. [CrossRef] [PubMed]

14. Bakaletz, L.O. Viral–bacterial co-infections in the respiratory tract. *Curr. Opin. Microbiol.* **2017**, *35*, 30–35. [CrossRef] [PubMed]

15. Warnking, K.; Klemm, C.; Löffler, B.; Niemann, S.; van Krüchten, A.; Peters, G.; Ludwig, S.; Ehrhardt, C. Super-infection with Staphylococcus aureus inhibits influenza virus-induced type I IFN signalling through impaired STAT1-STAT2 dimerization. *Cell. Microbiol.* **2015**, *17*, 303–317. [CrossRef]

16. LeMessurier, K.S.; Tiwary, M.; Morin, N.P.; Samarasinghe, A.E. Respiratory Barrier as a Safeguard and Regulator of Defense Against Influenza A Virus and Streptococcus pneumoniae. *Front. Immunol.* **2020**, *11*, 1–15. [CrossRef]

17. Morris, D.E.; Cleary, D.W.; Clarke, S.C. Secondary Bacterial Infections Associated with Influenza Pandemics. *Front. Microbiol.* **2017**, *8*, 1–17. [CrossRef]

18. Yu, D.; Wei, L.; Zhengxiu, L.; Jian, L.; Lijia, W.; Wei, L.; Xiqiang, Y.; Xiaodong, Z.; Zhou, F.; Enmei, L. Impact of bacterial colonization on the severity, and accompanying airway inflammation, of virus-induced wheezing in children. *Clin. Microbiol. Infect.* **2010**, *16*, 1399–1404. [CrossRef]

19. Von Eiff, C.; Becker, K.; Metze, D.; Lubritz, G.; Hockmann, J.; Schwarz, T.; Peters, G. Intracellular Persistence of Staphylococcus aureus Small-Colony Variants within Keratinocytes: A Cause for Antibiotic Treatment Failure in a Patient with Darier's Disease. *Clin. Infect. Dis.* **2001**, *32*, 1643–1647. [CrossRef]

20. von Eiff, C.; Bettin, D.; Proctor, R.A.; Rolauffs, B.; Lindner, N.; Winkelmann, W.; Peters, G. Recovery of Small Colony Variants of Staphylococcus aureus Following Gentamicin Bead Placement for Osteomyelitis. *Clin. Infect. Dis.* **1997**, *25*, 1250–1251. [CrossRef]

21. Abu-Qatouseh, L.F.; Chinni, S.V.; Seggewiß, J.; Proctor, R.A.; Brosius, J.; Rozhdestvensky, T.S.; Peters, G.; von Eiff, C.; Becker, K. Identification of differentially expressed small non-protein-coding RNAs in Staphylococcus aureus displaying both the normal and the small-colony variant phenotype. *J. Mol. Med.* **2010**, *88*, 565–575. [CrossRef]

22. Flory, E.; Kunz, M.; Scheller, C.; Jassoy, C.; Stauber, R.; Rapp, U.R.; Ludwig, S. Influenza Virus-induced NF-κB-dependent Gene Expression Is Mediated by Overexpression of Viral Proteins and Involves Oxidative Radicals and Activation of IκB Kinase. *J. Biol. Chem.* **2000**, *275*, 8307–8314. [CrossRef] [PubMed]

23. Ludwig, S.; Ehrhardt, C.; Neumeier, E.R.; Kracht, M.; Rapp, U.R.; Pleschka, S. Influenza Virus-induced AP-1-dependent Gene Expression Requires Activation of the JNK Signaling Pathway. *J. Biol. Chem.* **2001**, *276*, 10990–10998. [CrossRef]

24. Mazur, I.; Wurzer, W.J.; Ehrhardt, C.; Pleschka, S.; Puthavathana, P.; Silberzahn, T.; Wolff, T.; Planz, O.; Ludwig, S. Acetylsalicylic acid (ASA) blocks influenza virus propagation via its NF-kB-inhibiting activity. *Cell. Microbiol.* **2007**, *9*, 1683–1694. [CrossRef] [PubMed]

25. Livak, K.J.; Schmittgen, T.D. Analysis of Relative Gene Expression Data Using Real-Time Quantitative PCR and the 2−ΔΔCT Method. *Methods* **2001**, *25*, 402–408. [CrossRef] [PubMed]

26. Qiagen GeneGlobe RT2 Profiler PCR Data Analysis. Available online: https://geneglobe.qiagen.com/de/analyze/ (accessed on 16 June 2020).

27. Van Krüchten, A.; Wilden, J.J.; Niemann, S.; Peters, G.; Löffler, B.; Ludwig, S.; Ehrhardt, C. Staphylococcus aureus triggers a shift from influenza virus-induced apoptosis to necrotic cell death. *FASEB J.* **2018**, *32*, 2779–2793. [CrossRef] [PubMed]

28. Wurzer, W.J.; Planz, O.; Ehrhardt, C.; Giner, M.; Silberzahn, T.; Pleschka, S.; Ludwig, S. Caspase 3 activation is essential for effcient infuenza virus propagation. *EMBO J.* **2003**, *22*, 2717–2728. [CrossRef]

29. Wurzer, W.J.; Ehrhardt, C.; Pleschka, S.; Berberich-Siebelt, F.; Wolff, T.; Walczak, H.; Planz, O.; Ludwig, S. NF-κB-dependent Induction of Tumor Necrosis Factor-related Apoptosis-inducing Ligand (TRAIL) and Fas/FasL Is Crucial for Efficient Influenza Virus Propagation. *J. Biol. Chem.* **2004**, *279*, 30931–30937. [CrossRef]

30. Ehrhardt, C.; Wolff, T.; Ludwig, S. Activation of phosphatidylinositol 3-kinase signaling by the nonstructural NS1 protein is not conserved among type A and B influenza viruses. *J. Virol.* **2007**, *81*, 12097–12100. [CrossRef]

31. Korea, C.G.; Balsamo, G.; Pezzicoli, A.; Merakou, C.; Tavarini, S.; Bagnoli, F.; Serruto, D.; Unnikrishnan, M. Staphylococcal Esx Proteins Modulate Apoptosis and Release of Intracellular Staphylococcus aureus during Infection in Epithelial Cells. *Infect. Immun.* **2014**, *82*, 4144–4153. [CrossRef]

32. Chi, C.-Y.; Lin, C.-C.; Liao, I.-C.; Yao, Y.-C.; Shen, F.-C.; Liu, C.-C.; Lin, C.-F. Panton-Valentine Leukocidin Facilitates the Escape of Staphylococcus aureus From Human Keratinocyte Endosomes and Induces Apoptosis. *J. Infect. Dis.* **2014**, *209*, 224–235. [CrossRef]

33. Kitur, K.; Parker, D.; Nieto, P.; Ahn, D.S.; Cohen, T.S.; Chung, S.; Wachtel, S.; Bueno, S.; Prince, A. Toxin-Induced Necroptosis Is a Major Mechanism of Staphylococcus aureus Lung Damage. *PLoS Pathog.* **2015**, *11*, e1004820. [CrossRef] [PubMed]

34. McCullers, J.A. Preventing and treating secondary bacterial infections with antiviral agents. *Antivir. Ther.* **2011**, *16*, 123–135. [CrossRef] [PubMed]

35. Klemm, C.; Bruchhagen, C.; van Krüchten, A.; Niemann, S.; Löffler, B.; Peters, G.; Ludwig, S.; Ehrhardt, C. Mitogen-activated protein kinases (MAPKs) regulate IL-6 over-production during concomitant influenza virus and Staphylococcus aureus infection. *Sci. Rep.* **2017**, *7*, 42473. [CrossRef] [PubMed]

36. Tse, L.V.; Whittaker, G.R. Modification of the hemagglutinin cleavage site allows indirect activation of avian influenza virus H9N2 by bacterial staphylokinase. *Virology* **2015**, *482*, 1–8. [CrossRef] [PubMed]

37. Kriegeskorte, A.; Grubmüller, S.; Huber, C.; Kahl, B.C.; von Eiff, C.; Proctor, R.A.; Peters, G.; Eisenreich, W.; Becker, K. Staphylococcus aureus small colony variants show common metabolic features in central metabolism irrespective of the underlying auxotrophism. *Front. Cell. Infect. Microbiol.* **2014**, *4*, 1–8. [CrossRef] [PubMed]

38. Kanehisa, M. Toward understanding the origin and evolution of cellular organisms. *Protein Sci.* **2019**, *28*, 1947–1951. [CrossRef]

39. Kanehisa, M.; Sato, Y.; Furumichi, M.; Morishima, K.; Tanabe, M. New approach for understanding genome variations in KEGG. *Nucleic Acids Res.* **2019**, *47*, D590–D595. [CrossRef]

40. Kanehisa, M. KEGG: Kyoto Encyclopedia of Genes and Genomes. *Nucleic Acids Res.* **2000**, *28*, 27–30. [CrossRef]

41. Li, W.; Moltedo, B.; Moran, T.M. Type I Interferon Induction during Influenza Virus Infection Increases Susceptibility to Secondary Streptococcus pneumoniae Infection by Negative Regulation of T Cells. *J. Virol.* **2012**, *86*, 12304–12312. [CrossRef]

42. Liu, T.; Zhang, L.; Joo, D.; Sun, S.-C. NF-κB signaling in inflammation. *Signal Transduct. Target. Ther.* **2017**, *2*, 17023. [CrossRef]

43. Claro, T.; Widaa, A.; McDonnell, C.; Foster, T.J.; O'Brien, F.J.; Kerrigan, S.W. Staphylococcus aureus protein A binding to osteoblast tumour necrosis factor receptor 1 results in activation of nuclear factor kappa B and release of interleukin-6 in bone infection. *Microbiology* **2013**, *159*, 147–154. [CrossRef] [PubMed]

44. Chiaretti, A.; Pulitanò, S.; Barone, G.; Ferrara, P.; Romano, V.; Capozzi, D.; Riccardi, R. IL-1 β and IL-6 Upregulation in Children with H1N1 Influenza Virus Infection. *Mediat. Inflamm.* **2013**, *2013*, 1–8. [CrossRef] [PubMed]

45. Chan, M.; Cheung, C.; Chui, W.; Tsao, S.; Nicholls, J.; Chan, Y.; Chan, R.; Long, H.; Poon, L.; Guan, Y.; et al. Proinflammatory cytokine responses induced by influenza A (H5N1) viruses in primary human alveolar and bronchial epithelial cells. *Respir. Res.* **2005**, *6*, 135. [CrossRef] [PubMed]

46. Yoneyama, M.; Fujita, T. RIG-I family RNA helicases: Cytoplasmic sensor for antiviral innate immunity. *Cytokine Growth Factor Rev.* **2007**, *18*, 545–551. [CrossRef]

47. Fournier, B.; Philpott, D.J. Recognition of Staphylococcus aureus by the Innate Immune System. *Clin. Microbiol. Rev.* **2005**, *18*, 521–540. [CrossRef]

48. Holler, N.; Zaru, R.; Micheau, O.; Thome, M.; Attinger, A.; Valitutti, S.; Bodmer, J.-L.; Schneider, P.; Seed, B.; Tschopp, J. Fas triggers an alternative, caspase-8–independent cell death pathway using the kinase RIP as effector molecule. *Nat. Immunol.* **2000**, *1*, 489–495. [CrossRef]

49. Berghe, T.V.; Linkermann, A.; Jouan-Lanhouet, S.; Walczak, H.; Vandenabeele, P. Regulated necrosis: The expanding network of non-apoptotic cell death pathways. *Nat. Rev. Mol. Cell Biol.* **2014**, *15*, 135–147. [CrossRef]

50. Kearney, C.J.; Martin, S.J. An Inflammatory Perspective on Necroptosis. *Mol. Cell* **2017**, *65*, 965–973. [CrossRef]

51. Wong Fok Lung, T.; Monk, I.R.; Acker, K.P.; Mu, A.; Wang, N.; Riquelme, S.A.; Pires, S.; Noguera, L.P.; Dach, F.; Gabryszewski, S.J.; et al. Staphylococcus aureus small colony variants impair host immunity by activating host cell glycolysis and inducing necroptosis. *Nat. Microbiol.* **2020**, *5*, 141–153. [CrossRef]

52. Tewari, M.; Quan, L.T.; O'Rourke, K.; Desnoyers, S.; Zeng, Z.; Beidler, D.R.; Poirier, G.G.; Salvesen, G.S.; Dixit, V.M. Yama/CPP32β, a mammalian homolog of CED-3, is a CrmA-inhibitable protease that cleaves the death substrate poly(ADP-ribose) polymerase. *Cell* **1995**, *81*, 801–809. [CrossRef]

53. Jacobsen, K.A. Mitteilungen über einen variablen Typhusstamm (Bacterium typhi mutabile), sowie über eine eigentümliche hemmende Wirkung des gewöhnlichen Agar, verursacht durch Autoklavierung. *Zentralbl. Bakteriol.* **1910**, *56*, 208–2016.

54. Tuchscherr, L.; Heitmann, V.; Hussain, M.; Viemann, D.; Roth, J.; von Eiff, C.; Peters, G.; Becker, K.; Löffler, B. Staphylococcus aureus Small-Colony Variants Are Adapted Phenotypes for Intracellular Persistence. *J. Infect. Dis.* **2010**, *202*, 1031–1040. [CrossRef] [PubMed]

55. Betáková, T.; Kostrábová, A.; Lachová, V.; Turianová, L. Cytokines Induced During Influenza Virus Infection. *Curr. Pharm. Des.* **2017**, *23*, 2616–2622. [CrossRef] [PubMed]

56. Demine, S.; Schiavo, A.A.; Marín-Cañas, S.; Marchetti, P.; Cnop, M.; Eizirik, D.L. Pro-inflammatory cytokines induce cell death, inflammatory responses, and endoplasmic reticulum stress in human iPSC-derived beta cells. *Stem Cell Res. Ther.* **2020**, *11*, 7. [CrossRef] [PubMed]

57. Vanlangenakker, N.; Bertrand, M.J.M.; Bogaert, P.; Vandenabeele, P.; Vanden Berghe, T. TNF-induced necroptosis in L929 cells is tightly regulated by multiple TNFR1 complex I and II members. *Cell Death Dis.* **2011**, *2*, e230. [CrossRef]

58. Riquelme, S.A.; Wong, T.F.L.; Prince, A. Pulmonary Pathogens Adapt to Immune Signaling Metabolites in the Airway. *Front. Immunol.* **2020**, *11*, 1–14. [CrossRef]

59. Surewaard, B.G.J.; de Haas, C.J.C.; Vervoort, F.; Rigby, K.M.; DeLeo, F.R.; Otto, M.; van Strijp, J.A.G.; Nijland, R. Staphylococcal alpha-phenol soluble modulins contribute to neutrophil lysis after phagocytosis. *Cell. Microbiol.* **2013**, *15*, 1427–1437. [CrossRef]

60. Yang, D.; Elner, S.G.; Bian, Z.-M.; Till, G.O.; Petty, H.R.; Elner, V.M. Pro-inflammatory cytokines increase reactive oxygen species through mitochondria and NADPH oxidase in cultured RPE cells. *Exp. Eye Res.* **2007**, *85*, 462–472. [CrossRef]

microorganisms

Communication

Mild Lactic Acid Stress Causes Strain-Dependent Reduction in SEC Protein Levels

Danai Etter [1,2], Céline Jenni [2], Taurai Tasara [1] and Sophia Johler [1,*]

[1] Institute for Food Safety and Hygiene, University of Zurich, 8057 Zurich, Switzerland; danai.etter@uzh.ch (D.E.); taurai.tasara@uzh.ch (T.T.)
[2] Laboratory of Food Microbiology, Institute for Food, Nutrition and Health (IFNH), ETH Zurich, 8092 Zurich, Switzerland; jennic@student.ethz.ch
* Correspondence: sophia.johler@uzh.ch

Abstract: Staphylococcal enterotoxin C (SEC) is a major cause of staphylococcal food poisoning in humans and plays a role in bovine mastitis. *Staphylococcus aureus* (*S. aureus*) benefits from a competitive growth advantage under stress conditions encountered in foods such as a low pH. Therefore, understanding the role of stressors such as lactic acid on SEC production is of pivotal relevance to food safety. However, stress-dependent cues and their effects on enterotoxin expression are still poorly understood. In this study, we used human and animal strains harboring different SEC variants in order to evaluate the influence of mild lactic acid stress (pH 6.0) on SEC expression both on transcriptional and translational level. Although only a modest decrease in *sec* mRNA levels was observed under lactic acid stress, protein levels showed a significant decrease in SEC levels for some strains. These findings indicate that post-transcriptional modifications can act in SEC expression under lactic acid stress.

Keywords: superantigen; mastitis; food intoxication; regulation; sec variants

Citation: Etter, D.; Jenni, C.; Tasara, T.; Johler, S. Mild Lactic Acid Stress Causes Strain-Dependent Reduction in SEC Protein Levels. *Microorganisms* **2021**, *9*, 1014. https://doi.org/10.3390/microorganisms9051014

Academic Editor: Rajan P. Adhikari

Received: 24 March 2021
Accepted: 5 May 2021
Published: 8 May 2021

Publisher's Note: MDPI stays neutral with regard to jurisdictional claims in published maps and institutional affiliations.

1. Introduction

Staphylococcus aureus (*S. aureus*) is of major relevance in food intoxications and infectious diseases of humans and animals [1,2]. *S. aureus* employs a plethora of virulence factors including secreted enterotoxins (SEs). They lead to an emetic response when ingested and act as superantigens [3,4]. The consumption of one or several preformed enterotoxins produced by *S. aureus* causes staphylococcal food poisoning (SFP). SFP symptoms include nausea, vomiting, and abdominal pain, followed by diarrhea [5]. It is one of the most common causes of foodborne intoxications worldwide [6]. The European Food Safety Authority (EFSA) reported 393 SFP outbreaks in 2014 and an increase in cases with 434 outbreaks in 2016 [7,8]. Moreover, in the United States, the Centers for Disease Control and Prevention (CDC) reported 17 SFP outbreaks and 566 cases in 2014. The most common contributing factors in these outbreaks were improper maintenance of the cold chain and inadequate food preparation practices, leading to the proliferation of pathogens and the concomitant production of enterotoxins [9].

Of the currently 25 known SEs, SEC is of particular interest since various, often host-specific variants have been reported [10–13]. SEC plays a crucial role in SFP and the development of atopic dermatitis [14]. SEC is also frequently found in milk and milk products [15–24] and represents a key driver of the inflammatory response in bovine mastitis in dairy cattle [25–28]. We recently provided a comprehensive review of SEC variants and their role in SFP, as well as their structure and properties [10].

To reduce the burden of *S. aureus* in the dairy value chain, lactic acid bacteria (LAB) have been used as an intervention in mastitis treatment and as starter cultures in cheese production. For instance, LAB were successfully used to alleviate mastitis symptoms in cows [29] and to inhibit mastitis-causing pathogens including *S. aureus* [30]. The use of

177

Weissella paramesenteroides GIR16L4 and *Lactobacillus rhamnosus* D1 as starter cultures was shown to decrease SEC expression in several *S. aureus* strains [31]. Another study showed decreased expression of *sec* and SEC in co-cultures with LAB compared with pure *S. aureus* cultures by up to 331-fold in TSB and milk [32]. Metabolites of LAB, in particular lactic acid, might present an additional metabolic burden to *S. aureus* and therefore interfere with toxin expression.

Lactic acid is a lipophilic weak organic acid that freely dissociates through the bacterial membrane. Inside the cell, lactic acid dissociates and thereby releases protons that acidify the cytoplasm. Additional energy is required to maintain internal pH, leading to adaptations in bacterial metabolic processes [33,34]. It can therefore alter toxin regulation by interfering with *S. aureus* regulatory systems such as the accessory gene regulator (*agr*) [35]. The *agr* regulon employs a multicomponent system that is activated by autoinducing peptides (AIP). Upon activation, RNAIII is transcribed, which deactivates the repressor of toxins (*rot*) [36]. The activity of *agr* may be influenced by other regulatory elements that react to changes in the microenvironment or external stressors.

External stressors that have been shown to influence SE production include NaCl [37], low pH [38], nitrite [39], and others [40,41]. Although previous studies demonstrated the substantial resilience of *S. aureus* against low pH, with growth being observed at pH 4 [42–44], pH stress can still influence toxin expression. Lactic acid was previously shown to affect SEA and SED production [45–47]. In the presence of LAB, *agrA*, *sarA*, and *sigB* are typically downregulated, while *rot* is upregulated [48]. However, it remains unclear how external stressors affect the complex regulatory network and whether the presence of lactic acid influences SEC expression. Therefore, we investigated the role of lactic acid in SEC production on mRNA and protein level.

2. Materials and Methods

2.1. Bacterial Strains, Growth Conditions, and Sample Collection for sec mRNA and SEC Protein Quantification

All *S. aureus* strains used in this study are listed in Table 1. The strains were grown in LB medium (non-stress control conditions) and in LB supplemented with lactic acid. Mild acidic stress conditions encountered in food were mimicked by adjusting the LB medium (BD, France) to pH 6.0 using ~0.6 mL 90% (*v/v*) lactic acid (Merck, Darmstadt, Germany). The medium was buffered using 19.52 g 2-(N-Morpholino) ethanesulfonic acid-hydrate (MES hydrate, Sigma-Aldrich, Switzerland) per 1000 mL LB. pH was monitored for 2 representative strains and remained unchanged over the course of the experiment (Supplementary Figure S2). All media were sterile filtered and stored at 4 °C.

Table 1. Overview of *S. aureus* strains used in this study including their SEC variants, origin, and clonal complex. [1] Medical Department of the German Federal Armed Forces, Germany. [2] Bavarian State Office of Health and Food Safety, Germany.

Strain	Protein Variant	Promotor Variant	Origin	Clonal Complex	Reference
BW10	SEC_2	sec_p v1	SFP	CC45	[1]
NB6	SEC_2	sec_p v1	SFP	CC45	[2]
SAI3	SEC_1	sec_p v3 (H-EMRSA-15)	Human infection	CC8	[49]
SAI48	SEC_2	sec_p v1 (79_S10)	Human infection	CC5	[49]
SAR1	SEC_{bovine}	sec_p v2	bovine mastitis milk	CC151	[50]
SAR38	SEC_{bovine}	sec_p v2	bovine mastitis milk	CC151	[50]
OV20	SEC_{ovine}	sec_p v4	ovine	CC133	[51]

Single colonies from each strain were transferred from 5% sheep blood agar to 5 mL LB broth and grown overnight (37 °C, 125 rpm). Overnight cultures were centrifuged (5000× *g*, 2 min) and washed twice with 0.85% NaCl, before resuspending the pellet in 0.85% saline solution. We adjusted 50 mL of medium (LB and LB + lactic acid) to an OD_{600} of 0.05 using the washed bacteria. The culture was incubated at 37 °C at 125 rpm and harvested after 4, 10, and 24 h during exponential, early stationary, and late stationary growth phase, respectively. Three independent biological replicates were collected.

Growth curves were evaluated by plating serial dilutions on plate count agar (Oxoid, Pratteln, Switzerland) as previously described [37] with minor modifications. Namely, the culture was adjusted to a final OD of 0.05 in 50 mL medium in 250 mL Erlenmeyer flasks. Two independent replicates were assessed.

For *sec* mRNA quantification, 1 mL of sample was added to 3 mL RNAprotect®®Tissue Reagent (Qiagen, Hilden, Germany) and processed according to manufacturer's instructions. The cell pellets were stored at −20 °C until further processing. For SEC protein quantification, 1 mL of sample was collected in low protein binding micro-centrifuge tubes (Thermo Scientific, Waltham, MA, USA) and stored at −20 °C until further processing.

2.2. RNA Extraction

RNA extraction was performed with the RNeasy mini Kit Plus (Qiagen, Hilden, Germany) as previously described [52]. All RNA samples were quantified using QuantusFluorometer (Promega, Dübendorf, Switzerland) instruments. Quality control was performed by the Agilent 2100 Bioanalyzer (Agilent Technologies, Waldbronn, Germany) instrument using Agilent RNA 6000 PicoReagents according to the manufacturer's instructions. Samples were included in the study if they met the inclusion criteria of RNA integrity number > 6. RNA integrity numbers ranged from 6.18.5.

2.3. Reverse-Transcription and Quantitative Real-Time PCR

All RNA samples were diluted to 40 ng/µL in RNase-free water. A total of 480 ng was converted to cDNA using the QuantiTect®®Reverse Transcription Kit (Qiagen, Germany) according to the manufacturer's instructions. A no-RT control and a negative control were included in every run. The final cDNA was diluted 1:10 with DNase-free water (Promega, Madison, WI, USA) and stored at −20 °C. The following primers sequences were used for real-time qPCR: forward 5′TAA CGG CAA TAC TTT TTG GT3′ and reverse primer 5′AGG TGG ACT TCT ATC TTC AC3′. A 5 µL template was added to 10 µL LightCycler®® 480 SYBR Green I Master, 2 µL of each primer (5 µM), and 1 µL nuclease-free water in LightCycler®®480 Mulitwell Plate 96white (Roche, Basel, Germany). The plate was centrifuged for 2 min at 1500× *g*. Quantification was performed on the Lightcycler®®96 Instrument (Roche, Basel, Switzerland) as previously described [52]. The relative expression of the target gene *sec* was normalized using the housekeeping genes *rho* and *rplD* [52]. Ct values were determined using the Lightcycler®®Software v. 1.1.0.1320 (Roche). The influence of lactic acid stress on *sec* expression in each strain is expressed as Δct values and relative expression ($2^{-\Delta\Delta ct}$). The following formula was used to calculate expression values: $2^{-(\text{ref control-sec control})-(\text{ref lactic acid-sec lactic acid})}$. Statistical analysis was performed with RStudio 1.3.1093 and GraphPad Prism 9.0.0. For RNA analysis a mixed-effect linear model was fitted on the fold change, with a full three-way interaction between reference gene, strain, and time effects. Fold change was $\log_{10}$-transformed to ensure normal distribution. To determine whether individual mRNA levels were increased or decreased (indicated by a fold change significantly larger than 1), lsmeans was used to perform a two-sided effect test, with Holm–Bonferroni-corrected *p*-values. The results were regarded as significant if $p < 0.05$.

2.4. Protein Quantification

An enzyme-linked immunosorbent assay (ELISA) was developed to quantify the effect of mild lactic acid stress on SEC protein levels. The protocol was based on [53] with some modifications according to [41]. Sheep Anti-SEC IgG (Toxin Technology Inc., Sarasota, FL, USA) was used to measure SEC concentrations. For quantification, a standard curve was obtained using SEC_2 (Toxin Technology, Inc., USA). Absorbance was measured at 405 nm in a Synergy HT plate reader (BioTek, Sursee, Switzerland). Absorbance values were plotted against toxin concentrations, and values were determined from linear regression in Excel (version 16.44). ELISA measurements were performed in duplicates. Statistical analysis was performed with RStudio 1.3.1093 and GraphPad Prism 9.0.0. Protein data were analyzed via two-way ANOVA and post hoc Tukey's multiple comparisons. Results were regarded as significant if $p < 0.05$.

3. Results

3.1. Effect of Mild Lactic Acid Stress (pH 6.0) on Bacterial Growth and sec mRNA Levels

The bacterial growth of the seven *S. aureus* strains was compared in LB and LB supplemented with lactic acid by plate counting. The growth behavior was similar for all strains under control and stress conditions. Growth was not impaired by lactic acid in any of the investigated strains (Supplementary Figures S1 and S2).

Under lactic acid stress, we observed a trend toward decreased *sec* expression (Figure 1) compared with control conditions. However, the reduction in *sec* expression was strain-dependent, being only significant in strains BW10 (10 and 24 h), NB6 (24 h), SAR1 (4 h), SAR38 (4 and 10 h), and OV20 (10 h).

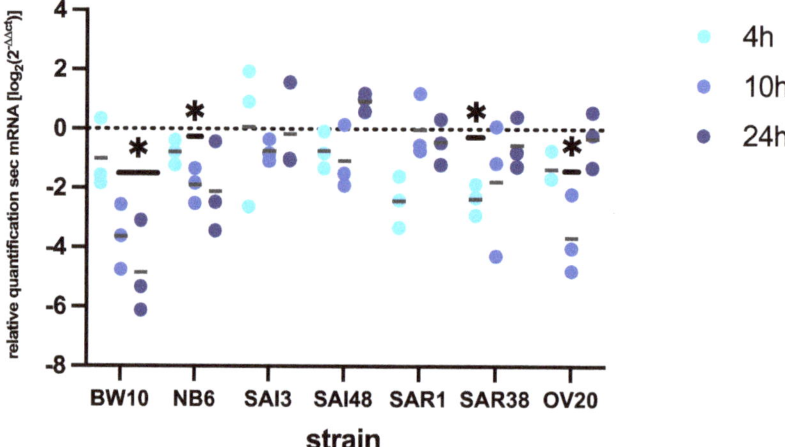

Figure 1. Effect of lactic acid stress on *sec* mRNA levels in *Staphylococcus aureus* strains BW10, NB6, SAI3, SAI48, SAR1, SAR38, and OV20 in exponential (4 h), early stationary (10 h), and late stationary (24 h) phase. *sec* mRNA levels are expressed as relative quantification values. *sec* mRNA expression was normalized to reference genes *rho* and *rplD* [52]. Replicates are shown as single data points; horizontal grey lines indicate means. Timepoints are signified by fill color (4 h, light blue; 10 h, medium blue; 24 h, dark blue). Significant differences in *sec* mRNA levels in LB compared with LB + lactic acid are marked by asterisks ($p < 0.05$).

3.2. SEC Protein Levels under Lactic Acid Stress

SEC concentrations under lactic acid stress and control conditions were assessed by ELISA at 4, 10, and 24 h (Figure 2, Table 2). Under control conditions, two different expression levels were observed. BW10 and SAI48 were classified as SEC over-producers (>1000 ng/mL) with concentrations ranging from 3410 ng/mL to 9867 ng/mL after 24 h, respectively. NB6, SAI3, SAR1, SAR38, and OV20 were classified as low to moderate level SEC producers (<1000 ng/mL) with concentrations from 54.5 to 344.9 ng/mL after 24 h (Table 2). Expression levels were lowest after 4 h and highest after 24 h for all strains.

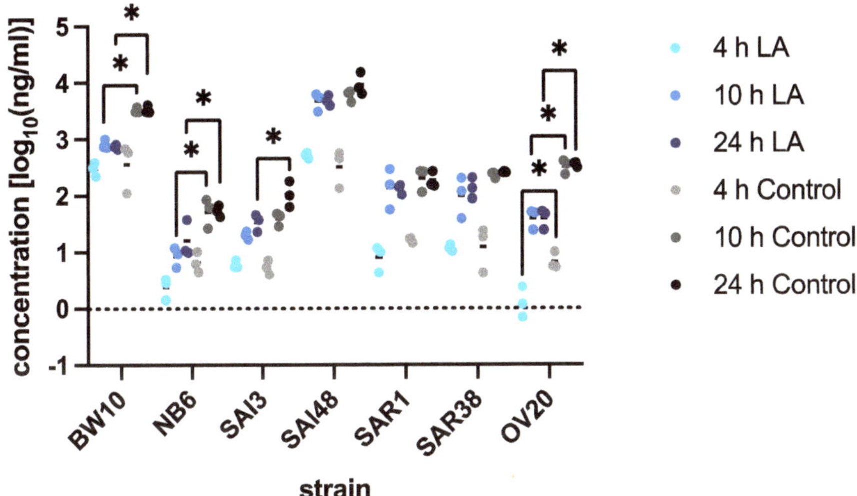

Figure 2. SEC concentration in $\log_{10}$ ng/mL under lactic acid stress (pH 6.0) compared with control conditions in *Staphylococcus aureus* strains BW10, NB6, SAI3, SAI48, SAR1, SAR38, and OV20 in exponential (4 h), early stationary (10 h), and late stationary (24 h) phase. SEC levels under control conditions are shown in grey; pH stress levels are shown in blue. Replicates are shown as single data points; horizontal lines indicate means. Darkening fill colors indicate progressing time points. Statistically significant differences between conditions are indicated by lines with asterisks (* $p < 0.05$).

SEC concentrations under lactic acid stress were generally lower than those under non-stress control conditions, with the exception of SAI48, SAR1, and SAR38 (Figure 2). BW10 and SAI48 again showed the highest expression levels with 740 ± 69 and 4887 ± 1027 ng/mL at the late exponential phase, respectively. The low-to-moderate SEC producers NB6, SAI3, SAR1, SAR38, and OV20 ranged from 1 ng/mL (OV20) to 142 ng/mL (OV20) over all time points. SEC production in the human infection isolate SAI48 was the least impaired by lactic acid stress (-29%) while the ovine mastitis isolate OV20 was affected the most (-255%) (Table 2).

Table 2. Effect of pH stress on SEC expression. Absolute values in ng/mL including standard deviation. Effect is shown as a percentage difference under lactic acid stress (pH 6.0) compared with non-stress control conditions.

Strain	SEC Produced under pH Stress (ng/mL)									Effect of pH Stress (%)			
	4 h			10 h			24 h			4 h	10 h	24 h	Sum
BW10	307	±	81	832	±	161	740	±	69	−33	−75	−78	−186
NB6	3	±	1	9	±	3	20	±	16	−63	−85	−64	−212
SAI3	6	±	1	20	±	3	35	±	11	10	−50	−69	−109
SAI48	523	±	63	4968	±	1641	4887	±	1027	38	−17	−50	−29
SAR1	9	±	4	169	±	119	132	±	25	−46	−21	−30	−97
SAR38	12	±	2	120	±	82	143	±	59	−22	−47	−44	−113
OV20	1	±	1	41	±	15	41	±	14	−80	−87	−88	−255
SEC produced under control conditions (ng/mL)													
BW10	462	±	307	3325	±	360	3410	±	562	-	-	-	-
NB6	7	±	3	58	±	29	54	±	12	-	-	-	-
SAI3	6	±	2	40	±	10	114	±	59	-	-	-	-
SAI48	380	±	220	5988	±	1317	9868	±	4688	-	-	-	-
SAR1	16	±	2	213	±	87	189	±	65	-	-	-	-
SAR38	15	±	10	228	±	27	254	±	7	-	-	-	-
OV20	7	±	3	328	±	87	345	±	34	-	-	-	-

4. Discussion

Since *S. aureus* possesses a competitive growth advantage in many food matrices, it is crucial to identify compounds that interfere with SE production. Here, we used lactic acid and pH 6.0 to mimic conditions comparable to food matrices such as ham, cheese, and other fermented products [47]. LB medium was used to ensure reproducibility and to allow observation of the effect of lactic acid as an individual constituent. Experiments were performed at 37 °C to provide optimal growth conditions for *S. aureus*.

In the investigated strains, lactic acid stress resulted in a trend toward lower *sec* transcription, although the results were not significant for all strains. Significant reductions in mRNA levels were observed for strains isolated from food and bovine mastitis, whereas human infection isolates did not alter *sec* transcription levels. The influence of a complex food matrix containing lactic acid such as milk was demonstrated to reduce *sec* expression, especially in late stationary phase after 48 h [54]. Other SEs, for instance *sed*, were not significantly altered under lactic acid stress [47]. How lactic acid influences toxin transcription likely depends on the genetic background of a strain. We could, however, not observe any correlation between transcription levels and factors such as clonal complex or the SEC toxin variant of the respective strain. Possibly, a larger strain set may provide further insights.

SEC protein data only partly correlated with the *sec* transcriptional patterns. Whereas BW10 showed a significant reduction in mRNA and protein levels after 10 and 24 h, SAI3 did not display any reduced transcriptional activity, but exhibited significantly reduced SEC concentrations. Especially under certain environmental stress situations such as NaCl, sorbic acid, or complex food environments, mRNA levels do not always reflect protein levels, indicating that post-transcriptional modification might be at play [55–58]. Overall, lactic acid stress led to decreased levels of SEC in all strains. Since growth under lactic acid stress was similar to control conditions, factors independent of growth rates must be at play. Again, we did not observe any correlation between toxin production and clonal complex, source, or toxin variant. Still, SEC overproducers BW10 and SAI48 were also the highest SEC producers under lactic acid stress. Interestingly, previous studies reported substantially lower toxin concentrations of 1–70 ng/mL in milk and even under optimal growth conditions [55]. For SEA, an ingested amount of 60 ng was sufficient to reach SFP attack rates of almost 100%, highlighting the threat of overproducer strains such as BW10 and SAI48 [59]. Another study demonstrated even more pronounced reductions in SEA, SEB, SEC, and SED caused by lactic acid. This pronounced reduction could be expected, since higher concentrations (pH < 5) than in the present study were used [60]. In contrast, mild lactic acid stress was also reported to increase the formation of SEA [45]. It was suggested that *S. aureus* strains from human and food sources produce higher SEC levels in contrast with strains from animal sources [11]. The SEC overproducers in our study, BW10 and SAI48, originated from cases of SFP and human infection, respectively, and showed high SEC concentrations under control conditions and lactic acid stress. However, the moderate SEC producers SAR1, SAR38, and OV20 originating from bovine and ovine mastitis milk showed higher SEC concentrations than NB6 and SAI3, which originated from cases of SFP and human infection, respectively. This underlines the importance of investigating multiple strains for *S. aureus* enterotoxin analysis as postulated by previous studies [39,61–63]. The observed reduction in toxin expression may be magnified at lower temperatures as found in food environments.

Several SEs including SEC are regulated by the quorum-sensing system *agr*. The activity of *agr* may be influenced by other regulatory elements that react to changes in the microenvironment or external stressors. Such additional regulators may include, but are not limited to, *sigB*, *sarA*, *saeRS*, *srrAB*, *arlRS*, and *mgrA* [64,65]. pH stress was shown to affect regulatory elements in *S. aureus* such as *rot*, *agr*, and *sarA* [66]. In the presence of LAB, *agrA*, *sarA*, and *sigB* are typically downregulated, while *rot* is upregulated [48]. Lactic acid can influence toxin expression by interacting with DNA, enzymes, or structural proteins (Figure 3). Since the current data suggest a stronger decrease in SEC protein levels compared with mRNA, post-transcriptional events may be involved. The shift in internal pH may influence mRNA degradation susceptibility or interfere with translational events. For example, elongation factors were shown to be influenced by acid stress [67]. An overall decreased enzyme activity in the cytoplasm due to lower pH may impact SE handling after translation. In addition, folding and transport of SEs may be impaired by a lower pH, leading to less efficient excretion [68]. Which regulatory elements are affected remains unclear at this time.

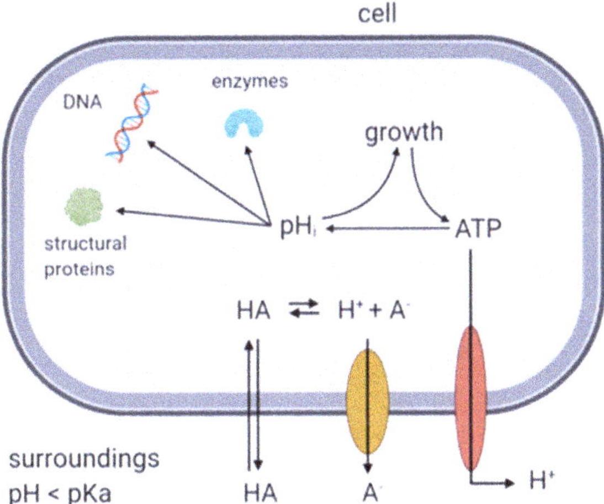

Figure 3. Mechanism of weak organic acids such as lactic acid on bacterial cells. The undissociated form of organic acids (HA) can cross cell membranes when the pH of the surroundings is lower than that of the cellular cytoplasm. Inside the cell, HA can dissociate and acidify the cytoplasm. Acidic pH damages or modifies internal structures such as enzymes, structural proteins, or DNA. In order to control internal pH, energy is required for active export of protons. The figure was adapted from [34] and created in BioRender.com.

In conclusion, our study demonstrated that lactic acid slightly decreases *sec* transcription and has a more pronounced effect on SEC protein levels. It can therefore be a useful tool in minimizing SEC synthesis during food production and preservation. Since transcription patterns and SEC concentrations are highly variable, it is of utmost importance to investigate several strains with different genomic backgrounds and isolated from different sources. We did not find any correlation between the observed data and any strain-specific properties such as toxin variant or clonal complex. Therefore, further research is needed to determine biomarkers associated with the toxicity of a strain. Future studies should also investigate the mechanistic action of lactic acid on the reduction in SEs to provide further insights into the regulation of SE production. Our research highlights the importance of food composition in mitigating SFP. Compounds such as lactic acid may be used as natural preservatives to minimize toxin production and alleviate the burden of staphylococcal intoxications.

Supplementary Materials: The following are available online at https://www.mdpi.com/article/10.3390/microorganisms9051014/s1: Figure S1: Growth curves under lactic acid stress and control conditions. Figure S2: pH values for 2 representative strains.

Author Contributions: Conceptualization, S.J. and D.E.; methodology, S.J., T.T. and D.E.; formal analysis, D.E. and C.J.; investigation, D.E. and C.J.; resources, S.J.; data curation, D.E. and C.J.; writing—original draft preparation, D.E. and C.J.; writing—review and editing, D.E., S.J. and T.T.; visualization, D.E. and C.J.; supervision, D.E. and S.J. All authors have read and agreed to the published version of the manuscript.

Funding: This research received no external funding.

Institutional Review Board Statement: Not applicable.

Informed Consent Statement: Not applicable.

Data Availability Statement: Not applicable.

Conflicts of Interest: The authors declare no conflict of interest.

References

1. Fetsch, A.; Johler, S. *Staphylococcus aureus* as a foodborne pathogen. *Curr. Clin. Microbiol.* **2018**, *9*, 1–8. [CrossRef]
2. Food-Borne Disease Burden Epidemiology Reference Group. *WHO Estimates of the Global Burden of Foodborne Diseases*; WHO Press: Geneva, Switzerland, 2015.
3. Langley, R.J.; Fraser, J.D.; Proft, T. *Bacterial Superantigens and Superantigen-Like Toxins*; Elsevier Ltd.: Amsterdam, The Netherlands, 2015; ISBN 9780128005897.
4. Johnson, H.M.; Russell, J.K.; Pontzer, C.H. Staphylococcal enterotoxin superantigens. *Proc. Soc. Exp. Biol. Med.* **1991**, *198*, 765–771. [CrossRef] [PubMed]
5. Balaban, N.; Rasooly, A. Staphylococcal enterotoxins. *Int. J. Food Microbiol.* **2000**, *61*, 1–10. [CrossRef]
6. Nia, Y.; Mutel, I.; Assere, A.; Lombard, B.; Auvray, F.; Hennekinne, J.A. Review over a 3-year period of european union proficiency tests for detection of staphylococcal enterotoxins in food matrices. *Toxins (Basel)* **2016**, *8*, 107. [CrossRef] [PubMed]
7. EFSA The Euroean Union summary report on trends and sources of zoonoses, zoonotic agents and food-borne outbreaks in 2015. *EFSA J.* **2016**, *13*. [CrossRef]
8. Food, E.; Authority, S.; Centre, E. The European Union summary report on trends and sources of zoonoses, zoonotic agents and food-borne outbreaks in 2014. *EFSA J.* **2015**, *13*. [CrossRef]
9. Gould, L.H.; Nisler, A.L.; Herman, K.M.; Cole, D.J.; Williams, I.T.; Mahon, B.E.; Griffin, P.M.; Hall, A.J. Surveillance for foodborne disease outbreaks—United States, 2008. *Mmwr. Morb. Mortal. Wkly. Rep.* **2011**, *60*, 1197–1197202.
10. Etter, D.; Schelin, J.; Schuppler, M.; Johler, S. Staphylococcal Enterotoxin C—An Update on SEC Variants, Their Structure and Properties, and Their Role in Foodborne Intoxications. *Toxins* **2020**, *12*, 584. [CrossRef]
11. Marr, J.C.; Lyon, J.D.; Roberson, J.R.; Lupher, M.; Davis, W.C.; Bohach, G.A. Characterization of novel type C staphylococcal enterotoxins: Biological and evolutionary implications. *Infect. Immun.* **1993**, *61*, 4254–4262. [CrossRef] [PubMed]
12. Viana, D.; Blanco, J.; Tormo-Más, M.Á.; Selva, L.; Guinane, C.M.; Baselga, R.; Corpa, J.M.; Lasa, Í.; Novick, R.P.; Fitzgerald, J.R.; et al. Adaptation of *Staphylococcus aureus* to ruminant and equine hosts involves SaPI-carried variants of von Willebrand factor-binding protein. *Mol. Microbiol.* **2010**, *77*, 1583–1594. [CrossRef]
13. Hajek, V. Identification of enterotoxigenic *staphylococci* from sheep and sheep cheese. *Appl. Environ. Microbiol.* **1978**, *35*, 264–268. [CrossRef] [PubMed]
14. Bunikowski, R.; Mielke, M.E.A.; Skarabis, H.; Worm, M.; Anagnostopoulos, I.; Kolde, G.; Wahn, U.; Renz, H. Evidence for a disease-promoting effect of *Staphylococcus aureus*-derived exotoxins in atopic dermatitis. *J. Allergy Clin. Immunol.* **2000**, *105*, 814–819. [CrossRef]
15. Benkerroum, N. Staphylococcal enterotoxins and enterotoxin-like toxins with special reference to dairy products: An overview. *Crit. Rev. Food Sci. Nutr.* **2018**, *58*, 1943–1970. [CrossRef]
16. Aragon-Alegro, L.C.; Konta, E.M.; Suzuki, K.; Silva, M.G.; Júnior, A.F.; Rall, R.; Rall, V.L.M. Occurrence of coagulase-positive *Staphylococcus* in various food products commercialized in Botucatu, SP, Brazil and detection of toxins from food and isolated strains. *Food Control* **2007**, *18*, 630–634. [CrossRef]
17. Lindqvist, R.; Sylve, S. Quantitative microbial risk assessment exemplified by Staphylococcus aureus in unripened cheese made from raw milk. *Int. J. Food Microbiol.* **2002**, *78*, 155–170. [CrossRef]
18. Scherrer, D.; Corti, S.; Muehlherr, J.E.; Zweifel, C.; Stephan, R. Phenotypic and genotypic characteristics of *Staphylococcus aureus* isolates from raw bulk-tank milk samples of goats and sheep. *Vet. Microbiol.* **2004**, *101*, 101–107. [CrossRef] [PubMed]
19. Rosengren, Å.; Fabricius, A.; Guss, B.; Sylvén, S.; Lindqvist, R. Occurrence of foodborne pathogens and characterization of *Staphylococcus aureus* in cheese produced on farm-dairies. *Int. J. Food Microbiol.* **2010**, *144*, 263–269. [CrossRef] [PubMed]
20. Wikler, M.; Cockerill, F.; Craig, W.; Dudley, M.; Eliopoulos, G.; Hecht, D. *Performance Standards for Antimicrobial Susceptibility Testing*; Seventeenth Informational Supplement; Clinical and Laboratory Standards Institute: Wayne, PA, USA, 2007; Volume 27, ISBN 1562386255.
21. Moser, A.; Stephan, R.; Corti, S.; Johler, S. Comparison of genomic and antimicrobial resistance features of latex agglutination test-positive and latex agglutination test-negative *Staphylococcus aureus* isolates causing bovine mastitis. *J. Dairy Sci.* **2013**, *96*, 329–334. [CrossRef] [PubMed]
22. Baumgartner, A.; Niederhauser, I.; Johler, S. Virulence and resistance gene profiles of staphylococcus aureus strains isolated from ready-to-eat foods. *J. Food Prot.* **2014**, *77*, 1232–1236. [CrossRef] [PubMed]
23. Stalder, U.; Stephan, R.; Corti, S.; Bludau, M.; Maeschli, A.; Klocke, P.; Johler, S. Short communication: Staphylococcus aureus isolated from colostrum of dairy heifers represent a closely related group exhibiting highly homogeneous genomic and antimicrobial resistance features. *J. Dairy Sci.* **2014**, *97*, 4997–5000. [CrossRef]
24. EFSA. The European Union One Health 2018 Zoonoses Report. *Sci. Rep.* **2019**, *17*. [CrossRef]

25. Fang, R.; Cui, J.; Cui, T.; Guo, H.; Ono, H.; Park, C.-H.; Okamura, M.; Nakane, A.; Hu, D.-L. Staphylococcal Enterotoxin C Is an Important Virulence Factor for Mastitis. *Toxins* **2019**, *11*, 141. [CrossRef]

26. Niskanen, A.; Koranen, L.; Roine, K. Staphylococcal enterotoxin and thermonuclease production during induced bovine mastitis and the clinical reaction of enterotoxin in udders. *Infect. Immun.* **1978**, *19*, 493–498. [CrossRef] [PubMed]

27. Fitzgerald, J.R.; Monday, S.R.; Foster, T.J.; Bohach, G.A.; Hartigan, P.J.; Meaney, W.J.; Smyth, C.J. Characterization of a putative pathogenicity island from bovine *Staphylococcus aureus* encoding multiple superantigens. *J. Bacteriol.* **2001**, *183*, 63–70. [CrossRef] [PubMed]

28. Orden, J.A.; Goyache, J.; Hernández, J.; Doménech, A.; Suárez, G.; Gómez-Lucia, E. Detection of enterotoxins and TSST-1 secreted by *Staphylococcus aureus* isolated from ruminant mastitis. Comparison of ELISA and immunoblot. *J. Appl. Bacteriol.* **1992**, *72*, 486–489. [CrossRef] [PubMed]

29. Gao, J.; Liu, Y.C.; Wang, Y.; Li, H.; Wang, X.M.; Wu, Y.; Zhang, D.R.; Gao, S.; Qi, Z.L. Impact of yeast and lactic acid bacteria on mastitis and milk microbiota composition of dairy cows. *AMB Express* **2020**, *10*. [CrossRef]

30. Diepers, A.C.; Krömker, V.; Zinke, C.; Wente, N.; Pan, L.; Paulsen, K.; Paduch, J.H. In vitro ability of lactic acid bacteria to inhibit mastitis-causing pathogens. *Sustain. Chem. Pharm.* **2017**, *5*, 84–92. [CrossRef]

31. Silva, G.O.; Castro, R.D.; Oliveira, L.G.; Sant'Anna, F.M.; Barbosa, C.D.; Sandes, S.H.C.; Silva, R.S.; Resende, M.F.S.; Lana, A.M.Q.; Nunes, A.C.; et al. Viability of *Staphylococcus aureus* and expression of its toxins (SEC and TSST-1) in cheeses using *Lactobacillus rhamnosus* D1 or *Weissella paramesenteroides* GIR16L4 or both as starter cultures. *J. Dairy Sci.* **2020**, *103*, 4100–4108. [CrossRef]

32. Zdenkova, K.; Alibayov, B.; Karamonova, L.; Purkrtova, S.; Karpiskova, R.; Demnerova, K. Transcriptomic and metabolic responses of Staphylococcus aureus in mixed culture with Lactobacillus plantarum, Streptococcus thermophilus and Enterococcus durans in milk. *J. Ind. Microbiol. Biotechnol.* **2016**, *43*, 1237–1247. [CrossRef]

33. Guan, N.; Liu, L. Microbial response to acid stress: Mechanisms and applications. *Appl. Microbiol. Biotechnol.* **2020**, *104*, 51–65. [CrossRef]

34. Mani-López, E.; García, H.S.; López-Malo, A. Organic acids as antimicrobials to control Salmonella in meat and poultry products. *Food Res. Int.* **2012**, *45*, 713–721. [CrossRef]

35. Le Marc, Y.; Valík, L.; Medved'ová, A. Modelling the effect of the starter culture on the growth of *Staphylococcus aureus* in milk. *Int. J. Food Microbiol.* **2009**, *129*, 306–311. [CrossRef] [PubMed]

36. Novick, R.P.; Geisinger, E. Quorum Sensing in Staphylococci. *Annu. Rev. Genet.* **2008**, *42*, 541–564. [CrossRef] [PubMed]

37. Sihto, H.-M.; Tasara, T.; Stephan, R.; Johler, S. Temporal expression of the staphylococcal enterotoxin D gene under NaCl stress conditions. *FEMS Microbiol. Lett.* **2015**, *362*. [CrossRef] [PubMed]

38. Genigeorgis, C.; Foda, M.S.; Mantis, A.; Sadler, W.W. Effect of sodium chloride and pH on enterotoxin C production. *Appl. Microbiol.* **1971**, *21*, 862–866. [CrossRef] [PubMed]

39. Sihto, H.-M.; Budi Susilo, Y.; Tasara, T.; Rådström, P.; Stephan, R.; Schelin, J.; Johler, S. Effect of sodium nitrite and regulatory mutations Δagr, ΔsarA, and ΔsigB on the mRNA and protein levels of staphylococcal enterotoxin D. *Food Control* **2016**, *65*, 37–45. [CrossRef]

40. Schelin, J.; Susilo, Y.; Johler, S. Expression of Staphylococcal Enterotoxins under Stress Encountered during Food Production and Preservation. *Toxins (Basel)* **2017**, *9*, 401. [CrossRef] [PubMed]

41. Wallin-Carlquist, N.; Cao, R.; Márta, D.; Da Silva, A.S.; Schelin, J.; Rådström, P. Acetic acid increases the phage-encoded enterotoxin A expression in *Staphylococcus aureus*. *BMC Microbiol.* **2010**, *10*. [CrossRef]

42. Barber, L.E.; Deibel, R.H. Effect of pH and oxygen tension on staphylococcal growth and enterotoxin formation in fermented sausage. *Appl. Microbiol.* **1972**, *24*, 891–898. [CrossRef]

43. Stewart, C.M. Staphylococcus aureus and Staphylococcal Enterotoxins. In *Foodborne Microorganisms of Public Health Significance*; Hocking, A.D., Ed.; Australian Institute of Food Science and Technology: North Ryde, Australia, 2003; pp. 359–380.

44. Rode, T.M.; Møretrø, T.; Langsrud, S.; Langsrud, Ø.; Vogt, G.; Holck, A. Responses of *Staphylococcus aureus* exposed to HCl and organic acid stress. *Can. J. Microbiol.* **2010**, *56*, 777–792. [CrossRef]

45. Rosengren, Å.; Lindblad, M.; Lindqvist, R. The effect of undissociated lactic acid on Staphylococcus aureus growth and enterotoxin A production. *Int. J. Food Microbiol.* **2013**, *162*, 159–166. [CrossRef]

46. Elahi, S.; Fujikawa, H. Effects of Lactic Acid and Salt on Enterotoxin A Production and Growth of Staphylococcus aureus. *J. Food Sci.* **2019**, *84*, 3233–3240. [CrossRef]

47. Sihto, H.-M.; Tasara, T.; Stephan, R.; Johler, S. Growth behavior and temporal enterotoxin D expression of *Staphylococcus aureus* strains under glucose and lactic acid stress. *Food Control* **2016**, *62*, 69–73. [CrossRef]

48. Cretenet, M.; Nouaille, S.; Thouin, J.; Rault, L.; Stenz, L.; François, P.; Hennekinne, J.A.; Piot, M.; Maillard, M.B.; Fauquant, J.; et al. *Staphylococcus aureus* virulence and metabolism are dramatically affected by *Lactococcus lactis* in cheese matrix. *Environ. Microbiol. Rep.* **2011**, *3*, 340–351. [CrossRef]

49. Wattinger, L.; Stephan, R.; Layer, F.; Johler, S. Comparison of *Staphylococcus aureus* isolates associated with food intoxication with isolates from human nasal carriers and human infections. *Eur. J. Clin. Microbiol. Infect. Dis.* **2012**, *31*, 455–464. [CrossRef] [PubMed]

50. Johler, S.; Layer, F.; Stephan, R. Comparison of virulence and antibiotic resistance genes of food poisoning outbreak isolates of *Staphylococcus aureus* with isolates obtained from bovine mastitis milk and pig carcasses. *J. Food Prot.* **2011**, *74*, 1852–1859. [CrossRef] [PubMed]

51. Guinane, C.M.; Zakour, N.L.B.; Tormo-Mas, M.A.; Weinert, L.A.; Lowder, B.V.; Cartwright, R.A.; Smyth, D.S.; Smyth, C.J.; Lindsay, J.A.; Gould, K.A.; et al. Evolutionary genomics of *Staphylococcus aureus* reveals insights into the origin and molecular basis of ruminant host adaptation. *Genome Biol. Evol.* **2010**, *2*, 454–466. [CrossRef]
52. Sihto, H.M.; Tasara, T.; Stephan, R.; Johler, S. Validation of reference genes for normalization of qPCR mRNA expression levels in *Staphylococcus aureus* exposed to osmotic and lactic acid stress conditions encountered during food production and preservation. *FEMS Microbiol. Lett.* **2014**, *356*, 134–140. [CrossRef]
53. Poli, M.A.; Rivera, V.R.; Neal, D. Sensitive and specific colorimetric ELISAs for *Staphylococcus aureus* enterotoxins A and B in urine and buffer. *Toxicon* **2002**, *40*, 1723–1726. [CrossRef]
54. Valihrach, L.; Alibayov, B.; Zdenkova, K.; Demnerova, K. Expression and production of staphylococcal enterotoxin C is substantially reduced in milk. *Food Microbiol.* **2014**, *44*, 54–59. [CrossRef] [PubMed]
55. Valihrach, L.; Alibayov, B.; Demnerova, K. Production of staphylococcal enterotoxin C in milk. *Int. Dairy J.* **2013**, *30*, 103–107. [CrossRef]
56. Zeaki, N.; Cao, R.; Skandamis, P.N.; Rådström, P.; Schelin, J. Assessment of high and low enterotoxin A producing *Staphylococcus aureus* strains on pork sausage. *Int. J. Food Microbiol.* **2014**, *182–183*, 44–50. [CrossRef]
57. Zeaki, N.; Rådström, P.; Schelin, J. Evaluation of Potential Effects of NaCl and Sorbic Acid on Staphylococcal Enterotoxin A Formation. *Microorganisms* **2015**, *3*, 551–566. [CrossRef]
58. Birbaum-Sihto, H.M. *Stress Response in* Staphylococcus aureus: *Regulatory Mechanisms Influencing Enterotoxin Gene Expression under Stress Relevant to Food Production and Preservation*; University of Zurich: Zurich, Switzerland, 2016.
59. Guillier, L.; Bergis, H.; Guillier, F.; Noel, V.; Auvray, F.; Hennekinne, J.-A. Dose-response Modelling of Staphylococcal Enterotoxins Using Outbreak Data. *Procedia Food Sci.* **2016**, *7*, 129–132. [CrossRef]
60. Domenech, A.; Hernandez, F.J.; Orden, J.A.; Goyache, J.; Lopez, B.; Suarez, G.; Gomez-Lucia, E. Effect of six organic acids on staphylococcal growth and enterotoxin production. *Z. Lebensm. Unters.* **1992**, *194*, 124–128. [CrossRef]
61. Blevins, J.S.; Beenken, K.E.; Elasri, M.O.; Hurlburt, B.K.; Smeltzer, M.S. Strain-dependent differences in the regulatory roles of *sarA* and *agr* in *Staphylococcus aureus*. *Infect. Immun.* **2002**, *70*, 470–480. [CrossRef] [PubMed]
62. Cassat, J.; Dunman, P.M.; Murphy, E.; Projan, S.J.; Beenken, K.E.; Palm, K.J.; Yang, S.J.; Rice, K.C.; Bayles, K.W.; Smeltzer, M.S. Transcriptional profiling of a *Staphylococcus aureus* clinical isolate and its isogenic *agr* and *sarA* mutants reveals global differences in comparison to the laboratory strain RN6390. *Microbiology* **2006**, *152*, 3075–3090. [CrossRef] [PubMed]
63. Nagarajan, V.; Smeltzer, M.S.; Elasri, M.O. Genome-scale transcriptional profiling in *Staphylococcus aureus*: Bringing order out of chaos. *FEMS Microbiol. Lett.* **2009**, *295*, 204–210. [CrossRef] [PubMed]
64. Hsieh, H.Y.; Ching, W.T.; Stewart, G.C. Regulation of rot expression in *Staphylococcus aureus*. *J. Bacteriol.* **2008**, *190*, 546–554. [CrossRef]
65. Jenul, C.; Horswill, A.R. Regulation of *Staphylococcus aureus* Virulence. *Microbiol. Spectr.* **2019**, *7*, 1–21. [CrossRef]
66. Weinrick, B.; Dunman, P.M.; McAleese, F.; Murphy, E.; Projan, S.J.; Fang, Y.; Novick, R.P. Effect of Mild Acid on Gene Expression in *Staphylococcus aureus*. *J. Bacteriol.* **2004**, *186*, 8407–8423. [CrossRef] [PubMed]
67. Starosta, A.L.; Lassak, J.; Jung, K.; Wilson, D.N. The bacterial translation stress response. *FEMS Microbiol. Rev.* **2014**, *38*, 1172–1201. [CrossRef] [PubMed]
68. Lund, P.; Tramonti, A.; De Biase, D. Coping with low pH: Molecular strategies in neutralophilic bacteria. *FEMS Microbiol. Rev.* **2014**, *38*, 1091–1125. [CrossRef] [PubMed]

microorganisms

Article

Cell Membrane Adaptations Mediate β-Lactam-Induced Resensitization of Daptomycin-Resistant (DAP-R) *Staphylococcus aureus* In Vitro

Nagendra N. Mishra [1,2,*], Arnold S. Bayer [1,2], Sarah L. Baines [3], Ashleigh S. Hayes [3], Benjamin P. Howden [3], Christian K. Lapitan [1], Cassandra Lew [4] and Warren E. Rose [4]

[1] Division of Infectious Diseases, The Lundquist Institute at Harbor-UCLA Medical Center, Torrance, CA 90502, USA; abayer@lundquist.org (A.S.B.); christian.lapitan@lundquist.org (C.K.L.)
[2] David Geffen School of Medicine, University of California (UCLA), Los Angeles, CA 90024, USA
[3] Doherty Applied Microbial Genomics, Department of Microbiology and Immunology, The University of Melbourne at the Peter Doherty Institute for Infection and Immunity, Melbourne, VIC 3004, Australia; bainess@unimelb.edu.au (S.L.B.); ashleigh.hayes@unimelb.edu.au (A.S.H.); bhowden@unimelb.edu.au (B.P.H.)
[4] School of Pharmacy, University of Wisconsin-Madison, Madison, WI 53705, USA; clew@wisc.edu (C.L.); warren.rose@wisc.edu (W.E.R.)
*   Correspondence: nmishra@lundquist.org or nnmishra@ucla.edu; Tel.: +1-310-222-4013; Fax: +1-310-803-5620

check for **updates**

Citation: Mishra, N.N.; Bayer, A.S.; Baines, S.L.; Hayes, A.S.; Howden, B.P.; Lapitan, C.K.; Lew, C.; Rose, W.E. Cell Membrane Adaptations Mediate β-Lactam-Induced Resensitization of Daptomycin-Resistant (DAP-R) *Staphylococcus aureus* In Vitro. *Microorganisms* **2021**, *9*, 1028. https://doi.org/10.3390/microorganisms9051028

Academic Editor: Rajan P. Adhikari

Received: 23 April 2021
Accepted: 7 May 2021
Published: 11 May 2021

Publisher's Note: MDPI stays neutral with regard to jurisdictional claims in published maps and institutional affiliations.

Abstract: The reversal of daptomycin resistance in MRSA to a daptomycin-susceptible phenotype following prolonged passage in selected β-lactams occurs coincident with the accumulation of multiple point mutations in the *mprF* gene. MprF regulates surface charge by modulating the content and translocation of the positively charged cell membrane phospholipid, lysyl-phosphatidylglycerol (LPG). The precise cell membrane adaptations accompanying such β-lactam-induced *mprF* perturbations are unknown. This study examined key cell membrane metrics relevant to antimicrobial resistance among three daptomycin-resistant MRSA clinical strains, which became daptomycin-susceptible following prolonged exposure to cloxacillin ('daptomycin-resensitized'). The causal role of such secondary *mprF* mutations in mediating daptomycin resensitization was confirmed through allelic exchange strategies. The daptomycin-resensitized strains derived either post-cloxacillin passage or via allelic exchange (vs. their respective daptomycin-resistant strains) showed the following cell membrane changes: (i) enhanced BODIPY-DAP binding; (ii) significant reductions in LPG content, accompanied by significant increases in phosphatidylglycerol content ($p < 0.05$); (iii) no significant changes in positive cell surface charge; (iv) decreased cell membrane fluidity ($p < 0.05$); (v) enhanced carotenoid content ($p < 0.05$); and (vi) lower branched chain fatty acid profiles (antiso- vs. iso-), resulting in increases in saturated fatty acid composition ($p < 0.05$). Overall, the cell membrane characteristics of the daptomycin-resensitized strains resembled those of parental daptomycin-susceptible strains. Daptomycin resensitization with selected β-lactams results in both definable genetic changes (i.e., *mprF* mutations) and a number of key cell membrane phenotype modifications, which likely facilitate daptomycin activity.

Keywords: CM lipids; daptomycin resistance; resensitization; MRSA

1. Introduction

S. aureus is a leading cause of bacteremia and other endovascular infections including endocarditis, vascular catheter sepsis, and intracardiac device infections [1–3]. Of note, methicillin-resistant *S. aureus* (MRSA) comprise up to one-half of these cases [4]. Complications associated with the standard-of-care anti-MRSA antibiotic, vancomycin (e.g., poor overall clinical responses, persistent bacteremia, renal toxicity), has resulted in increased use of alternative MRSA therapies such as daptomycin (DAP), which are accompanied by excess health-care expenditures [5–7].

187

Daptomycin is an effective treatment against invasive *S. aureus* infections, including MRSA [8,9]. It remains the only rapidly bactericidal anti-MRSA antibiotic with consistent efficacy in bacteremia and infective endocarditis [7,10]. Due to the increasing usage of DAP, a number of reports describing DAP-resistant (R) MRSA strains emerging during DAP therapy, associated with treatment failures, have been cited, particularly in cases of osteomyelitis and endocarditis [11–14]. Since DAP has been considered a last resort antibiotic treatment option for severe MRSA infections, the evolution of DAP-R can be very problematic for patients. In addition, newer anti-MRSA antibiotics either have no proven efficacy in severe bacteremic syndromes (telavancin, tedizolid), have evoked documented clinical and microbiologic resistance (ceftaroline), and/or have issues regarding optimal dosing regimens in systemic MRSA infections (dalbavancin, oritavancin) [15–19]. Therefore, it is important to design strategies that can both salvage the bactericidal activity of DAP as well as prevent the development of DAP-R during treatment. One novel finding that has frequently accompanied the evolution of DAP-R in *S. aureus* is the so-called "see-saw" effect [20]; in this phenomenon, increasing DAP MICs are associated with a concomitant and significant enhancement in β-lactam susceptibility, despite retention of the *mecA* gene encoding for PBP2a-mediated β-lactam resistance [20–23]. This latter finding underscores both the likely complex adaptations that underlie the DAP-R phenotype and the possibility of modifying the DAP-R to a DAP-susceptible (S) phenotype pharmacologically [24].

Previous reports demonstrate that the DAP-R phenotype in MRSA features a number of mutations in global regulatory genes, as well as in genes involved in CM and/or cell wall homeostasis [25–27]. The most frequently cited mutations in MRSA associated with DAP-R are perturbations in conserved "hot spots" within the multipeptide resistance factor (*mprF*) gene [25–27]. MprF is responsible for lysinylating the anionic phospholipid, phosphatidylglycerol (PG), into the cationic lysyl-phosphatidylglycerol (LPG) species, by increasing LPG synthesis and/or its translocation into the outer CM leaflet [28–36]. This results in an enhancement of net positive surface charge and putative formation of a more charge-repulsive milieu against calcium-DAP oligomeric aggregates, which reduces DAP insertion into the target CM [28–37]. Of note, DAP-R MRSA strains often undergo several other phenotypic modifications in CM metrics including in CM order (fluidity/rigidity), carotenoid content, and fatty acid composition [32–36].

β-lactam antibiotics enhance the activity of DAP in vitro and in vivo against both DAP-S and DAP-R MRSA [38]. The mechanisms associated with this combinatorial interaction are incompletely understood but have been suggested to include (i) β-lactam-induced enhancement of DAP binding to the target bacterial surface [38] or (ii) more targeted binding to those CM regions where DAP is most effective (i.e., the cell divisome) [39–41]. In this regard, we have recently reported a novel genetic linkage related to β-lactam resensitization of DAP-R strains [41–43]. In those investigations, the penicillin-binding protein (PBP)-1-specific β-lactam, cloxacillin (LOX), was uniquely effective in restoring a DAP-S phenotype [41]; the genetic perturbations occurring in parental DAP-R strains (with single *mprF* mutations), when exposed to prolonged LOX passage, featured an accumulation of additional mutations in *mprF* and/or mutations in the divisome gene, *div1b* [41]. Of interest, previous in vitro studies determined that dual *mprF* mutations in DAP-R strains reversed this phenotype to DAP-S [29].

The current study examined a selected cadre of relevant CM phenotypic modifications that occurred during the transition from the DAP-R-to-DAP-S phenotype, which is induced by prolonged passage in vitro to LOX.

2. Materials and Methods

Bacterial strains and growth conditions. The bacterial strains used in this investigation are listed in Table 1. Three isogenic DAP-S parental (WT)/DAP-R MRSA strain pairs (J01/J03, D592/D712, and C24/C25) were utilized in this study, representing: (i) clinically derived DAP-S isolates and their respective DAP-R variants emerging during DAP therapy and (ii) the most common clonal complex (CC) types causing clinical infections in the

United States (USA100 and USA300, CC5 and CC8, respectively) [29]. In addition, for each strain-pair, we included a DAP-S variant selected by prolonged LOX passage (see below). Comparing each DAP-S parental strain with its DAP-R mutant revealed a single mutation within the *mprF* locus in this latter strain. In contrast, the post-LOX-passage DAP-S variant had accumulated an additional *mprF* mutation. These detailed genotypic characteristics are further described elsewhere [41].

Table 1. List of study strains and their DAP MIC.

Strain Set [a]	Strain Name	Strain Description	DAP MIC [b] (µg/mL)	LOX MIC [c] (µg/mL)	SNPs in *mprF* [d]
I	C24	DAP-S	0.5	8	WT
	C25	DAP-R	2	4	S295L
	C25-LOX	DAP-resensitized (LOX passaged)	<0.125	8	S295L + L84 (Translocase domain)
	C25, *mprF* DM	Secondary *mprF* mutation (L84 [e]) introduced into C25	0.125	16	S295L + L84 [e]
II	D592	DAP-S	0.5	512	WT
	D712	DAP-R	2	512	L341S
	D712-LOX	DAP-resensitized (LOX passaged)	0.5	1024	L341S + S136L (Translocase domain)
	D712, *mprF* DM	Secondary *mprF* mutation (S136L) introduced into D712	0.5	1024	L341S + S136L
III	J01	DAP-S	0.5	16	WT
	J03	DAP-R	2–4	2	T345I
	J03-LOX	DAP-resensitized (LOX passaged)	0.125	32	T345I + R788L Synthase domain
	J03, *mprF* DM	Secondary *mprF* mutation (R788L) introduced into J03	0.125	16	T345I + R788L

[a] Sets of isolates are represented by alternative shading and no shading, with the first strain in each set being the DAP-S parental strain, the second in each set being the DAP-R strain and the third and fourth being the DAP-resensitized strain generated by LOX passage or allelic exchange, respectively; [b,c,d] Data in this table have been previously published (41); [e] nonsense mutation (41).

For most experiments in this investigation, Mueller–Hinton broth (MHB; Difco, Rock Island, IL, USA) was utilized [41]. However, for studies quantifying surface charge, as well as CM carotenoid and fatty acid contents and phospholipids, an enriched media was required (brain heart infusion broth (BHI; Bacto, Mount Pritchard, NSW, Australia)), with all cultures being aerated at 37 °C. Carotenoid and fatty acid are major determinants of CM biophysical properties (e.g., CM rigidity/fluidity); in turn, these characteristics impact key events such as susceptibility to antimicrobials (e.g., DAP and β-lactams), staphylococcal pathogenesis, and organism responses to environmental stressors [34–36].

In vitro passaging of DAP-R strains in LOX. As described before [41], DAP-R isolates, J03, D712, and C25 were passaged in LOX for 28 days, with the post-passage strains now exhibiting DAP-S by MIC testing. The concentrations of LOX used for serial passage (1.4 mg/L) represented sub-MIC free average levels achieved in human serum (fC_{avg}) for each DAP-R isolate. Cultures were grown overnight, diluted, and resuspended in fresh media (MHB supplemented with 25 mg/L calcium, 12.5 mg/L magnesium and 2% NaCl) to a total volume of one ml for daily passage as previously described [41]. All experimental passage experiments were performed in triplicate. The previously published mutations acquired during prolonged LOX passage, as well as DAP and LOX susceptibilities of the strain-sets are described in Table 1 [41].

Construction of *mprF*-mutants by allelic exchange. Introduction of a secondary *mprF* mutation into the three DAP-R backgrounds (which contain a single *mprF* mutation) was conducted using the allelic exchange protocol developed by Monk and Stinear [44] with modification (full detail provided in supplementary methods). Oligonucleotides tailed with sequence complementary to pIMAY-Z were designed to amplify a ≈ 1.2 kb region surrounding the secondary *mprF* mutation (Table S1) [45], in which the LOX passaged DAP-resensitized strains served as a donor for the sequence. *E. coli* strain IM08B was used for electrocompetent transformation [46].

Suspected *mprF* double mutant (DM) colonies and cultures of the parent DAP-R strains used for allelic exchange underwent whole genome sequencing (WGS) to confirm their genotype (NextSeq; Illumina, San Diego, CA, USA), as previously described [41]. Both the primary (from the DAP-R parent) and secondary (introduced) *mprF* mutations were confirmed in all three backgrounds. Only one off-target missense mutation was identified: a A308V amino acid change in predicted gene FFX42_RS09315 in the D712 *mprF* DM (reference D592, CC5 background).

Surface charge. The relative positive cell surface charge of the three DAP-S/DAP-R/LOX-DAP-resensitized strain-sets was assayed using the standard polycationic cytochrome C (Cyt C) binding assay as described elsewhere [32–34]. Briefly, *S. aureus* strains were grown in BHI broth to stationary phase, washed with MOPS (3-morpholinopropane-1-sulfonic acid) buffer (pH 7.0), resuspended in the same buffer at OD578 ≈ 1.0, and then incubated with 0.5 mg/mL of Cyt C for 10 min. Then, the residual quantity of Cyt C remaining in the bacterial supernatant was measured spectrophotometrically at OD530 nm, as described previously [32–34]. A decrease in the quantity of Cyt C binding (i.e., more cation in the supernate) reflects a greater positively charged bacterial surface [32–34]. The data are presented as mean (±SD) of bound Cyt C. A minimum of three independent experiments was performed on separate days.

CM phospholipid (PL) composition and amino-PL (LPG) asymmetry. The lipid extraction methodology has been described before [32–36]. *S. aureus'* major PLs (lysyl-phosphatidylglycerol (LPG); phosphatidylglycerol (PG); and cardiolipin (CL)) were separated using two-dimensional thin-layer chromatography (2D-TLC), using a unique solvent system as previously described [32–36]. The isolated PL spots on TLC plates were scraped, digested at 180 °C for 3 h with 0.3 mL 70% perchloric acid to convert into the inorganic form of phosphate, and quantified spectrophotometrically at OD_{660} by phosphate estimation assay. The identification of all spots on the TLC plate were carried out by exposure to iodine vapors and by spraying with $CuSO_4$ (100 mg/mL) containing 8% phosphoric acid (*v/v*) and heated at 180 °C.

Fluorescamine labeling (a CM-impermeant UV fluorophore that binds to amino PLs, such as LPG, in the outer CM leaflet and is a measure of LPG translocation) was performed, using the same 2D TLC plates [32–36]. The percentage of fluorescamine-labeled LPG was calculated from the phosphorus data relative to total PLs. In general, LPG resides predominantly in the inner leaflet of the *S. aureus* CM; however, variable amounts of LPG can be translocated from the inner-to-outer CM leaflet to maintain lipid homeostasis. Fluorescamine labeling of outer CM (O)-LPG was detected by using a UV detector (365 nm). Fluorescamine-labeled LPG alters its mobility characteristics, and its ability to be detected by ninhydrin staining is attenuated. Unlabeled LPG (inner CM [I]-LPG), was visualized by ninhydrin staining. The identity of each of the major TLC spots was made in relation to known positive control PLs. Data were presented as the mean (±SD) percentages of the three major PLs (Total LPG + PG + CL = 100%).

BODIPY-DAP fluorescence microscopy. DAP binding was performed using confocal microscopy with BODIPY-labeled DAP. Cells were incubated with BODIPY-labeled DAP as previously described [37]. The cells were concentrated 20-fold, and 3 μL was placed on a glass slide. Slides were set with prolonged diamond antifade mountant and a #1.5 glass coverslip. Images were collected using a Leica SP8 3X STED Super-Resolution Confocal Microscope using a 489 nm laser line and 510–579 nm emission with 660 nm depletion. ImageJ was utilized to measure integrated fluorescence density of 30 cells, and corrected cell total fluorescence was calculated.

Quantification of carotenoids: The extraction and quantification of CM carotenoids was performed as described previously [33–36]. Strains were grown at the late stationary phase in BHI broth at 37 °C for 18–24 h and then harvested, washed, and pelleted in PBS by centrifugation. The post-removal of extra liquid from final pellets occurred by inversion for 2 min; then, pellet wet-weights were recorded. After 0.5 mL methanol was added to 0.1 g of the bacterial pellet, cells were vortexed vigorously and centrifuged. The upper layer of methanol extract was collected for the quantification of overall carotenoid content [33–36], which was determined spectrophotometrically at OD_{450} [33–36]. These assays were performed a minimum of five times for individual strains on different days.

Fatty acid content. *S. aureus* cells were saponified, methylated, and fatty acid esters were extracted into hexane as detailed [43]. The resulting methyl ester mixtures were separated by gas chromatography [43]. Fatty acids were identified by a well-characterized microbial identification system. The external calibration standards (a mixture of the straight-chained saturated fatty acids from 9 to 20 carbons in length and five hydroxy acids) and individual known fatty acids were used to calibrate equivalent chain length (ECL) data for fatty acid identification [43]. The ECL value for each fatty acid are represented as a function of its elution time in reference to the elution time of known standards of straight-chain fatty acids [43]. Short, medium, and long-chain saturated fatty acids were grouped per carbon number [43]. ECLx = (Rtx − Rtn/Rt (n + 1) − Rtn) + n, where Rtx is the retention time of x, Rtn is the retention time of the saturated fatty acid methyl ester preceding x, and Rt(n + 1) is the retention time of the saturated fatty acid methyl ester eluting after x. FA data represent the means (±SD) from a minimum of two independent determinations on different days. Data were expressed as the percentage of the major FAs (branch chain [BC] FA + saturated [S]FA + unsaturated [U]FA = 100%). FAs present representing <1% of the total content were not included in the data analysis.

CM order (fluidity/rigidity). CM order profoundly impacts the interactions of DAP with the *S. aureus* CM [33–36]. MRSA strain-sets were grown overnight in BHI broth at 37 °C, harvested by centrifugation, and then washed with PBS. A whole-cell suspension of the MRSA strains was prepared at an OD_{600} = 1.0 ($\approx 10^8$ CFU mL^{-1}). CM fluidity was measured by polarizing spectrofluorometry utilizing the fluorescent probe, 1,6-diphenyl-1,3,5-hexatriene (DPH) (excitation and emission wavelengths of 360 and 426 nm). The detailed methods for quantifying DPH incorporation into CMs and the calculations of the degree of fluorescence polarization (polarization index (PI)) are described elsewhere [21,22]. An inverse relationship occurs between PI values and CM fluidity (i.e., lower PI equates to a greater extent of CM fluidity) [32–36]. These experiments were carried out a minimum of four times for each strain-set on different days.

Statistical Analyses. The two-tailed Student's t-test was used for statistical analysis of all quantitative data. *p* values of ≤0.05 were considered "significant".

3. Results

Resensitization of DAP-R MRSA to DAP-S. As previously published, LOX resensitized all DAP-R strains (J03, D712, and C25) to DAP (DAP-resensitized) after serial passage for 28 days in vitro (Table 1) [41]. The secondary *mprF* mutations derived from LOX passage strains were reintroduced (*mprF* DM) into the respective DAP-R parental strains. This allelic replacement resulted in a reduction of DAP MICs to levels similar to the post-LOX passage strains. This enhanced DAP resensitization correlated with increased BODIPY-DAP binding by confocal microscopy (Figure 1).

CM phospholipid (PL) content. As expected, the DAP-R variants showed CM PL profiles featuring increased total LPG content vs. the respective DAP-S parental straincon-sistent with other previously described DAP-R strains (Table 2); this reflects the gain-in-function impacts typical of single *mprF* mutations [42]. Of interest, the increased total LPG content in this mutant was associated with enhanced synthesis but not increased outer CM translocation (data not shown). In contrast, the DAP-resensitized variants, either derived post-LOX passage or via allelic replacement, demonstrated reductions in overall synthesis of LPG to levels compatible with the DAP-S parental strain (Table 2). This CM PL profile in the DAP-resensitized strains is consistent with the documented accumulation of an additional *mprF* mutation in these strains [41], resulting in a decrease-of-function phenotype.

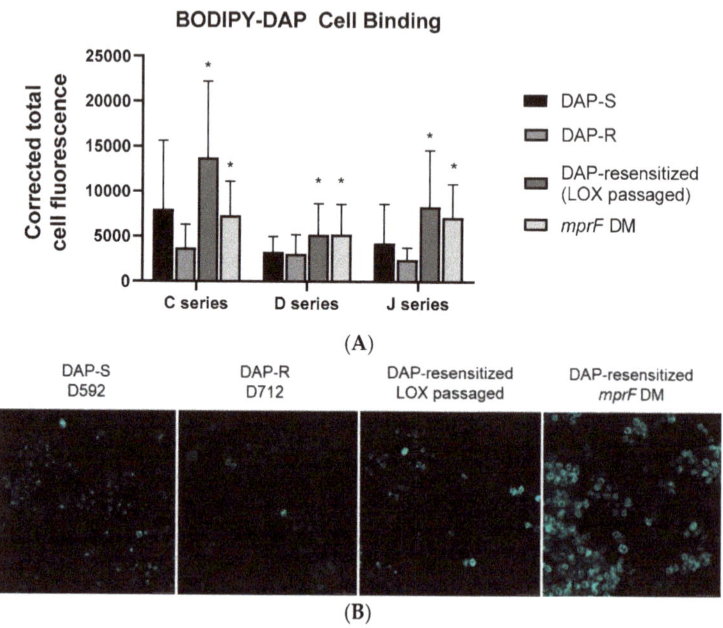

Figure 1. Binding of BODIPY-labeled daptomycin (BODIPY-DAP) to *S. aureus* study strains. (**A**) Corrected total cell fluoresence of BODIPY-DAP binding. (**B**) Representative confocal microscopy images of BODPY-DAP binding in the well-characterized D series strains. * $p < 0.01$ vs. DAP R-strain.

Table 2. Phospholipid composition (%) of LOX passaged strain vs. DAP-R/DAP-S.

Strains	Total LPG	PG	CL
C24	12 ± 3	80 ± 6	8 ± 5
C25	25 ± 5 [a]	70 ± 5 [a]	6 ± 3
C25-LOX	5 ± 1 [b]	94 ± 1 [b]	2 ± 1 [b]
C25, *mprF* DM	11 ± 2 [c]	83 ± 4 [c]	6 ± 6
D592	20 ± 3	77 ± 3	2 ± 3
D712	23 ± 2 [a]	74 ± 4 [a]	3 ± 2
D712-LOX	16 ± 4 [b]	81 ± 5 [b]	3 ± 2
D712, *mprF* DM	21 ± 2	75 ± 2	4 ± 1
J01	22 ± 2	70 ± 2	8 ± 2
J03	31 ± 7 [a]	66 ± 6 [a]	3 ± 1 [a]
J03-LOX	20 ± 1 [b]	78 ± 3 [b]	3 ± 2 [b]
J03, *mprF* DM	16 ± 3 [c]	78 ± 4 [c]	6 ± 1 [c]

[a] p-value < 0.05; DAP-R vs. DAP-S; [b] p-value < 0.05; LOX passaged strains vs. DAP-R; [c] p-value < 0.05; *mprF* DM strains vs. DAP-R.

Cell surface charge. As shown in Table 3, in two of the three DAP-R strains, more unbound cytochrome C (reflecting a more positive cell surface charge) was observed vs. their respective DAP-S parental strains. In strains with secondary *mprF* mutation derived post-LOX passage as well as in the allelic reintroduction strains (*mprF* DM), a more negative surface charge was observed as compared to respective DAP-R strains ($p < 0.05$) and similar to the DAP-S parental strains.

Table 3. Surface charge of study strains.

Strains	% Cytochrome C Unbound
C24	53 ± 1
C25	62 ± 0 [a]
C25-LOX	54 ± 0 [b]
C25, *mprF* DM	45 ± 0 [c]
D592	56 ± 0
D712	85 ± 3 [a]
D712-LOX	46 ± 1 [b]
D712, *mprF* DM	57 ± 1
J01	58 ± 0
J03	48 ± 0 [a]
J03-LOX	55 ± 0 [b]
J03, *mprF* DM	44 ± 0 [c]

[a] *p*-value < 0.05; DAP-R vs. DAP-S; [b] *p*-value < 0.05; LOX passaged strains vs. DAP-R; [c] *p*-value < 0.05; *mprF* DM strains vs. DAP-R.

Fatty acid content. We observed considerable shifts in the patterns of saturated fatty acids (SFAs) as well as iso- and anteiso-branched chain fatty acid (BCFA) profiles among the DAP-resensitized strains in comparison to both the parental DAP-S and their respective DAP-R mutant strains (Tables 4 and S2). Interestingly, the DAP-resensitized strains had elevated SFA content as compared to their respective DAP-R mutants. However, changes in SFA content in DAP-resensitized strains vs. respective DAP-S parental strains were minimal. A significant decrease in the proportion of total anteiso-BCFAs was noted in the DAP-resensitized strains versus the corresponding DAP-R mutants (*p* < 0.01). The DAP-resensitized strains and their DAP-S parental strains had similar levels of anteiso-BCFAs. In addition, no consistent pattern of iso-BCFAs differences was observed among the strain-sets (Tables 4 and S2). It should be noted that increased anteiso-SFA, reduced iso-SFA, and SFA content correlates with more fluid CM in *S. aureus* [34–36].

Table 4. Fatty acids (%) composition of LOX passaged strain vs. DAP-R/DAP-S.

Strain Set [a]	Iso FA	Anteiso FA	SFA
C24	30 ± 0.1	41 ± 0.12	25 ± 0.03
C25	27 ± 0.01 [a]	44 ± 0.02 [a]	22 ± 0.03 [a]
C25-LOX	25 ± 0.07 [b,c]	41 ± 0.1 [b,c]	29 ± 0.3 [b,c]
D592	24 ± 0.6	40 ± 0.03	31 ± 0.4
D712	25 ± 0.01 [a]	40 ± 0.03	26 ± 0.2 [a]
D712-LOX	25 ± 0.04 [b]	45 ± 0.2 [b,c]	31 ± 0.2 [c]
J01	31 ± 0.03	40 ± 0.03	25 ± 0.02
J03	29 ± 0.1 [a]	45 ± 0.2 [a]	22 ± 0.11 [a]
J03-LOX	27 ± 0.1 [b,c]	43 ± 0.11 [b,c]	26 ± 0.04 [b,c]

[a] *p*-value < 0.05; DAP-R vs. DAP = S; [b] *p*-value < 0.05; LOX passaged strains vs. DAP-S; [c] *p*-value < 0.05; LOX passaged strains vs. DAP-R; SFA = Saturated/Straight Chain FAs.

CM carotenoids. *S. aureus* utilizes its carotenoid content to modulate its CM order (i.e., the more carotenoid content, the more rigid the CM) to resist the microbicidal action of DAP and other cationic peptides [33–35]. The DAP-R strains had lower carotenoid content vs. their corresponding DAP-S strains (Table 5), paralleling our prior data [36]. This is also consistent with the more fluid CMs noted below in the DAP-R vs. their respective DAP-S parental strains. There was a significant enhancement in CM carotenoid content among the DAP-resensitized variants, either derived from LOX passage or allelic exchange vs. their respective DAP-R strains (Table 5) (*p* < 0.05). These data in line with the re-establishing of a more rigid, parental-level CM.

Table 5. CM fluidity and carotenoid content of study strains.

Strains	CM Fluidity (PI Value)	Carotenoids (OD_{450nm})
C24	0.389 ± 0.01	0.685 ± 0.04
C25	0.370 ± 0.01 [a]	0.261 ± 0.03 [a]
C25-LOX	0.413 ± 0.00 [b]	1.129 ± 0.24 [b]
C25, *mprF* DM	0.368 ± 0.00	0.627 ± 0.02 [c]
D592	0.408 ± 0.01	1.338 ± 0.01
D712	0.372 ± 0.01 [a]	0.878 ± 0.01 [a]
D712-LOX	0.389 ± 0.00 [b]	1.037 ± 0.02 [b]
C25, *mprF* DM	0.395 ± 0.00 [c]	0.971 ± 0.03 [c]
J01	0.381 ± 0.01	1.121 ± 0.04
J03	0.359 ± 0.01	0.697 ± 0.12 [a]
J03-LOX	0.395 ± 0.01 [b]	1.518 ± 0.27 [b]
C25, *mprF* DM	0.430 ± 0.03 [c]	1.545 ± 0.37 [c]

[a] *p*-value < 0.05; DAP-R vs. DAP-S; [b] *p*-value < 0.05; LOX passaged strains vs. DAP-R; [c] *p*-value < 0.05; *mprF* DM strains vs. DAP-R.

CM order (fluidity/rigidity). There appears to be an optimal degree of CM order for the interaction of most CM-targeting cationic peptides, including calcium-complexed DAP [34–36]. As seen with other DAP-R mutants of MRSA [47], the three DAP-R mutants in our study displayed more fluid CMs vs. each respective parental DAP-S strain (Table 5). The DAP-resensitized strains overall had a shift in CM order toward a less fluid (more rigid) CM, similar to the pattern of their respective DAP-S parental strains (Table 5).

4. Discussion

Our recent studies suggest that β-lactams can synergize with DAP against DAP-R MRSA through the inhibition of PBP-1 without necessarily enhancing DAP binding [38]. Furthermore, distinct β-lactam antibiotics (such as LOX) can uniquely resensitize DAP-R MRSA strains to a DAP-S phenotype [41], which is a phenomenon that was only previously found through in vitro genetic manipulation of *mprF* [28–30]. Of note, WGS confirmed that DAP-resensitized strains frequently acquire additional *mprF* mutations in either their translocase or synthase domains [41]. The accumulation of such secondary *mprF* mutations in our LOX-post passage variants correlated with a significant drop in DAP MICs, which is often below that of the DAP-S parental strains [41]. The emergence of DAP-R in MRSA and mutations in *mprF* are often associated with distinct compensatory changes in PL composition, surface charge, and CM biophysics (e.g., fluidity profiles). Understanding the CM-associated adaptations in DAP-R mutants when exposed to selected β-lactams has important therapeutic implications, as infections with such strains are now being more frequently treated with DAP-β-lactam combinations [34–36,38,42].

Several interesting themes emerged from our study. First, MprF mutations among DAP-R clinical strains have been noted in a variety of "hot spot" loci within this protein, most frequently (>50%) in its central bifunctional domain [28,30]. Of interest, recent in vitro studies [29] have confirmed that the presence of secondary *mprF* mutations can yield a decrease-of-function paradigm, featuring both reversal of DAP MICs, as well as a shift in PL profiles, paralleling DAP-S parental phenotypes. As noted above, prolonged LOX passage induces the accumulation of secondary mutations in *mprF*; however, the causal nature of this event for daptomycin resensitization was unclear, since other genetic mutations were also observed, most notably in *div1b* and *rpoC*. In this present study, we confirmed that these secondary *mprF* mutations (either via LOX passage or allelic exchange) are sufficient to restore parental level daptomycin MICs, as well as induce prototypical modifications in CM phenotypes. Studies are in progress to understand the mechanism(s) by which β-lactams can trigger the accumulation of secondary *mprF* mutations.

Second, DAP-R MRSA strains can modulate their surface charge toward a more relatively positive charge phenotype, potentially creating a "charge-repulsive" surface milieu against cationic molecules such as calcium-complexed DAP [26–28]. However, altered positive surface charge was only noted in two of our three strain-sets, and thus, it is not likely being determinative of the DAP-resensitization phenotype. Of note, β-lactams with PBP-1-targeting specificity may, in fact, not significantly alter surface charge, making their DAP-β-lactam synergy impacts independent of such events [38]. These data suggest that charge repulsion itself is not sufficient to fully explain the β-lactam resensitization of DAP-R strains.

Third, all three DAP-resensitized strains demonstrated substantially decreased CM fluidity as compared to their respective DAP-R strains and similar to their respective DAP-S parental strains. There are optimal biophysical metrics within the CM microenvironment that appear to maximize interactions of cationic peptides with the CM of MRSA [20–22,24]. Therefore, MRSA CMs containing extremes of rigidity/fluidity are comparatively resistant to interactions with such peptides [34,36,47]. The similar patterns of CM order of DAP-S parental and DAP-resensitized strains, on the one hand, vs. the distinctly different patterns of CM order in their respective DAP-R variants, underscored the likelihood that this phenotype is playing a relevant role in the overall differences in ultimate DAP-induced killing.

Finally, the above differences in CM order within the strain-sets were well correlated with changes in both carotenoid content and the patterns of anteiso-BCFAs; thus, changes in CM carotenoid content can lead to a buildup of C30 precursor species that shuttle into the menaquinone-fatty acid oxidation pathways [34]. In addition, CM–carotenoid interaction with the scaffold protein, flotillin, leads to CM microdomain formation, which is an important scaffolding for PBP maturation [48]; thus, the disassembly of these microdomains can have a major impact on MRSA antibiotic resistance [45]. Furthermore, it should be underscored that DAP inhibits the net synthesis of the cell envelope by interfering with such microdomains of DAP-S bacteria, leading to reorganization of the overall CM architecture, followed by the delocalization of essential CM proteins, such as the lipid II synthase, MurG [39,40]. Moreover, it is well known that various "stressors" can evoke a pattern of enhanced production of anteiso-BCFAs, with resultant increases in CM fluidity as a protective survival mechanism. Thus, Bacillus species can elicit such anteiso-BCFA shifts during 'cold shock' [49]. The shift we observed in anteiso-BCFAs and other FAs in the DAP-R mutants may represent the equivalent of a "cold shock" response at 37 °C. These same iso-to-anteiso-BCFA shifts, linked with perturbed CM order, have also been documented in previously studied *S. aureus* strains [35].

We recognize that the current investigation has several key limitations: (i) only three DAP S/DAP-R/DAP-resensitized strain-sets were studied; (ii) only a relatively focused cadre of phenotypic characteristics were interrogated in comparing the strain-sets; (iii) only a single β lactam antibiotic was used for prolonged passage, leaving unresolved whether other PBP-specific or PBP-promiscuous β-lactams can elicit the same adaptations; (iv) the linkage of our CM perturbations with specific metabolomics modifications was not explored [50]; and (vi) additional mutations documented previously in prolonged LOX-passaged strains (41) were not systematically investigated (e.g., via allelic exchange) to determine their impacts on the above CM parameters. It should be noted that DAP-resensitization did not occur post-LOX passage in DAP-R strains lacking a primary *mprF* mutation [41]; thus, it is highly likely that *mprF* and its associated CM changes play a critical role in the DAP-resensitization phenomenon.

Supplementary Materials: The following are available online at https://www.mdpi.com/article/10.3390/microorganisms9051028/s1.

Author Contributions: Conceptualization, N.N.M., W.E.R., A.S.B., C.L.; methodology, N.N.M., W.E.R., S.L.B., A.S.H., B.P.H., C.K.L.; validation, N.N.M., A.S.B., W.E.R., S.L.B., A.S.H., B.P.H.; writing—original draft preparation, N.N.M., A.S.B., W.E.R., writing—review and editing, N.N.M., A.S.B., W.E.R., B.P.H.; visualization, N.N.M., A.S.B., W.E.R., B.P.H. All authors have read and agreed to the published version of the manuscript.

Funding: This research was supported by grants from NIH/NIAID: R01 AI132627 (to WER) and R01AI130056 (to ASB). NNM was supported by Lundquist Institute at Harbor UCLA intramural research grant (#531604-01-01).

Acknowledgments: We thank to Sabrina Farah for technical assistance. This work was presented in part at the 29th European Congress of Clinical Microbiology and Infectious Diseases (ECCMID) Meeting, Amsterdam, Netherlands, 13–16 April 2019, poster # P2843.

Conflicts of Interest: The authors declare no conflict of interest. The funders had no role in the design of the study; in the collection, analyses, or interpretation of data; in the writing of the manuscript, or in the decision to publish the results.

References

1. Fowler, V.G., Jr.; Miro, J.M.; Hoen, B.; Cabell, C.H.; Abrutyn, E.; Rubinstein, E.; Corey, G.R.; Spelman, D.; Bradley, S.F.; Barsic, B.; et al. *Staphylococcus aureus* endocarditis: A consequence of medical progress. *JAMA* **2005**, *293*, 3012–3021. [CrossRef] [PubMed]
2. Van Hal, S.J.; Jensen, S.O.; Vaska, V.L.; Espedido, B.A.; Paterson, D.L.; Gosbell, I.B. Predictors of Mortality in *Staphylococcus aureus* Bacteremia. *Clin. Microbiol. Rev.* **2012**, *25*, 362–386. [CrossRef]
3. Kullar, R.; Sakoulas, G.; Deresinski, S.; Van Hal, S.J. When sepsis persists: A review of MRSA bacteraemia salvage therapy. *J. Antimicrob. Chemother.* **2016**, *71*, 576–586. [CrossRef]
4. Gerber, J.S.; Coffin, S.E.; Smathers, S.A.; Zaoutis, T.E. Trends in the Incidence of Methicillin-Resistant *Staphylococcus aureus* Infection in Children's Hospitals in the United States. *Clin. Infect. Dis.* **2009**, *49*, 65–71. [CrossRef]
5. Guimaraes, A.O.; Cao, Y.; Hong, K.; Mayba, O.; Peck, M.C.; Gutierrez, J.; Ruffin, F.; Carrasco-Triguero, M.; Dinoso, J.B.; Clemenzi-Allen, A.; et al. A prognostic model of persistent bacteremia and mortality in complicated *Staphylococcus aureus* bloodstream infection. *Clin. Infect. Dis.* **2019**, *68*, 1502–1511. [CrossRef] [PubMed]
6. Rose, W.E.; Shukla, S.K.; Berti, A.D.; Hayney, M.S.; Henriquez, K.M.; Ranzoni, A.; Cooper, M.A.; Proctor, R.A.; Nizet, V.; Sakoulas, G. Increased Endovascular *Staphylococcus aureus* Inoculum Is the Link Between Elevated Serum Interleukin 10 Concentrations and Mortality in Patients with Bacteremia. *Clin. Infect. Dis.* **2017**, *64*, 1406–1412. [CrossRef] [PubMed]
7. Liu, C.; Bayer, A.; Cosgrove, S.E.; Daum, R.S.; Fridkin, S.K.; Gorwitz, R.J.; Kaplan, S.L.; Karchmer, A.W.; Levine, D.P.; Murray, B.E.; et al. Clinical Practice Guidelines by the Infectious Diseases Society of America for the Treatment of Methicillin-Resistant *Staphylococcus aureus* Infections in Adults and Children: Executive Summary. *Clin. Infect. Dis.* **2011**, *52*, 285–292. [CrossRef] [PubMed]
8. Claeys, K.C.; Zasowski, E.J.; Casapao, A.M.; Lagnf, A.M.; Nagel, J.L.; Nguyen, C.T.; Hallesy, J.A.; Compton, M.T.; Kaye, K.S.; Levine, D.P.; et al. Daptomycin improves outcomes regardless of vancomycin MIC in a pro-pensity-matched analysis of methicillin-resistant *Staphylococcus aureus* bloodstream infections. *Antimicrob. Agents Chemother.* **2016**, *60*, 5841–5848. [CrossRef]
9. Moise, P.A.; Culshaw, D.L.; Wong-Beringer, A.; Bensman, J.; Lamp, K.C.; Smith, W.J.; Bauer, K.; Goff, D.A.; Adamson, R.; Leuthner, K.; et al. Comparative effectiveness of vancomycin versus daptomycin for MRSA bacteremia with vancomycin MIC >1 mg/L: A multicenter evaluation. *Clin. Ther.* **2016**, *38*, 16–30. [CrossRef]
10. Fowler, V.G., Jr.; Boucher, H.W.; Corey, G.R.; Abrutyn, E.; Karchmer, A.W.; Rupp, M.E.; Levine, D.P.; Chambers, H.F.; Tally, F.P.; Vigliani, G.A.; et al. Daptomycin versus standard therapy for bacteremia and endocarditis caused by *Staphylococcus aureus*. *N. Engl. J. Med.* **2006**, *355*, 653–665. [CrossRef]
11. Capone, A.; Cafiso, V.; Campanile, F.; Parisi, G.; Mariani, B.; Petrosillo, N.; Stefani, S. In vivo development of daptomycin resistance in vancomycin-susceptible methicillin-resistant *Staphylococcus aureus* severe infections previously treated with glycopeptides. *Eur. J. Clin. Microbiol. Infect. Dis.* **2016**, *35*, 625–631. [CrossRef]
12. Marty, F.M.; Yeh, W.W.; Wennersten, C.B.; Venkataraman, L.; Albano, E.; Alyea, E.P.; Gold, H.S.; Baden, L.R.; Pillai, S.K. Emergence of a clinical daptomycin-resistant *Staphylococcus aureus* isolate during treatment of methicillin-resistant Staphylococcus aureus bacteremia and osteomyelitis. *J. Clin. Microbiol.* **2006**, *44*, 595–597. [CrossRef] [PubMed]
13. Stefani, S.; Campanile, F.; Santagati, M.; Mezzatesta, M.L.; Cafiso, V.; Pacini, G. Insights and clinical perspectives of daptomycin resistance in *Staphylococcus aureus*: A review of the available evidence. *Int. J. Antimicrob. Agents* **2015**, *46*, 278–289. [CrossRef]
14. Skiest, D.J. Treatment Failure Resulting from Resistance of *Staphylococcus aureus* to Daptomycin. *J. Clin. Microbiol.* **2006**, *44*, 655–656. [CrossRef]

15. Chan, L.C.; Basuino, L.; Dip, E.C.; Chambers, H.F. Comparative Efficacies of Tedizolid Phosphate, Vancomycin, and Daptomycin in a Rabbit Model of Methicillin-Resistant *Staphylococcus aureus* Endocarditis. *Antimicrob. Agents Chemother.* **2015**, *59*, 3252–3256. [CrossRef] [PubMed]

16. Roberts, K.D.; Sulaiman, R.M.; Rybak, M.J. Dalbavancin and Oritavancin: An Innovative Approach to the Treatment of Gram-Positive Infections. *Pharmacotherapy* **2015**, *35*, 935–948. [CrossRef] [PubMed]

17. Guskey, M.T.; Tsuji, B.T. A Comparative Review of the Lipoglycopeptides: Oritavancin, Dalbavancin, and Telavancin. *Pharmacotherapy* **2010**, *30*, 80–94. [CrossRef] [PubMed]

18. Long, S.W.; Olsen, R.J.; Mehta, S.C.; Palzkill, T.; Cernoch, P.L.; Perez, K.K.; Musick, W.L.; Rosato, A.E.; Musser, J.M. PBP2a Mutations Causing High-Level Ceftaroline Resistance in Clinical Methicillin-Resistant *Staphylococcus aureus* Isolates. *Antimicrob. Agents Chemother.* **2014**, *58*, 6668–6674. [CrossRef] [PubMed]

19. Chan, L.C.; Basuino, L.; Diep, B.; Hamilton, S.; Chatterjee, S.S.; Chambers, H.F. Ceftobiprole- and Ceftaroline-Resistant Methicillin-Resistant *Staphylococcus aureus*. *Antimicrob. Agents Chemother.* **2015**, *59*, 2960–2963. [CrossRef]

20. Yang, S.-J.; Xiong, Y.Q.; Boyle-Vavra, S.; Daum, R.S.; Jones, T.; Bayer, A.S. Daptomycin-Oxacillin Combinations in Treatment of Experimental Endocarditis Caused by Daptomycin-Nonsusceptible Strains of Methicillin-Resistant *Staphylococcus aureus* with Evolving Oxacillin Susceptibility (the "Seesaw Effect"). *Antimicrob. Agents Chemother.* **2010**, *54*, 3161–3169. [CrossRef]

21. Ortwine, J.K.; Werth, B.J.; Sakoulas, G.; Rybak, M.J. Reduced glycopeptide and lipopeptide susceptibility in *Staphylococcus aureus* and the "seesaw effect": Taking advantage of the back door left open? *Drug Resist. Updates* **2013**, *16*, 73–79. [CrossRef] [PubMed]

22. Vignaroli, C.; Rinaldi, C.; Varaldo, P.E. Striking "seesaw effect" between daptomycin nonsusceptibility and beta-lactam susceptibility in *Staphylococcus haemolyticus*. *Antimicrob. Agents Chemother.* **2011**, *55*, 2495–2496. [CrossRef]

23. Patel, J.B.; Jevitt, L.A.; Hageman, J.; McDonald, L.C.; Tenover, F.C. An Association between Reduced Susceptibility to Daptomycin and Reduced Susceptibility to Vancomycin in *Staphylococcus aureus*. *Clin. Infect. Dis.* **2006**, *42*, 1652–1653. [CrossRef]

24. Fischer, A.; Yang, S.-J.; Bayer, A.S.; Vaezzadeh, A.R.; Herzig, S.; Stenz, L.; Girard, M.; Sakoulas, G.; Scherl, A.; Yeaman, M.R.; et al. Daptomycin resistance mechanisms in clinically derived *Staphylococcus aureus* strains assessed by a combined transcriptomics and proteomics approach. *J. Antimicrob. Chemother.* **2011**, *66*, 1696–1711. [CrossRef]

25. Bayer, A.S.; Mishra, N.N.; Chen, L.; Kreiswirth, B.N.; Rubio, A.; Yang, S.J. Frequency and distribution of single-nucleotide polymorphisms within *mprF* in methicillin-resistant *Staphylococcus aureus* clinical isolates and their role in cross-resistance to daptomycin and host defense antimicrobial peptides. *Antimicrob. Agents Chemother.* **2015**, *59*, 4930–4937. [CrossRef] [PubMed]

26. Bayer, A.S.; Mishra, N.N.; Cheung, A.L.; Rubio, A.; Yang, S.J. Dysregulation of *mprF* and *dltABCD* expression among dap-tomycin-non-susceptible MRSA clinical isolates. *J. Antimicrob. Chemother.* **2016**, *71*, 2100–2104. [CrossRef] [PubMed]

27. Bayer, A.S.; Schneider, T.; Sahl, H.-G. Mechanisms of daptomycin resistance in *Staphylococcus aureus*: Role of the cell membrane and cell wall. *Ann. N. Y. Acad. Sci.* **2013**, *1277*, 139–158. [CrossRef] [PubMed]

28. Ernst, C.M.; Slavetinsky, C.J.; Kuhn, S.; Hauser, J.N.; Nega, M.; Mishra, N.N.; Gekeler, C.; Bayer, A.S.; Peschel, A. Gain-of-Function Mutations in the Phospholipid Flippase *MprF* Confer Specific Daptomycin Resistance. *mBio* **2018**, *9*, e0659-18. [CrossRef] [PubMed]

29. Yang, S.J.; Mishra, N.N.; Kang, K.M.; Lee, G.Y.; Park, J.H.; Bayer, A.S. Impact of multiple single-nucleotide polymorphisms within *mprF* on daptomycin resistance in *Staphylococcus aureus*. *Microb. Drug Resist.* **2018**, *24*, 1075–1081. [CrossRef]

30. Ernst, C.M.; Staubitz, P.; Mishra, N.N.; Yang, S.-J.; Hornig, G.; Kalbacher, H.; Bayer, A.S.; Kraus, D.; Peschel, A. The Bacterial Defensin Resistance Protein *MprF* Consists of Separable Domains for Lipid Lysinylation and Antimicrobial Peptide Repulsion. *PLOS Pathog.* **2009**, *5*, e1000660. [CrossRef]

31. Mishra, N.N.; Yang, S.J.; Chen, L.; Muller, C.; Saleh-Mghir, A.; Kuhn, S.; Peschel, A.; Yeaman, M.R.; Nast, C.C.; Kreiswirt, B.N.; et al. Emergence of daptomycin resistance in daptomycin-naïve rabbits with methicil-lin-resistant *Staphylococcus aureus* prosthetic joint infection is associated with resistance to host defense cationic peptides and *mprF* polymorphisms. *PLoS ONE* **2013**, *8*, e71151. [CrossRef]

32. Mishra, N.N.; Bayer, A.S.; Weidenmaier, C.; Grau, T.; Wanner, S.; Stefani, S.; Cafiso, V.; Bertuccio, T.; Yeaman, M.R.; Nast, C.C.; et al. Phenotypic and Genotypic Characterization of Daptomycin-Resistant Methicillin-Resistant *Staphylococcus aureus* Strains: Relative Roles of *mprF* and *dlt* Operons. *PLoS ONE* **2014**, *9*, e107426. [CrossRef] [PubMed]

33. Mishra, N.N.; Rubio, A.; Nast, C.C.; Bayer, A.S. Differential adaptations of methicillin-resistant *Staphylococcus aureus* to serial in vitro passage in daptomycin: Evolution of daptomycin resistance and role of membrane carotenoid content and fluidity. *Int. J. Microbiol.* **2012**, *2012*, 683450. [CrossRef]

34. Mishra, N.N.; Liu, G.Y.; Yeaman, M.R.; Nast, C.C.; Proctor, R.A.; McKinnell, J.; Bayer, A.S. Carotenoid-Related Alteration of Cell Membrane Fluidity Impacts *Staphylococcus aureus* Susceptibility to Host Defense Peptides. *Antimicrob. Agents Chemother.* **2011**, *55*, 526–531. [CrossRef] [PubMed]

35. Mishra, N.N.; Bayer, A.S. Correlation of cell membrane lipid profiles with daptomycin resistance in methicillin-resistant *Staphylococcus aureus*. *Antimicrob. Agents Chemother.* **2013**, *57*, 1082–1085. [CrossRef]

36. Mishra, N.N.; Yang, S.-J.; Sawa, A.; Rubio, A.; Nast, C.C.; Yeaman, M.R.; Bayer, A.S. Analysis of Cell Membrane Characteristics of In vitro-Selected Daptomycin-Resistant Strains of Methicillin-Resistant *Staphylococcus aureus*. *Antimicrob. Agents Chemother.* **2009**, *53*, 2312–2318. [CrossRef] [PubMed]

37. Dhand, A.; Bayer, A.S.; Pogliano, J.; Yang, S.J.; Bolaris, M.; Nizet, V.; Wang, G.; Sakoulas, G. Use of antistaphylococcal beta-lactams to increase daptomycin activity in eradicating persistent bacteremia due to methicillin-resistant *Staphylococcus aureus*: Role of enhanced daptomycin binding. *Clin. Infect. Dis.* **2011**, *53*, 158–163. [CrossRef] [PubMed]

38. Berti, A.D.; Theisen, E.; Sauer, J.-D.; Nonejuie, P.; Olson, J.; Pogliano, J.; Sakoulas, G.; Nizet, V.; Proctor, R.A.; Rose, W.E. Penicillin Binding Protein 1 Is Important in the Compensatory Response of *Staphylococcus aureus* to Daptomycin-Induced Membrane Damage and Is a Potential Target for β-Lactam–Daptomycin Synergy. *Antimicrob. Agents Chemother.* **2015**, *60*, 451–458. [CrossRef] [PubMed]

39. Müller, A.; Wenzel, M.; Strahl, H.; Grein, F.; Saaki, T.N.; Kohl, B.; Siersma, T.; Bandow, J.E.; Sahl, H.-G.; Schneider, T.; et al. Daptomycin inhibits cell envelope synthesis by interfering with fluid membrane mi-crodomains. *Proc. Natl. Acad. Sci. USA* **2016**, *113*, E7077–E7086. [CrossRef]

40. Grein, F.; Müller, A.; Scherer, K.M.; Liu, X.; Ludwig, K.C.; Klöckner, A.; Strach, M.; Sahl, H.-G.; Kubitscheck, U.; Schneider, T. Ca^{2+}-Daptomycin targets cell wall biosynthesis by forming a tripartite complex with undecaprenyl-coupled intermediates and membrane lipids. *Nat. Commun.* **2020**, *11*, 1455. [CrossRef]

41. Jenson, R.E.; Baines, S.L.; Howden, B.P.; Mishra, N.N.; Farah, S.; Lew, C.; Berti, A.D.; Shukla, S.K.; Bayer, A.S.; Rose, W.E. Prolonged exposure to β-lactam antibiotics reestablishes susceptibility of dap-tomycin-nonsusceptible *Staphylococcus aureus* to daptomycin. *Antimicrob. Agents Chemother.* **2020**, *64*, e00890-20. [CrossRef]

42. Berti, A.D.; Sakoulas, G.; Nizet, V.; Tewhey, R.; Rose, W.E. β-Lactam Antibiotics Targeting PBP1 Selectively Enhance Daptomycin Activity against Methicillin-Resistant *Staphylococcus aureus*. *Antimicrob. Agents Chemother.* **2013**, *57*, 5005–5012. [CrossRef] [PubMed]

43. Sasser, M. *Identification of Bacteria by Gas Chromatography of Cellular Fatty Acids*; MIDI Technical Note 101; MIDI Inc.: Newark, NJ, USA, 1990.

44. Monk, I.R.; Stinear, T.P. From cloning to mutant in 5 days: Rapid allelic exchange in *Staphylococcus aureus*. *Access Microbiol.* **2021**, *3*, 193. [CrossRef]

45. Zhang, Y.; Werling, U.; Edelmann, W. SLiCE: A novel bacterial cell extract-based DNA cloning method. *Nucleic Acids Res.* **2012**, *40*, e55. [CrossRef] [PubMed]

46. Monk, I.R.; Tree, J.J.; Howden, B.P.; Stinear, T.P.; Foster, T.J. Complete bypass of restriction systems for major *Staphylococcus aureus* lineages. *mBio* **2015**, *26*, e00308-15. [CrossRef] [PubMed]

47. Mishra, N.N.; McKinnell, J.; Yeaman, M.R.; Rubio, A.; Nast, C.C.; Chen, L.; Kreiswirth, B.N.; Bayer, A.S. In vitro Cross-Resistance to Daptomycin and Host Defense Cationic Antimicrobial Peptides in Clinical Methicillin-Resistant *Staphylococcus aureus* Isolates. *Antimicrob. Agents Chemother.* **2011**, *55*, 4012–4018. [CrossRef]

48. García-Fernández, E.; Koch, G.; Wagner, R.M.; Fekete, A.; Stengel, S.T.; Schneider, J.; Mielich-Süss, B.; Geibel, S.; Markert, S.M.; Stigloher, C.; et al. Membrane Microdomain Disassembly Inhibits MRSA Antibiotic Resistance. *Cell* **2017**, *171*, 1354–1367.e20. [CrossRef] [PubMed]

49. Klein, W.; Weber, M.H.W.; Marahiel, M.A. Cold Shock Response of *Bacillus subtilis*: Isoleucine-Dependent Switch in the Fatty Acid Branching Pattern for Membrane Adaptation to Low Temperatures. *J. Bacteriol.* **1999**, *181*, 5341–5349. [CrossRef] [PubMed]

50. Gaupp, R.; Lei, S.; Reed, J.M.; Peisker, H.; Boyle-Vavra, S.; Bayer, A.S.; Bischoff, M.; Herrmann, M.; Daum, R.S.; Powers, R.; et al. *Staphylococcus aureus* metabolic adaptations during the transition from a daptomycin susceptibility phenotype to a daptomycin nonsusceptibility phenotype. *Antimicrob. Agents Chemother.* **2015**, *59*, 4226–4238. [CrossRef]

Review

Human *mecC*-Carrying MRSA: Clinical Implications and Risk Factors

Carmen Lozano *, Rosa Fernández-Fernández, Laura Ruiz-Ripa, Paula Gómez, Myriam Zarazaga and Carmen Torres

Area of Biochemistry and Molecular Biology, University of La Rioja, 26006 Logroño, Spain; rosa.fernandez.1995@gmail.com (R.F.-F.); laura_ruiz_10@hotmail.com (L.R.-R.); paula_gv83@hotmail.com (P.G.); myriam.zarazaga@unirioja.es (M.Z.); carmen.torres@unirioja.es (C.T.)
* Correspondence: carmen.lozano@unirioja.es; Tel.: +34-941-299-752

Received: 18 September 2020; Accepted: 19 October 2020; Published: 20 October 2020

Abstract: A new methicillin resistance gene, named *mecC*, was first described in 2011 in both humans and animals. Since then, this gene has been detected in different production and free-living animals and as an agent causing infections in some humans. The possible impact that these isolates can have in clinical settings remains unknown. The current available information about *mecC*-carrying methicillin resistant *S. aureus* (MRSA) isolates obtained from human samples was analyzed in order to establish its possible clinical implications as well as to determine the infection types associated with this resistance mechanism, the characteristics of these *mecC*-carrying isolates, their possible relation with animals and the presence of other risk factors. Until now, most human *mecC*-MRSA infections have been reported in Europe and *mecC*-MRSA isolates have been identified belonging to a small number of clonal complexes. Although the prevalence of *mecC*-MRSA human infections is very low and isolates usually contain few resistance (except for beta-lactams) and virulence genes, first isolates harboring important virulence genes or that are resistant to non-beta lactams have already been described. Moreover, severe and even fatal human infection cases have been detected. *mecC*-carrying MRSA should be taken into consideration in hospital, veterinary and food safety laboratories and in prevention strategies in order to avoid possible emerging health problems.

Keywords: *Staphylococcus aureus*; methicillin resistance; human infection; CC130

---

1. Introduction

Staphylococcus aureus is an opportunistic pathogen that causes high morbidity and mortality. This microorganism is able to cause diverse diseases that range from having a relatively minor impact, such as a skin infection, to serious and life-threatening episodes, such as endocarditis, pneumonia or sepsis. The impact of *S. aureus* is enhanced by its great capacity to develop and acquire resistance to various antimicrobial agents. Among the antibiotic resistance of *S. aureus*, methicillin resistance mediated by the *mecA* gene is highly relevant as this mechanism provides this bacterium resistance to almost all beta-lactam antibiotics, seriously limiting therapeutic options [1,2]. Recently, the World Health Organization (WHO) outlined the greatest threats in terms of antimicrobial resistance and methicillin-resistant *S. aureus* (MRSA) was classified as a high-priority microorganism. For many years, MRSA infections were only reported in hospitals, with it being considered to be a nosocomial pathogen (hospital-associated MRSA or HA-MRSA). In the 1990s, community-associated MRSA (CA-MRSA) cases in healthy humans without any connection to healthcare settings started to be described and, nowadays, the distinction between CA-MRSA and HA-MRSA seems to be disappearing [3,4].

For the last two decades, a third epidemiological group known as livestock-associated MRSA (LA-MRSA) has been described. *S. aureus* has been considered to be an important zoonotic agent with a great capacity to cause infections in different animal species and in humans. Various studies have suggested that there is a high specificity of the different genetic lineages of *S. aureus* for the host [5]. However, many cases of clones related to animals have been detected and have caused infections in humans [6,7]. Presently, different clonal lineages associated with LA-MRSA have been described and, among these, the clonal complex (CC) CC398 stands out (Table S1). CC398 is related to production animals, mainly pigs, and has been detected worldwide [8]. Infection cases have been identified in humans, both in contact and without contact with animals [9–11]. In addition to CC398, there are other clonal complexes associated with animals such as CC5 in birds, CC9 in pigs, CC97 in cattle or CC133 in small ruminants [12–15].

Remarkably, a new methicillin resistance gene ($mecA_{LGA251}$, which shares only 70% similarity to *mecA* (Figure S1)) was first described in 2011 in both humans and animals [16,17]. Initially these strains were associated with dairy cows and these animals were considered to be a possible reservoir [16]. Since then, this gene has been detected in different production and free-living animals and as an agent causing infection in some humans [8,18]. This new gene was named *mecC* since *mecB* had previously been described in macrococci, but not in staphylococcal species [19]. Worryingly, *mecB* has been recently identified in *S. aureus* and future studies should determine the potential risk that this entails [20]. In the case of MRSA isolates carrying the *mecC* gene (*mecC*-MRSA isolates), these isolates have already been identified as belonging to diverse clonal lineages such as CC130, CC49, sequence type (ST) 151, ST425, CC599 or CC1943 and in very different hosts, including its detection in environmental samples [8,21–23]. There are different theories about the origin of the *mecC* gene and the possible impact that these isolates can have in clinical settings. In this review, the objective was to describe current knowledge about *mecC* detection in humans and its possible clinical implications, as well as to determine the infection types associated with this resistance mechanism, the characteristics of these *mecC*-carrying isolates, their possible relationship with animals and the presence of other risk factors.

2. Detection of *mecC*-MRSA Isolates in Humans

2.1. Human Studies Related to mecC-MRSA

Although the *mecC* gene was initially discovered in an isolate from bulk milk in England, the first human *mecC*-MRSA isolates were also identified in that same study [16]. These human isolates were obtained from patients from the United Kingdom and Denmark. Moreover, in a publication from the same year, two human MRSA isolates carrying this new resistance gene were independently identified in Ireland [17].

Since then, several retrospective and prospective studies using human *S. aureus* isolates/samples were carried out in order to search for *mecC*-MRSA isolates (Tables 1 and 2) [16,18,24–74]. Most of these studies were performed in European countries (Tables 1 and 2), and the UK and Denmark were the countries in which the highest levels of *mecC*-MRSA isolates were detected [16,24,25,39,41].

Table 1. Human studies related to *mecC*-MRSA isolates in which prevalence can be estimated [1]

Reference	Country	Sampling Date	Prevalence: *mecC* Positive Isolates/*S. aureus* or Methicillin Resistant *S. aureus* (MRSA) (%)	Type of Sample/Infection (Number of Isolates)	Clonal Complex: Sequence Type [2] (Number of Isolates)	IEC [3] (Number of Isolates)	Non-beta lactam Resistance (Number of Isolates) [4]	Possible Relationship with Animals
[18,24]	Denmark	1960–2011	112 (0.21%)/53746 MRSA	Wound (37), skin (26), blood (8), post-operative wound (5), urine (4), eye/ear (2), impetigo (2), unknown (28)	CC130 (98)/CC2361 (14): ST2173, ST2174	Negative (2)	Q (NOR) (1), S (111)	Most were from rural areas (106): 4 with contact with animals
[16]	United Kingdom (UK) and Denmark	1975–2011	51 (0.04%)/approximately 120500 *S. aureus*	Screen swab (10), Skin and soft tissue infections (7), nose (5), wound (5), blood (4), skin (4), nose/mouth (2), ear (1), eye/ear (1), finger (1), fluid (1), hand (1), PEG site(1), sputum (1), toe (1), unknown (6)	CC130: ST130 (18), ST1245 (3), ST1764 (3), ST1945 (3), ST1526 (1), ST1944 (1), n/d (17)/CC1943: ST1943 (1), ST1946 (1)/CC425: ST425 (3)	-	S (51)	-
[25]	Denmark	1975–011	127/-: in routine testing 12 (5.91%)/203 MRSA	-	CC130 (107): ST130, ST1245, ST1526, ST1945/CC1943 (14): ST1943, ST1946, ST2173, ST2174/CC425 (6): ST425	-	-	-
[26]	Ireland	2000–2012	1 (1.14%)/88 MRSA	-	CC130 (1)	Negative (1)	Q (NOR) (1)	Patient lived on a Farm
[27]	Germany	2000–2016	2 (0.16%)/1277 MRSA	-	CC130 (2)	-	-	-
[28]	Austria	2002–2012	1 (0.31%)/327 MRSA	-	CC130 (1)	-	-	-
[29]	Belgium	2003–2012	9 (0.18%)/4869 *S. aureus*	Screen swab (4), urine (2), wound (2), sputum (1)	CC130 (4)/CC49 (3)/CC1943 (2)	-	S (9)	Most were from a rural area with a high density of cattle farms
[30,31]	Germany and The Netherlands	2004–2011	16/–	nasal swab (11), wound (2), joint aspirate (1), mouth swab (1), sputum (1)	CC130 (14)/CC1943: ST2361 (1)/CC599: ST599 (1)	Negative (16)	S (1)	-
[32]	Germany	2004–2005 2010-2011	2 (0.06%)/3207 MRSA	Screen swab (1), sputum (1)	-	-	-	-
[33]	Germany	2006–2011	11 (0.09%)/12691 MRSA	Wound (8), dermatitis (1), nasal swab (1), nosocomial pneumonia (1)	CC130 (11)	Negative (11)	Q [CIP (1), MFL (1)] (2), S (9)	Veterinarian (1)
[34,35]	UK	2006–2012	2/–	Screen swab (2)	CC130 (2)	-	-	-
[36]	Slovenia	2006–2013	6 (1.52%)/395 MRSA	Wound (4), Screen swab (2)	CC130: ST130 (6)	Negative (6)	S (6)	Most were from rural areas
[37]	Spain	2008–2013	2 (0.04%)/5505 *S. aureus*	Joint fluid (1), wound (1)	CC130: ST1945 (2)	-	S (2)	-.
[38]	Austria	2009-2013	6 (2%)/301 *S. aureus*	blood (2), screen swab (2), wound (2)	CC130: ST130 (3), new SLV (1)/ CC599: ST599 (2)	-	S (6)	Contact with pet rabbit (1), unknown (5)

Table 1. *Cont.*

Reference	Country	Sampling Date	Prevalence: *mecC* Positive Isolates/*S. aureus* or Methicillin Resistant *S. aureus* (MRSA) (%)	Type of Sample/Infection (Number of Isolates)	Clonal Complex: Sequence Type [2] (Number of Isolates)	IEC [3] (Number of Isolates)	Non-beta lactam Resistance (Number of Isolates) [4]	Possible Relationship with Animals
[39]	Denmark	2010-2011	6 (6.32%)/95 MRSA	-	-	-	-	-
[40]	England	2011-2012	9 (0.45%)/2010 MRSA	Screen swab (6), wound (2), leg ulcer (1)	CC130: ST130 (2), ST1245(4), ST2573 (1), ST2574 (1)/CC425: ST425 (1)	-	L (CLI) (1) -M (ERY) (1), S (8)	-
[41]	UK	2012-2013	12 (0.53%)/2282 MRSA	Screen swab (9), SSTI (3)	CC130: ST1245 (6), ST130 (2), ST1945 (1), ST2574 (1)/CC425: ST425 (1)/CC1943: ST1943 (1)	Negative (11)/type E (1)	M (ERY) (1), S (11)	-
[42]	England	2015	1 (0.08%)/1242 MRSA	Screen swab (1)	CC130: ST130 (1)	Negative (1)	S (1)	-
[43]	England	2018-2019	1 (0.7%)/142 *S. aureus*	-	-	-	-	-
[44]	Germany, UK, Belgium	-	80/-	-	-	-	-	-

[1] Case reports were also analyzed in some other studies but, in this table, only results from prevalence studies are included. [2] CC, clonal complex; ST, sequence type; [3] IEC, immune evasion cluster; [4] L, resistant to lincosamides (CLI, clindamycin); M, resistant to macrolides (ERY, erythromycin); Q, resistant to fluoroquinolones (CIP, ciprofloxacin, MFL, moxifloxacin, NOR, norfloxacin); S, susceptible to all non-beta lactam agents tested. UK, United Kingdom.

Table 2. Studies performed on humans, in which *mec*C-MRSA isolates were sought but not detected.

Reference	Country	Sampling Date	Type of Samples [1]	Number of Samples or (*S. aureus* or MRSA) Isolates
[45]	Switzerland	2005–2012	Clinical/screening	1695 *S. aureus* isolates
[46]	Ghana	2007–2012	Clinical	9834 blood samples
[47]	Turkey	2007–2014	Clinical	1700 *S. aureus* isolates
[48]	Belgium	2009–2011	Screening	149 farmers and family members (41 MRSA isolates)
[49]	Hungary	2009–2011	Screening	878 children
[50]	United States	2009–2011	Clinical/screening	364 *S. aureus* isolates (102 MRSA isolates)
[51]	Ireland	2011	Screening	64 residents
[52]	UK	2011	Screening	307 cattle veterinarians
[53]	Jordan	2011–2012	Screening	716 humans (56 MRSA isolates)
[54]	Germany	2011–2013	Screening	1878 non-hospitalized adults
[55]	Belgium	2012–2013	Clinical	510 cystic fibrosis patients
[56]	Greece	2012–2013	Screening	18 veterinary personnel
[57]	The Netherlands	2012–2013	Clinical/screening	13,387 samples
[58]	UK	2012–2013	Clinical	500 *S. aureus* isolates
[59]	Egypt	2013	Clinical/screening	1300 dental patients
[60]	Taiwan	2013–2014	Clinical/screening	3717 *S. aureus* isolates
[61]	Turkey	2013–2014	Screening	7 MRSA isolates
[62]	Turkey	2013–2016	Clinical/screening	494 MRSA isolates
[63]	Spain	2014	Screening	15 humans in contact with animals
[64]	Poland	2014–2016	Screening	955 students (only one MRSA isolate)
[65]	Germany	2015	Clinical	140 Gram-positive isolates
[66]	UK	2015	Clinical	520 *S. aureus* isolates
[67]	United States	2015	Screening	479 patients
[68]	India	2015–2017	Screening	32 animal handlers
[69]	Spain	2016	Clinical/screening	45 non-beta-lactam susceptible MRSA isolates
[70]	Greece	2016–2017	Screening	68 farmers
[71]	Denmark	2017	Screening	16 workers at wildlife rehabilitation centres
[72]	Italy	2017–2018	Clinical/screening	102 MRSA isolates
[73]	Egypt	-	Screening	223 health care personnel
[74]	Madagascar	-	Screening	1548 students and healthcare workers

[1] Screening: isolates obtained in epidemiological studies for colonization detection.

Unfortunately, the design of these studies was very different, which complicates any comparison of the data obtained. Importantly, the criteria chosen for the selection of the initial isolates/samples varied significantly. While all *S. aureus* isolates were collected in some studies [29,38,45], only MRSA isolates were included in others [24,26–28,32,33,36,41,42]. Moreover, several studies were more restrictive and only used isolates that showed characteristics suspected of carrying the *mecC* gene such as *spa* types associated with *mecC*-positive clonal lineages previously described, *mecA*-negative MRSA isolates, isolates with antimicrobial susceptibility suspected to be *mecC*-positive or *pvl*-negative MRSA isolates [25,26,37,69] (Table S1). In any case, the *mecC*-MRSA human prevalence detected in most of the studies was very low. Several studies did not identify any *mecC*-positive *S. aureus* among included human isolates/samples (Table 2) [45–74]. In studies in which this gene was detected (Table 1), the prevalence identified, considering the total number of isolates/samples included, was < 1% in most of the cases [24,27–29,32,33,37,40–43], similar to that identified in the first study in which *mecC* was discovered (approximately 0.04%) [16]. In a few studies, the prevalence was > 1% but, in all of these, only a small number of initial isolates (<400 isolates) was used; this may be the reason for the high prevalence value obtained (up to 6.3%) [25,26,36,38,39]. Recently, a meta-analysis of the prevalence of *mecC*-MRSA, based on previously published results, estimated the prevalence of *mecC*-MRSA in the human subgroup at 0.004% (95% CI = 0.002–0.007), and the prevalence in the animal subgroup to be 0.098% (95% CI = 0.033–0.174) [75].

2.2. mecC-MRSA Human Case Reports

A total of 61 human case reports associated with *mecC*-MRSA isolates has been described (Table 3) [17,36,37,45,76–81]. Although *mecC*-positive isolates have been identified in Asia, Europe, and Oceania [21,82,83] in different hosts, all human case reports were described in European countries (Table 3). This was to be expected considering that the majority of the papers in which *mecC*-MRSA has been detected in both animals and humans, as well as in environmental samples, have been focused on countries on this continent [8,21–23].

In 4 of the 61 human case reports, *mecC*-MRSA was only identified in screen swabs (for colonization detection), with it not being related to the cause of the patient's admission [36,37,45], and the clinical information was not indicated in another two case reports [17]. In the remaining 56 studies, *mecC*-MRSA isolates were related to (number of cases): skin and wound infections (47 cases) [37,76,79,81], joint and bone infections (3 cases) [37,77,78], respiratory infections (2 cases) [76] and bacteremia (2 cases) [37,80]. Taking into consideration the type of samples in which *mecC*-positive isolates have been detected in humans (Tables 1–4), most *mecC* human cases were implicated in skin or wound infections. However, the detection of *mecC*-MRSA isolates in other types of samples such as blood, sputum or urine is remarkable (Table 4). Pertinently, some serious infections have been described, such as severe bone infections [78], nosocomial pneumonia [33] and bacteremia [16,24,80], in some cases ending with the death of the patient [37].

Table 3. Human *mecC* MRSA case reports.

Reference [1]	Country	Sampling Date	Number of Described Case Reports	Year-Old Patient	Type of Sample/Infection [2]	Clonal Complex: Sequence Type [3]	IEC [4]	Non-Beta Lactam Resistance [5]	Possible Relationship with Animals
[76]	Sweden	2005–2014	45	Median age (range) 60 (2–86)	Wound, sputum, nasopharynx	CC130/CC2361	-	L-M (1 isolate), S (44 isolates)	Most were from a rural area: farmer (1), patients lived on farms (4)
[77]	France	2007	1	67	Fluid of lesion heel	CC130: ST1945	-	-	-
[37]	Spain	2008–2013	7	3, 50, 63, 64, 76, 80, 85	Blood, joint fluid, nasal screen swab, urine, wound	CC130: ST130, ST1945	-	S	No epidemiological data were available except for one patient who did not have any contact with animals
[78]	France	2010	1	48	Blood, ear fluid, retrosternal abscess	CC130	Negative	S	Contact with cows
[17]	Ireland	2010	2	64, 85	-	CC130: ST130, ST1764	Negative	S (but detection of *tet* efflux)	-
[45]	Switzerland	2011	1	59	Groin, nose, and throat screen swab	CC130: ST130	-	-	Contact with a cat
[79]	Spain	2012	1	46	Skin lesion swab	CC130	-	S	Patient lived in rural area with high density of livestock animals
[36]	Slovenia	2013	1	86	Nose and skin screen swabs	CC130: ST130	Negative	S	Patient lived on a farm and had contact with pigs, cats and dogs
[80]	Spain	2013	1	76	Blood	CC130: ST1945	-	S	No contact with livestock
[81]	Spain	2013–2014	1	34	Superficial skin lesion swab	CC130: ST130	Negative	S	Contact with livestock animals

[1] Prevalence studies were also included in some papers but, in this table, only case report results are present. [2] Screen swab: sample for colonization detection. [3] CC, clonal complex; ST, sequence type; [4] IEC, immune evasion cluster. [5] L, resistant to lincosamides; M, resistant to macrolides; S, susceptible to all non-beta lactam agents tested.

Table 4. Type of sample/infection in which *mecC*-MRSA isolates have been identified among human patients.

Type of Sample/Infection	Number of Isolates [2]	References
Screen swab [1]	54	[16,29–34,36,38,40–42,45,76]
Skin lesion/dermatitis/impetigo wound/post-operative wound/skin and soft tissue infections	158	16,24,29,30,33,36-38,40,41,76,79,81
Blood	16	[16,24,37,38,80]
Urine	7	[24,29,37]
Nosocomial pneumonia/sputum/ Tracheal aspirate	7	[29,30,32,33,76]
Nose	5	[16]
Eye/ear	3	[16,24]
Fluid of heel/joint fluid	3	[30,37,77]
Mouth/Nose	2	[16]
Ear	1	[16]
Finger	1	[16]
Fluid	1	[16]
Hand	1	[16]
Percutaneous endoscopic gastrostomy site	1	[16]
Retrosternal abscess	1	[78]
Toe	1	[16]
Unknown	34	[16,24]

[1] In screen swab: all samples in which was clearly indicated that they did not cause infection were included. However, in several studies it was not indicated whether samples were screen samples or if these samples were taken in infection sites. [2] In human case reports, only one isolate from the most representative infection sample was considered.

3. Risk Factors for *mecC*-MRSA Infection

3.1. Contact with Animals

Since the first description of the *mecC* gene, contact with animals has been considered to be a risk factor for *mecC*-MRSA infection or carriage for several reasons [16,17]. This gene was identified in isolates belonging to CC130, and this clonal complex was predominantly detected among methicillin-susceptible *S. aureus* (MSSA) isolates from bovine sources [17]. Moreover, the discovery of this gene in isolates obtained from dairy cows suggested that these animals might provide a reservoir of this resistance mechanism [16]. Thereby, in some of the studies carried out since then, information about the possible contact of patients with animals was indicated (Tables 1 and 3). Many studies found out that most of the patients lived in rural areas or areas with a high density of farms [18,24,26,29,36,76,79]. In this sense, four studies indicated patient contact with livestock or farm animals [18,24,76,78,81], two referred to only contact with pets [38,45], two patients had no contact with animals and the authors did not have a plausible explanation for the detection of these isolates [37,80], one patient was a veterinarian [33], and in several studies this information was not indicated [16,17,27,30,31,41,77]. Interestingly, *mecC*-MRSA transmission between animals and humans was demonstrated in two human infection cases by whole genome sequencing. Specific clusters including isolates from each human infection case and their own livestock were detected. Thus, human and animal isolates from the same farm only differed by a small number of SNPs [18]. These findings highlight the role of livestock as a potential reservoir for *mecC*-MRSA.

3.2. mecC-MRSA Carriage in Humans

S. aureus shows a great capacity to colonize the skin and nares of hosts, being able to last over time and cause opportunistic infections [84,85]. *mecC*-MRSA isolates were identified as commensals in several prevalence and case report studies (see screen swab in Tables 1–4). At least 54 *mecC*-MRSA positive isolates were obtained from screen swabs, mainly from the nose, but also from throat and groin sites. Moreover, isolates obtained from other types of samples could also be considered as commensals, as in one human case report in Spain in which the isolate recovered from the urine of one patient was considered as a colonizer since the patient did not present urinary symptoms [37].

mecC-MRSA isolates implicated in both colonization and infection were obtained from the same patient in some studies [18,37]. Indeed, one patient with bacteremia due to an *mecC*-MRSA isolate also presented nasal colonization by the same *mecC*-MRSA isolate (with the same genetic characteristics) [18]. These results corroborated the importance of colonization being the previous step, which enables isolates causing severe disease. Interestingly, in another bacteremia case in which the patient died, a household transmission between grandfather and grandson was detected, with the grandson being colonized by the same isolate [37]. Nevertheless, in other studies, *mecC*-MRSA isolates were not identified as colonizers from patients with *mecC* infections [81], and it has been suggested that *mecC*-MRSA isolates might be worse colonizers and less contagious in humans than *mecA*-MRSA isolates [76]. In the study carried out in Sweden, only two out of the patient's 27 family members were positive for *mecC*-MRSA isolates and the median time for *mecC* carriage was 21 days [76].

3.3. Patient Age

Most of the patients described in case reports (Table 3) were middle-aged or elderly [17,36,37,45,77–80], except two patients: one of them was a 34 year-old farm worker with high contact with animals who presented a superficial skin lesion [81], and the other was a healthy 3 year-old child [37]. The average age of patients with *mecC*-MRSA detected in Denmark during 2007–2011 was 51 [24] and the average detected in Sweden in 2005–2014 was 60 [76]. In the Danish study, CA-MRSA *mecC* patients were significantly older than other CA-MRSA cases, indicating that *mecC*-MRSA seems to have a different origin and epidemiology to typical CA-MRSA [24].

3.4. Underlying Chronic Disease

Remarkably, in the 45 human cases detected in Sweden, most patients had some kind of underlying chronic disease (diabetes mellitus, cancer, autoimmune diseases or atherosclerotic diseases), or an existing skin lesion [76]. Infection by *mecC*-MRSA of wounds has also been suggested by others [79]. Moreover, *mecC*-MRSA infections were identified in patients with primary pathologies (diabetes, myelodysplastic syndrome, peripheral arterial occlusion disease, etc.) in one study in Austria [38], and in a patient with an urothelial carcinoma in Spain [80]. Unfortunately, information about other underlying diseases of *mecC*-MRSA positive patients is missing in most of the papers.

4. Characterization of *mecC*-MRSA Human Isolates

4.1. Clonal Lineages of mecC-MRSA of Human Origin

As in other hosts, most of the *mecC*-MRSA isolates obtained from human samples belonged to CC130 (Tables 1 and 3) (Figure 1). Other clonal complexes identified were CC49, CC425, CC599, CC1943 and CC2361 [16,24,25,29–31,38,40,41,76] (Table 1) (Figure 1). Worryingly, it has been hypothesized that SCC*mec* XI (the SCC element that contains the *mecC* gene) might have the potential to be transferred to other *S. aureus* clonal lineages due to the fact that it is bounded by integration site sequence repeats and that it has intact site specific recombination components [16] (Table S1 and S2). Until now, *mecC*-MRSA CC130 isolates have been identified in all countries in which clonal lineages were determined and it was the unique CC detected in Spain, France, Ireland, Slovenia and Switzerland [17,37,45,77–79,81] (Figure 1). Remarkably, in France and Spain there were several human infection reports, but in all of them the *mecC*-MRSA isolates belonged to CC130 (Table 3). After CC130, the clonal complexes CC1943 and CC599 were the most widely detected in humans, being identified in four and three countries respectively [16,25,29,31,38,41] (Figure 1). Conversely, CC49 was only described in one study in Belgium [29]. While CC49, CC130, CC425, CC599 and CC1943 were also identified in *mecC*-MRSA isolates from a non-human origin, CC2361 has been only described in humans so far [24,76]. Thus, CC130 was described in farm, domestic and wild animals and in food samples; CC49 in horses and small mammals, CC425 in wild animals and food, CC599 in pets and farm animals and CC1943 in pets [8].

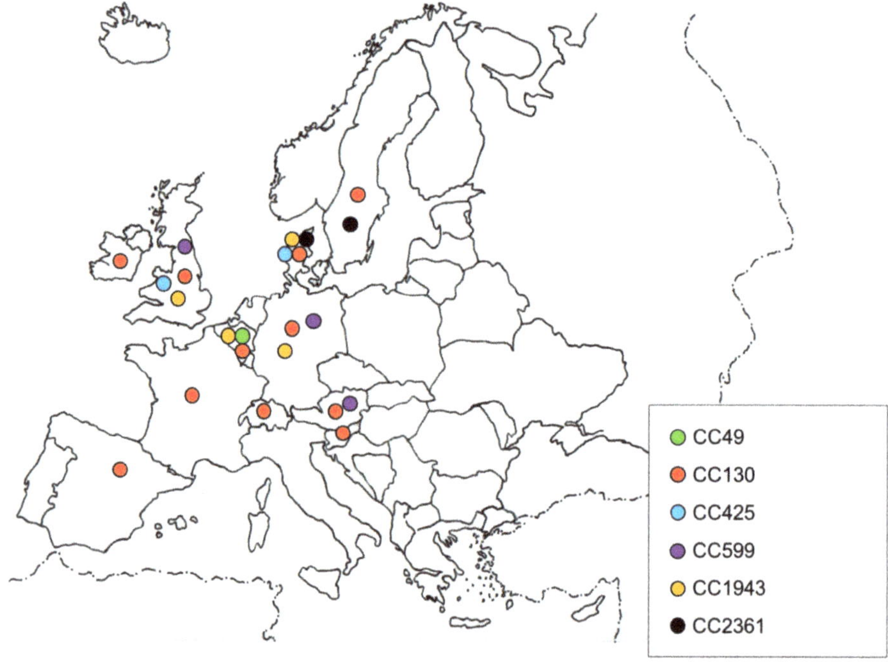

Figure 1. Clonal complexes (CCs) detected in *mecC*-MRSA human isolates.

A large variety of *spa* types was detected among the human *mecC*-MRSA isolates (Figure 2). The most predominant *spa* type was t843, which is associated with CC130 and was identified in a total of 260 human isolates. This *spa* type was detected in all countries in which human *mecC*-MRSA isolates were detected, except in Switzerland [45]. Other *spa* types were also described in several countries. Some of them were identified only in two countries, this is the case of t792, t1773, t5930, t6293, t6386, t7485, t7734, t7945, t7946, t7947 and t9397, but others were more widely spread as t978, t1535, t1736, t3391 or t6220 (Figure 2). Although there is a strong association among *spa* types and MLST clonal complexes [86], some *spa* types were associated with different clonal complexes. Two isolates obtained in screen swabs from two patients in two different hospitals from England presented the *spa* type t11706 [40]; one of these isolates belonged to ST1245 (CC130) and the other one to ST425 (CC425). Moreover, the *spa* types t978, t2345, t3391 and t8835 were associated in some studies with CC1943 [16,25,29], and in others with CC2361 [24,76]. Nevertheless, the founders of both clonal complexes, ST1943 and ST2361, are Single Locus Variant (SLV) of each other (and only differ at the *aroE* allele), which could explain these results (Table S1).

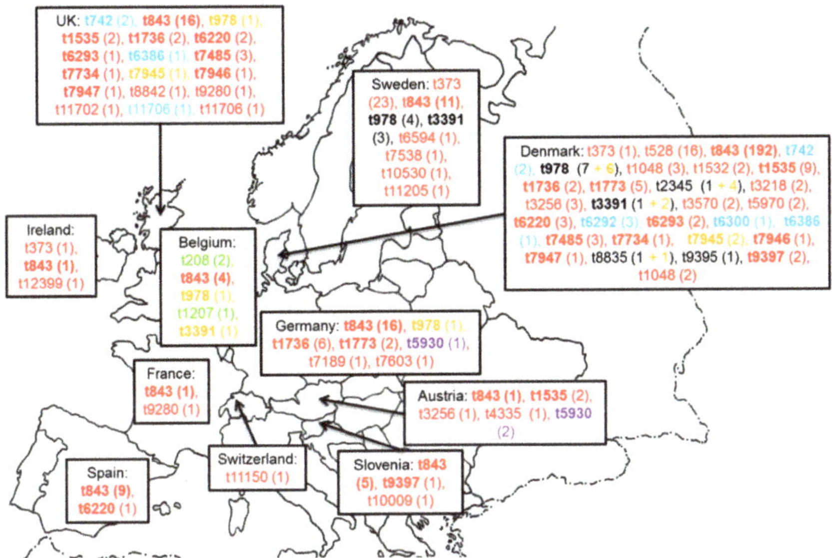

Figure 2. *spa* types detected in *mecC*-MRSA isolates of humans. Colors indicate the clonal complexes associated with each *spa* type: green CC49, red CC130, blue CC425, purple CC599, orange CC1943, black CC2361. The number of isolates of each *spa* type is indicated in parentheses (to calculate the number of isolates in human case reports, only one isolate from each *spa* type and each patient was considered)

4.2. Treatment and Antimicrobial Resistance Profile of mecC-MRSA Human Isolates

Most of the human *mecC*-MRSA isolates detected were susceptible to all non-beta-lactam antimicrobials tested (Tables 1 and 3). This is in accordance with results obtained in *mecC*-MRSA isolates from other origins [21]. In one study performed in Spain, isolates using this criterion were selected in order to identify *mecC*-MRSA or CA-MRSA isolates [69]. Although *mecC*-MRSA was not detected, this resistance phenotype was a valuable marker for Panton-Valentine Leukocidin (PVL)-producer isolates (Table S1). Nevertheless, the low prevalence of *mecC*-MRSA isolates in this region could be responsible for this result and the use of this phenotype to suspect the presence of the *mecC* mechanism should not be discarded.

The most important problem for treating *mecC*-MRSA infections is that these isolates must be correctly identified. Although *mecC* isolates are considered to be MRSA, these isolates sometimes show borderline susceptibility results for oxacillin or cefoxitin, being identified as MSSA if the *mecA* gene is only tested [44]. This could lead to the implementation of inappropriate therapies. Resistance development to other antimicrobial agents could be added to this, considering the capacity of *S. aureus* to acquire resistance to various antimicrobial agents. Some *mecC*-MRSA isolates detected in humans have shown resistance to non-beta-lactam antimicrobials [24,33,40,41,76] (Table 1). Fluoroquinolone resistance was identified in two isolates in Germany [33] and in one isolate in Denmark [24]. Macrolide and lincosamide resistance was also detected in some studies: one erythromycin resistant isolated in the UK [41] and one erythromycin and clindamycin resistant isolate in Sweden [76] and England [40]. Regarding resistance mechanisms, in two studies carried out in Ireland, the gene *sdrM*, which encodes a multidrug efflux pump related to norfloxacin resistance and *tet* efflux related to tetracycline resistance, were identified in one and two *mecC*-MRSA CC130 isolates, respectively [17,26]. Although there has only been one study, whose objective was to compare different diagnostic tests, human *mecC*-MRSA isolates resistant to several antimicrobial families have been detected [87]. The presence of resistance to non beta-lactam agents in *mecC*-MRSA isolates significantly limits our therapeutic options.

4.3. Virulence of mecC-MRSA Human Isolates

The search for virulence genes in human *mecC*-MRSA isolates has been highly variable from one study to another. In any case, for the moment, the most common virulence genes detected have been *hla, hld, hlb, edinB, lukED, cap8* or *ica*, with these genes being highly associated with CC130 [17,26,31,33,36,38,41]. Fortunately, no *mecC*-MRSA isolates carrying the PVL genes were detected. However, other clonal lineages associated with animals have been able to acquire this virulence factor [88]. For this, their detection in *mecC*-MRSA isolates cannot be ruled out in the future. Significantly, some pyrogenic toxin superantigen (PTSAg) genes have been detected in *mecC*-MRSA isolates [29,31,38,41]. These genes might be related to specific clonal lineages such as CC599, CC1943 or CC2361. Thus, the gene *tst* encoding the toxic shock syndrome toxin has been found in three CC1943 isolates (two harboring *sec* gene and one containing *seg* and *sei* genes) [29,41], in three CC599 isolates (two of them positive for *sel* gene and one for *sec* and *sel*) [31,38] and in one *sec, seg, sei, sel, sen, seo* and *seu* positive CC2361 isolate [31] (Table S1).

5. *mecC*-MRSA Problem: What is its Origin? Is It an Emerging Problem?

The oldest known *mecC*-MRSA isolate, dated in 1975, was detected in the retrospective study performed by García-Álvarez et al. [16] This isolate was identified in a human blood sample from Denmark and its detection suggested a possible human origin for the *mecC* gene [16]. Later, in two other retrospective studies also carried out in Denmark, two *mecC*-positive isolates were identified in samples dated in 1975 [24,25], both were also identified in human blood samples [24,25]. However, in two of these studies, the oldest sample studied was obtained in 1975, so the presence of older isolates cannot be ruled out [16,25]. With respect to the remaining retrospective studies in which the presence of the *mecC* gene was sought, the dates of the samples were much later than these three studies, with them being isolates obtained from the year 2000 and later (Table 1). Regarding the earliest reported *mecC*-MRSA isolate in other hosts, 1975 also seems to be the key date [89,90]. Therefore, this resistance mechanism might have been present for over 45 years.

Moreover, this resistance mechanism is highly associated with CC130 since most of the *mecC*-MRSA isolates belong to this clonal lineage. A human-to-bovine host-jump of CC130, which occurred ~5429 years ago, has been suggested [91]. The time and host in which CC130 MSSA isolates acquired the *mecC* resistance gene remain unknown today. In order to establish a possible human or animal origin for the detected *mecC*-MRSA isolates in human samples, several studies analyzed the presence of IEC (immune evasion cluster) genes [17,24,26,31,33,36,41,42,78,81] (Tables 1 and 3). In all cases, human *mecC*-MRSA isolates were negative for *sak, chp* and *scn* except for one ST1945 (CC130) isolate obtained from a screen swab from a patient in the UK that was positive for *sak* and *scn* (IEC type E), suggesting a possible human origin [41]. Nevertheless, it has been suggested that IEC type E might be a conserved part of ST1945 since *mecC* MRSA ST1945 isolates from wild animals also showed IEC type E in several studies in Spain [41,63,92].

The newest human *mecC*-MRSA isolates detected so far in Europe were obtained in 2015, one of them in Germany [27], and the other in England [42]. Both strains showed similarities to those identified in the first studies [16,17] and both belonged to CC130. Nevertheless, after phylogenetic analysis, the strain identified in England seemed not to be highly related to any of the published sequenced *mecC*-MRSA CC130 isolates [42]. Despite the non-description of *mecC*-MRSA isolates in humans in the last 5 years in Europe, data provided by previous studies have detected an increasing tendency in the prevalence of *mecC*-positive isolates [24], indicating that surveillance in detecting this resistance mechanism must be maintained. The lack of detection could be due to the low prevalence of this resistance mechanism and/or problems in *mecC* diagnostic methods. Important difficulties in the detection of *mecC*-MRSA isolates have been indicated [44,93]. It has been shown that various clinical tests used in hospital labs might have failed to identify 0 to 41% of *mecC*-MRSA isolates [93]. It is important to optimize and develop new testing protocols and redefine currently available phenotypic testing methods [44]. In this regard, several commercial PCR-based tests that detect *mecC* and *mecA* genes have been developed. Moreover, recently, *mecA/mecC* MRSA isolates from cattle have been described [83]. The possible clinical impact of isolates carrying both genes is currently unknown.

6. Implications in Veterinary and Food Safety

Although this review is focused on the human health implications of *mecC*-MRSA isolates, the effects that these isolates can have on veterinary medicine should not be forgotten. *mecC*-MRSA isolates causing infections in domestic animals have been identified in several studies [8,94,95]. However, this resistance gene seems to be most frequently detected in livestock animals including cattle, sheep and rabbits [8,21]. Although *mecC*-MRSA rarely causes clinical disease in these food-producing animals, there are reports of bovine mastitis in several countries [96,97]. As observed in humans, most of the *mecC*-MRSA isolates detected in other hosts also belong to CC130, with the characteristics of these animal *mecC*-MRSA isolates being very similar to those detected in humans [8,21].

On the other hand, the presence of *mecC*-MRSA in dairy animals is highly relevant since it could be a route of entry of these isolates into the food chain. Indeed, milk samples have been identified carrying *mecC*-MRSA [8] with the consequent risk of colonization for food handlers. In this case, it is worth highlighting one of the clinical cases included in this review in which the patient was a cheese producer [81]. *mecC*-MRSA zoonotic transmission has been demonstrated in some studies [18], with the correct prevention, detection and control measures in veterinary and food safety laboratories being necessary.

7. Conclusions and Future Problems Associated with *mecC*

Currently, the prevalence of human *mecC*-MRSA infections is very low. However, *mecC*-MRSA isolate transmission between different hosts indicates the great capacity of these isolates for spreading. There is a wide range of reservoirs in wild, livestock and companion animals and zoonotic transmission of these isolates could increase the number of *mecC*-MRSA human clinical cases. Moreover, SCC*mec* XI might have the potential to be transferred to other clonal lineages in the future. Their transfer to more virulent and better-adapted human clones would be deeply troubling. Worryingly, the *mecC* gene has already been detected in clonal lineages in which important virulence genes were identified (CC599, CC1943 or CC2361) or in which IEC was described (ST1945-CC130). Moreover, *mecC*-MRSA isolates resistant to non-beta lactams have been detected. Acquisition of non-beta lactam resistance by *mecC*-MRSA isolates significantly limits our therapeutic options. *mecC*-MRSA should be taken into consideration in hospital and veterinary laboratories and in food safety institutions, and prevention strategies must be implemented in order to avoid possible emerging health problems.

Supplementary Materials: Supplementary Materials are available online at http://www.mdpi.com/2076-2607/8/10/1615/s1.

Author Contributions: C.L. contributed to the search of articles and to their tabulation and classification into different categories. She also contributed to the design and the analysis of the review and to the writing of the paper. R.F.-F., L.R.-R., P.G. and M.Z. helped in the general review of the manuscript. C.T. contributed with project funding, and with the original idea and the design of the manuscript and reviewed the initial version of the manuscript. All authors have read and agreed to the published version of the manuscript.

Funding: This research was funded by project SAF2016-76571-R of the Agencia Estatal de Investigación (AEI) and the Fondo Europeo de Desarrollo Regional (FEDER) of EU. Laura Ruiz-Ripa has a pre-doctoral fellowship from the Universidad de La Rioja. Rosa Fernández-Fernández has a predoctoral fellowship from the Ministerio de Ciencia, Innovación y Universidades of Spain (FPU18/05438).

Conflicts of Interest: The authors declare no conflict of interest.

References

1. Cheng, M.; Antignac, A.; Kim, C.; Tomasz, A. Comparative study of the susceptibilities of major epidemic clones of methicillin-resistant *Staphylococcus aureus* to oxacillin and to the new broad-spectrum cephalosporin ceftobiprole. *Antimicrob. Agents Chemother.* **2008**, *52*, 2709–2717. [CrossRef] [PubMed]
2. Jacqueline, C.; Caillon, J.; Mabecque, V.L.; Miègeville, A.F.; Ge, Y.; Biek, D.; Batard, E.; Potel, G. In vivo activity of a novel anti-methicillin-resistant *Staphylococcus aureus* cephalosporin, ceftaroline, against vancomycin-susceptible and -resistant *Enterococcus faecalis* strains in a rabbit endocarditis model: A comparative study with linezolid and vancomycin. *Antimicrob. Agents Chemother.* **2009**, *53*, 5300–5302. [CrossRef] [PubMed]
3. Mediavilla, J.R.; Chen, L.; Mathema, B.; Kreiswirth, B.N. Global epidemiology of communityassociated methicillin resistant *Staphylococcus aureus* (CA-MRSA). *Curr. Opin. Microbiol.* **2012**, *15*, 588–595. [CrossRef] [PubMed]
4. Sung, J.Y.; Lee, J.; Choi, E.H.; Lee, H.J. Changes in molecular epidemiology of communityassociated and health care-associated methicillin-resistant *Staphylococcus aureus* in Korean children. *Diagn. Microbiol. Infect. Dis.* **2012**, *74*, 28–33. [CrossRef] [PubMed]
5. Herron-Olson, L.; Fitzgerald, J.R.; Musser, J.M.; Kapur, V. Molecular correlates of host specialization in *Staphylococcus aureus*. *PLoS ONE* **2007**, *2*, e1120. [CrossRef] [PubMed]
6. Aspiroz, C.; Lozano, C.; Vindel, A.; Lasarte, J.J.; Zarazaga, M.; Torres, C. Skin lesion caused by ST398 and ST1 MRSA, Spain. *Emerg. Infect. Dis.* **2010**, *16*, 157–159. [CrossRef] [PubMed]
7. Lozano, C.; Aspiroz, C.; Ezpeleta, A.I.; Gómez-Sanz, E.; Zarazaga, M.; Torres, C. Empyema caused by MRSA ST398 with atypical resistance profile, Spain. *Emerg Infect. Dis.* **2011**, *17*, 138–140. [CrossRef]
8. Zarazaga, M.; Gómez, P.; Ceballos, S.; Torres, C. Molecular epidemiology of *Staphylococcus aureus* lineages in the animal-human interface. In *Staphylococcus Aureus*; Fetsch, A., Ed.; Academic Press: Cambridge, MA, USA, 2018; pp. 189–214.
9. Benito, D.; Lozano, C.; Rezusta, A.; Ferrer, I.; Vasquez, M.A.; Ceballos, S.; Zarazaga, M.; Revillo, M.J.; Torres, C. Characterization of tetracycline and methicillin resistant *Staphylococcus aureus* Strains in a Spanish hospital: Is livestock-contact a risk factor in infections caused by MRSA CC398? *Int. J. Med. Microbiol.* **2014**, *304*, 1226–1232. [CrossRef]
10. Larsen, J.; Petersen, A.; Sørum, M.; Stegger, M.; van Alphen, L.; Valentiner-Branth, P.; Knudsen, L.K.; Larsen, L.S.; Feingold, B.; Price, L.B.; et al. Meticillin-resistant *Staphylococcus aureus* CC398 is an increasing cause of disease in people with no livestock contact in Denmark, 1999 to 2011. *Euro Surveill.* **2015**, *20*. [CrossRef]
11. Lekkerkerk, W.; van Wamel, J.B.; Snijders, S.B.; Willems, R.J.; van Duijkeren, E.; Broens, E.M.; Wagenaar, J.A.; Lindsay, J.A.; Vos, M.C. What is the origin of livestock-associated methicillin-resistant *Staphylococcus aureus* Clonal Complex 398 isolates from humans without livestock contact? an epidemiological and genetic analysis. *J. Clin. Microbiol.* **2015**, *53*, 1836–1841. [CrossRef]
12. Feltrin, F.; Alba, P.; Kraushaar, B.; Ianzano, A.; Argudín, M.A.; Di Matteo, P.; Porrero, M.C.; Aarestrup, F.M.; Butaye, P.; Franco, A.; et al. A Livestock-associated, multidrug-resistant, methicillin-resistant *Staphylococcus aureus* Clonal Complex 97 lineage spreading in dairy cattle and pigs in Italy. *Appl. Environ. Microbiol.* **2015**, *82*, 816–821. [CrossRef] [PubMed]
13. Lowder, B.V.; Guinane, C.M.; Ben Zakour, N.L.; Weinert, L.A.; Conway-Morris, A.; Cartwright, R.A.; Simpson, A.J.; Rambaut, A.; Nübel, U.; Fitzgerald, J.R. Recent human-to-poultry host jump, adaptation, and pandemic spread of *Staphylococcus aureus*. *Proc. Natl. Acad. Sci. USA* **2009**, *106*, 19545–19550. [CrossRef] [PubMed]
14. Monecke, S.; Gavier-Widén, D.; Hotzel, H.; Peters, M.; Guenther, S.; Lazaris, A.; Loncaric, I.; Müller, E.; Reissig, A.; Ruppelt-Lorz, A.; et al. Diversity of *Staphylococcus aureus* isolates in European wildlife. *PLoS ONE.* **2016**, *11*, e0168433. [CrossRef] [PubMed]
15. Ye, X.; Wang, X.; Fan, Y.; Peng, Y.; Li, L.; Li, S.; Huang, J.; Yao, Z.; Chen, S. Genotypic and phenotypic markers of livestock-associated methicillin-resistant *Staphylococcus aureus* CC9 in humans. *Appl. Environ. Microbiol.* **2016**, *82*, 3892–3899. [CrossRef]

16. García-Álvarez, L.; Holden, M.T.G.; Lindsay, H.; Webb, C.R.; Brown, D.F.J.; Curran, M.D.; Walpole, E.; Brooks, K.; Pickard, D.J.; Teale, C.; et al. Meticillin-resistant *Staphylococcus aureus* with a novel *mecA* homologue in human and bovine populations in the UK and Denmark: A descriptive study. *Lancet Infect. Dis.* **2011**, *11*, 595–603. [CrossRef]

17. Shore, A.C.; Deasy, E.C.; Slickers, P.; Brennan, G.; O'Connell, B.; Monecke, S.; Ehricht, R.; Coleman, D.C. Detection of staphylococcal cassette chromosome *mec* type XI carrying highly divergent *mecA, mecI, mecR1, blaZ,* and *ccr* genes in human clinical isolates of clonal complex 130 methicillin-resistant *Staphylococcus aureus*. *Antimicrob. Agents Chemother.* **2011**, *55*, 3765–3773. [CrossRef]

18. Harrison, E.M.; Paterson, G.K.; Holden, M.T.G.; Larsen, J.; Stegger, M.; Larsen, A.R.; Petersen, A.; Skov, R.L.; Christensen, J.M.; Zeuthen, A.B.; et al. Whole genome sequencing identifies zoonotic transmission of MRSA isolates with the novel *mecA* homologue *mecC*. *EMBO Mol. Med.* **2013**, *5*, 509–515. [CrossRef]

19. Becker, K.; Ballhausen, B.; Köck, R.; Kriegeskorte, A. Methicillin resistance in *Staphylococcus* isolates: The "*mec* alphabet*"* with specific consideration of *mecC*, a *mec* homolog associated with zoonotic *S. aureus* lineages. *Int. J. Med. Microbiol.* **2014**, *304*, 794–804. [CrossRef]

20. Becker, K.; van Alen, S.; Idelevich, E.A.; Schleimer, N.; Seggewiß, J.; Mellmann, A.; Kaspar, U.; Peters, G. Plasmid-Encoded Transferable *mecB*-Mediated Methicillin Resistance in *Staphylococcus aureus*. *Emerg. Infect. Dis.* **2018**, *24*, 242–248. [CrossRef]

21. Aires-de-Sousa, M. Methicillin-resistant *Staphylococcus aureus* among animals: Current overview. *Clin. Microbiol. Infect.* **2017**, *23*, 373–380. [CrossRef] [PubMed]

22. Gómez, P.; González-Barrio, D.; Benito, D.; García, J.T.; Viñuela, J.; Zarazaga, M.; Ruiz-Fons, F.; Torres, C. Detection of methicillin-resistant *Staphylococcus aureus* (MRSA) carrying the *mecC* gene in wild small mammals in Spain. *J. Antimicrobial Chemother.* **2014**, *69*, 2061–2064. [CrossRef] [PubMed]

23. Ruiz-Ripa, L.; Alcalá, L.; Simón, C.; Gómez, P.; Mama, O.M.; Rezusta, A.; Zarazaga, M.; Torres, C. Diversity of *Staphylococcus aureus* clones in wild mammals in Aragon, Spain, with detection of MRSA ST130-*mecC* in wild rabbits. *J. Appl. Microbiol.* **2019**, *127*, 284–291. [CrossRef] [PubMed]

24. Petersen, A.; Stegger, M.; Heltberg, O.; Christensen, J.; Zeuthen, A.; Knudsen, L.K.; Urth, T.; Sorum, M.; Schouls, L.; Larsen, J.; et al. Epidemiology of methicillin-resistant *Staphylococcus aureus* carrying the novel *mecC* gene in Denmark corroborates a zoonotic reservoir with transmission to humans. *Clin. Microbiol Infect.* **2013**, *19*, E16–E22. [CrossRef] [PubMed]

25. Stegger, M.; Andersen, P.S.; Kearns, A.; Pichon, B.; Holmes, M.A.; Edwards, G.; Laurent, F.; Teale, C.; Skov, R.; Larsen, A.R. Rapid detection, differentiation and typing of methicillin-resistant *Staphylococcus aureus* harbouring either *mecA* or the new *mecA* homologue *mecA*(LGA251). *Clin. Microbiol. Infect.* **2012**, *18*, 395–400. [CrossRef] [PubMed]

26. Kinnevey, P.M.; Shore, A.C.; Brennan, G.I.; Sullivan, D.J.; Ehricht, R.; Monecke, S.; Coleman, D.C. Extensive genetic diversity identified among sporadic methicillin-resistant *Staphylococcus aureus* isolates recovered in Irish hospitals between 2000 and 2012. *Antimicrob. Agents Chemother.* **2014**, *58*, 1907–1917. [CrossRef]

27. Monecke, S.; Jatzwauk, L.; Müller, E.; Nitschke, H.; Pfohl, K.; Slickers, P.; Reissig, A.; Ruppelt-Lorz, A.; Ehricht, R. Diversity of SCCmec elements in *Staphylococcus aureus* as observed in South-Eastern Germany. *PLoS ONE* **2016**, *11*, e0162654. [CrossRef]

28. Zarfel, G.; Luxner, J.; Folli, B.; Leitner, E.; Feierl, G.; Kittinger, C.; Grisold, A. Increase of genetic diversity and clonal replacement of epidemic methicillin-resistant *Staphylococcus aureus* strains in South-East Austria. *FEMS Microbiol Lett.* **2016**, *363*, fnw137. [CrossRef]

29. Deplano, A.; Vandendriessche, S.; Nonhoff, C.; Denis, O. Genetic diversity among methicillin-resistant *Staphylococcus aureus* isolates carrying the *mecC* gene in Belgium. *J. Antimicrob. Chemother.* **2014**, *69*, 1457–1460. [CrossRef]

30. Kriegeskorte, A.; Ballhausen, B.; Idelevich, E.A.; Köck, R.; Friedrich, A.W.; Karch, H.; Peters, G.; Becker, K. Human MRSA isolates with novel genetic homolog, Germany. *Emerg. Infect. Dis.* **2012**, *18*, 1016–1018. [CrossRef]

31. Sabat, A.J.; Koksal, M.; Akkerboom, V.; Monecke, S.; Kriegeskorte, A.; Hendrix, R.; Ehricht, R.; Köck, R.; Becker, K.; Friedrich, A.W. Detection of new methicillin-resistant *Staphylococcus aureus* strains that carry a novel genetic homologue and important virulence determinants. *J. Clin. Microbiol.* **2012**, *50*, 3374–3377. [CrossRef]

32. Schaumburg, F.; Köck, R.; Mellmann, A.; Richter, L.; Hasenberg, F.; Kriegeskorte, A.; Friedrich, A.W.; Gatermann, S.; Peters, G.; von Eiff, C.; et al. Population dynamics among methicillin-resistant *Staphylococcus aureus* isolates in Germany during a 6-year period. *J. Clin. Microbiol.* **2012**, *50*, 3186–3192. [CrossRef] [PubMed]

33. Cuny, C.; Layer, F.; Strommenger, B.; Witte, W. Rare occurrence of methicillin-resistant *Staphylococcus aureus* CC130 with a novel *mecA* homologue in humans in Germany. *PLoS ONE.* **2011**, *6*, e24360. [CrossRef] [PubMed]

34. Cartwright, E.J.P.; Paterson, G.K.; Raven, K.E.; Harrison, E.M.; Gouliouris, T.; Kearns, A.; Pichon, B.; Edwards, G.; Skov, R.L.; Larsen, A.; et al. Use of Vitek 2 antimicrobial susceptibility profile to identify *mecC* in methicillin-resistant *Staphylococcus aureus.* *J. Clin. Microbiol.* **2013**, *51*, 2732–2734. [CrossRef] [PubMed]

35. Petersdorf, S.; Herma, M.; Rosenblatt, M.; Layer, F.; Henrich, B. A Novel Staphylococcal Cassette Chromosome *mec* Type XI primer for detection of *mecC*-harboring methicillin-resistant *Staphylococcus aureus* directly from screening specimens. *J. Clin. Microbiol.* **2015**, *53*, 3938–3941. [CrossRef]

36. Dermota, U.; Zdovc, I.; Strumbelj, I.; Grmek-Kosnik, I.; Ribic, H.; Rupnik, M.; Golob, M.; Zajc, U.; Bes, M.; Laurent, F.; et al. Detection of methicillin-resistant *Staphylococcus aureus* carrying the *mecC* gene in, human samples in Slovenia. *Epidemiol. Infect.* **2015**, *143*, 1105–1108. [CrossRef]

37. García-Garrote, F.; Cercenado, E.; Marín, M.; Bal, M.; Trincado, P.; Corredoira, J.; Ballesteros, C.; Pita, J.; Alonso, P.; Vindel, A. Methicillin-resistant *Staphylococcus aureus* carrying the *mecC* gene: Emergence in Spain and report of a fatal case of bacteraemia. *J. Antimicrob. Chemother.* **2014**, *69*, 45–50. [CrossRef]

38. Kerschner, H.; Harrison, E.M.; Hartl, R.; Holmes, M.A.; Apfalter, P. First report of *mecC* MRSA in human samples from Austria: Molecular characteristics and clinical data. *New Microbes New Infect.* **2014**, *3*, 4–9. [CrossRef]

39. Petersen, A.; Medina, A.; Rhod Larsen, A. Ability of the GENSPEED(®) MRSA test kit to detect the novel *mecA* homologue *mecC* in *Staphylococcus aureus.* *APMIS.* **2015**, *123*, 478–481. [CrossRef]

40. Paterson, G.K.; Morgan, F.J.E.; Harrison, E.M.; Cartwright, E.J.P.; Török, M.E.; Zadoks, R.N.; Parkhill, J.; Peacock, S.J.; Holmes, M.A. Prevalence and characterization of human *mecC* methicillin-resistant *Staphylococcus aureus* isolates in England. *J. Antimicrob. Chemother.* **2014**, *69*, 907–910. [CrossRef]

41. Harrison, E.M.; Coll, F.; Toleman, M.S.; Blane, B.; Brown, N.B.; Török, M.E.; Parkhill, J.; Peacock, S.J. Genomic surveillance reveals low prevalence of livestock-associated methicillin-resistant *Staphylococcus aureus* in the East of England. *Sci. Rep.* **2017**, *7*, 7406. [CrossRef]

42. Paterson, G.K. Low prevalence of livestock-associated methicillin-resistant *Staphylococcus aureus* clonal complex 398 and *mecC* MRSA among human isolates in North-West England. *J. Appl. Microbiol.* **2020**, *128*, 1785–1792. [CrossRef] [PubMed]

43. Ciesielczuk, H.; Xenophontos, M.; Lambourne, J. Methicillin-resistant *Staphylococcus aureus* harboring *mecC* still eludes us in East London, United Kingdom. *J. Clin. Microbiol.* **2019**, *57*, e00020-19. [CrossRef] [PubMed]

44. Kriegeskorte, A.; Idelevich, E.A.; Schlattmann, A.; Layer, F.; Strommenger, B.; Denis, O.; Paterson, G.K.; Holmes, M.A.; Werner, G.; Becker, K. Comparison of different phenotypic approaches to screen and detect *mecC*-harboring methicillin-resistant *Staphylococcus aureus.* *J. Clin. Microbiol.* **2017**, *56*, e00826-17. [CrossRef] [PubMed]

45. Basset, P.; Prod'hom, G.; Senn, L.; Greub, G.; Blanc, D.S. Very low prevalence of meticillin-resistant *Staphylococcus aureus* carrying the *mecC* Gene in Western Switzerland. *J. Hosp. Infect.* **2013**, *83*, 257–259. [CrossRef]

46. Dekker, D.; Wolters, M.; Mertens, E.; Boahen, K.G.; Krumkamp, R.; Eibach, D.; Schwarz, N.G.; Adu-Sarkodie, Y.; Rohde, H.; Christner, M.; et al. Antibiotic resistance and clonal diversity of invasive *Staphylococcus aureus* in the rural Ashanti Region, Ghana. *BMC Infect. Dis.* **2016**, *16*, 720. [CrossRef]

47. Kılıç, A.; Doğan, E.; Kaya, S.; Baysallar, M. Investigation of the presence of *mecC* and Panton-Valentine leukocidin genes in *Staphylococcus aureus* strains isolated from clinical specimens during seven years period. *Mikrobiyol. Bul.* **2015**, *49*, 594–599. [CrossRef]

48. Vandendriessche, S.; Vanderhaeghen, W.; Valente Soares, F.; Hallin, M.; Catry, B.; Hermans, K.; Butaye, P.; Haesebrouck, F.; Struelens, M.J.; Denis, O. Prevalence, risk factors and genetic diversity of methicillin-resistant *Staphylococcus aureus* carried by humans and animals across livestock production sectors. *J. Antimicrob. Chemother.* **2013**, *68*, 1510–1516. [CrossRef]

49. Laub, K.; Tóthpál, A.; Kardos, S.; Dobay, O. Epidemiology and antibiotic sensitivity of *Staphylococcus aureus* nasal carriage in children in Hungary. *Acta Microbiol. Immunol. Hung.* **2017**, *64*, 51–62. [CrossRef]

50. Ganesan, A.; Crawford, K.; Mende, K.; Murray, C.K.; Lloyd, B.; Ellis, M.; Tribble, D.R.; Weintrob, A.C. Evaluation for a novel methicillin resistance (*mecC*) homologue in methicillin-resistant *Staphylococcus aureus* isolates obtained from injured military personnel. *J. Clin. Microbiol.* **2013**, *51*, 3073–3075. [CrossRef]

51. Ludden, C.; Brennan, G.; Morris, D.; Austin, B.; O'Connell, B.; Cormican, M. Characterization of methicillin-resistant *Staphylococcus aureus* from residents and the environment in a long-term care facility. *Epidemiol. Infect.* **2015**, *143*, 2985–2988. [CrossRef]

52. Paterson, G.K.; Harrison, E.M.; Craven, E.F.; Petersen, A.; Larsen, A.R.; Ellington, M.J.; Török, M.E.; Peacock, S.J.; Parkhill, J.; Zadoks, R.N.; et al. Incidence and characterisation of methicillin-resistant *Staphylococcus aureus* (MRSA) from nasal colonisation in participants attending a cattle veterinary conference in the UK. *PLoS ONE.* **2013**, *8*, e68463. [CrossRef] [PubMed]

53. Aqel, A.A.; Alzoubi, H.M.; Vickers, A.; Pichon, B.; Kearns, A.M. Molecular epidemiology of nasal isolates of methicillin-resistant *Staphylococcus aureus* from Jordan. *J. Infect. Public Health.* **2015**, *8*, 90–97. [CrossRef] [PubMed]

54. Becker, K.; Schaumburg, F.; Fegeler, C.; Friedrich, A.W.; Köck, R. Prevalence of Multiresistant Microorganisms PMM Study. *Staphylococcus aureus from the German general population is highly diverse. Int. J. Med. Microbiol.* **2017**, *307*, 21–27. [CrossRef] [PubMed]

55. Dodémont, M.; Argudín, M.A.; Willekens, J.; Vanderhelst, E.; Pierard, D.; Miendje Deyi, V.Y.; Hanssens, L.; Franckx, H.; Schelstraete, P.; Leroux-Roels, I.; et al. Emergence of livestock-associated MRSA isolated from cystic fibrosis patients: Result of a Belgian national survey. *J. Cyst. Fibros.* **2019**, *18*, 86–93. [CrossRef] [PubMed]

56. Drougka, E.; Foka, A.; Koutinas, C.K.; Jelastopulu, E.; Giormezis, N.; Farmaki, O.; Sarrou, S.; Anastassiou, E.D.; Petinaki, E.; Spiliopoulou, I. Interspecies spread of *Staphylococcus aureus* clones among companion animals and human close contacts in a veterinary teaching hospital. A cross-sectional study in Greece. *Prev. Vet. Med.* **2016**, *126*, 190–198. [CrossRef]

57. Nijhuis, R.H.; van Maarseveen, N.M.; van Hannen, E.J.; van Zwet, A.A.; Mascini, E.M. A rapid and high-throughput screening approach for methicillin-resistant *Staphylococcus aureus* based on the combination of two different real-time PCR assays. *J. Clin. Microbiol.* **2014**, *52*, 2861–2867. [CrossRef] [PubMed]

58. Saeed, K.; Ahmad, N.; Dryden, M.; Cortes, N.; Marsh, P.; Sitjar, A.; Wyllie, S.; Bourne, S.; Hemming, J.; Jeppesen, C.; et al. Oxacillin-susceptible methicillin-resistant *Staphylococcus aureus* (OS-MRSA), a hidden resistant mechanism among clinically significant isolates in the Wessex region/UK. *Infection* **2014**, *42*, 843–847. [CrossRef]

59. Khairalla, A.S.; Wasfi, R.; Ashour, H.M. Carriage frequency, phenotypic, and genotypic characteristics of methicillin-resistant *Staphylococcus aureus* isolated from dental health-care personnel, patients, and environment. *Sci. Rep.* **2017**, *7*, 7390. [CrossRef]

60. Ho, C.M.; Lin, C.Y.; Ho, M.W.; Lin, H.C.; Chen, C.J.; Lin, L.C.; Lu, J.J. Methicillin-resistant *Staphylococcus aureus* isolates with SCC*mec* type V and spa types t437 or t1081 associated to discordant susceptibility results between oxacillin and cefoxitin, Central Taiwan. *Diagn. Microbiol. Infect. Dis.* **2016**, *86*, 405–411. [CrossRef]

61. van Duijkeren, E.; Hengeveld, P.; Zomer, T.P.; Landman, F.; Bosch, T.; Haenen, A.; van de Giessen, A. Transmission of MRSA between humans and animals on duck and turkey farms. *J. Antimicrob. Chemother.* **2016**, *71*, 58–62. [CrossRef]

62. Cikman, A.; Aydin, M.; Gulhan, B.; Karakecili, F.; Kurtoglu, M.G.; Yuksekkaya, S.; Parlak, M.; Gultepe, B.S.; Cicek, A.C.; Bilman, F.B.; et al. Absence of the *mecC* gene in methicillin-resistant *Staphylococcus aureus* isolated from various clinical samples: The first multi-centered study in Turkey. Multicenter Study. *J. Infect. Public Health.* **2019**, *12*, 528–533. [CrossRef] [PubMed]

63. Gomez, P.; Lozano, C.; González-Barrio, D.; Zarazaga, M.; Ruiz-Fons, F.; Torres, C. High prevalence of methicillin-resistant *Staphylococcus aureus* (MRSA) carrying the *mecC* gene in a semi-extensive red deer (*Cervus elaphus hispanicus*) farm in Southern Spain. *Vet. Microbiology.* **2015**, *177*, 326–331. [CrossRef]

64. Szymanek-Majchrzak, K.; Kosiński, J.; Żak, K.; Sułek, K.; Młynarczyk, A.; Młynarczyk, G. Prevalence of methicillin resistant and mupirocin-resistant *Staphylococcus aureus* strains among medical students of Medical University in Warsaw. *Przegl Epidemiol.* **2019**, *73*, 39–48. [CrossRef] [PubMed]

65. Rödel, J.; Bohnert, J.A.; Stoll, S.; Wassill, L.; Edel, B.; Karrasch, M.; Löffler, B.; Pfister, W. Evaluation of loop-mediated isothermal amplification for the rapid identification of bacteria and resistance determinants in positive blood cultures. *Eur. J. Clin. Microbiol. Infect. Dis.* **2017**, *36*, 1033–1040. [CrossRef] [PubMed]

66. Horner, C.; Utsi, L.; Coole, L.; Denton, M. Epidemiology and microbiological characterization of clinical isolates of *Staphylococcus aureus* in a single healthcare region of the UK, 2015. *Epidemiol. Infect.* **2017**, *145*, 386–396. [CrossRef]

67. Mehta, S.R.; Estrada, J.; Ybarra, J.; Fierer, J. Comparison of the BD MAX MRSA XT to the Cepheid™ Xpert®MRSA assay for the molecular detection of methicillin-resistant *Staphylococcus aureus* from nasal swabs. *Diagn. Microbiol. Infect. Dis.* **2017**, *87*, 308–310. [CrossRef]

68. Venugopal, N.; Mitra, S.; Tewari, R.; Ganaie, F.; Shome, R.; Rahman, H.; Shome, B.R. Molecular detection and typing of methicillin-resistant *Staphylococcus aureus* and methicillin-resistant coagulase-negative staphylococci isolated from cattle, animal handlers, and their environment from Karnataka, Southern Province of India. *Vet. World.* **2019**, *12*, 1760–1768. [CrossRef]

69. Ceballos, S.; Aspiroz, C.; Ruiz-Ripa, l.; Azcona-Gutierrez, J.M.; López-Cerero, L.; López-Calleja, A.I.; Álvarez, L.; Gomáriz, M.; Fernández, M.; Torres, C.; et al. Multicenter study of clinical non-β-lactam-antibiotic susceptible MRSA strains: Genetic lineages and Panton-Valentine leukocidin (PVL) production. *Enferm. Infecc. Microbiol. Clin.* **2019**, *37*, 509–513. [CrossRef]

70. Papadopoulos, P.; Angelidis, A.S.; Papadopoulos, T.; Kotzamanidis, C.; Zdragas, A.; Papa, A.; Filioussis, G.; Sergelidis, D. *Staphylococcus aureus* and methicillin-resistant *S. aureus* (MRSA) in bulk tank milk, livestock and dairy-farm personnel in north-central and north-eastern Greece: Prevalence, characterization and genetic relatedness. *Food Microbiol.* **2019**, *84*, 103249. [CrossRef]

71. Rasmussen, S.L.; Larsen, J.; van Wijk, R.E.; Jones, O.R.; Bjørneboe Berg, T.; Angen, O.; Rhod Larsen, A. European hedgehogs (*Erinaceus europaeus*) as a natural reservoir of methicillin-resistant *Staphylococcus aureus* carrying *mecC* in Denmark. *PLoS ONE.* **2019**, *14*, e0222031. [CrossRef]

72. Morroni, G.; Brenciani, A.; Brescini, L.; Fioriti, S.; Simoni, S.; Pocognoli, A.; Mingoia, M.; Giovanetti, E.; Barchiesi, F.; Giacometti, A.; et al. High rate of ceftobiprole resistance among clinical methicillin-resistant *Staphylococcus aureus* isolates from a hospital in central Italy. *Antimicrob. Agents Chemother.* **2018**, *62*, e01663-18. [CrossRef]

73. Hefzy, E.M.; Hassan, G.M.; El Reheem, F.A. Detection of Panton-Valentine Leukocidin-positive methicillin-resistant *Staphylococcus aureus* nasal carriage among Egyptian health care workers. *Surg. Infect. (Larchmt).* **2016**, *17*, 369–375. [CrossRef] [PubMed]

74. Hogan, B.; Rakotozandrindrainy, R.; Al-Emran, H.; Dekker, D.; Hahn, A.; Jaeger, A.; Poppert, S.; Frickmann, H.; Hagen, R.M.; Micheel, V.; et al. Prevalence of nasal colonisation by methicillin-sensitive and methicillin-resistant *Staphylococcus aureus* among healthcare workers and students in Madagascar. *B.M.C. Infect. Dis.* **2016**, *16*, 420. [CrossRef] [PubMed]

75. Diaz, R.; Ramalheira, E.; Afreixo, V.; Gago, B. Methicillin-resistant *Staphylococcus aureus* carrying the new *mecC* gene–a meta-analysis. *Diagn. Microbiol. Infect. Dis.* **2016**, *84*, 135–140. [CrossRef] [PubMed]

76. Lindgren, A.K.; Gustafsson, E.; Petersson, A.C.; Melander, E. Methicillin-resistant *Staphylococcus aureus* with *mecC*: A description of 45 human cases in southern Sweden. *Eur. J. Clin. Microbiol. Infect. Dis.* **2016**, *35*, 971–975. [CrossRef] [PubMed]

77. Laurent, F.; Chardon, H.; Haenni, M.; Bes, M.; Reverdy, M.E.; Madec, J.V.; Lagier, E.; Vandenesch, F.; Tristan, A. MRSA harboring *mecA* variant gene *mecC*, France. *Emerg. Infect. Dis.* **2012**, *18*, 1465–1467. [CrossRef] [PubMed]

78. Barraud, O.; Laurent, F.; François, B.; Bes, M.; Vignon, P.; Ploy, M.C. Severe human bone infection due to methicillin-resistant *Staphylococcus aureus* carrying the novel *mecC* variant. *J. Antimicrob. Chemother.* **2013**, *68*, 2949–2950. [CrossRef]

79. Cano García, M.E.; Monteagudo Cimiano, I.; Mellado Encinas, P.; Ortega Álvarez, C. Methicillin-resistant *Staphylococcus aureus* carrying the *mecC* gene in a patient with a wound infection. *Enferm. Infecc. Microbiol. Clin.* **2015**, *33*, 287–288. [CrossRef]

80. Romero-Gómez, M.P.; Mora-Rillo, M.; Lázaro-Perona, F.; Gómez-Gil, M.R.; Mingorance, J. Bacteraemia due to meticillin-resistant *Staphylococcus aureus* carrying the *mecC* gene in a patient with urothelial carcinoma. *J. Med. Microbiol.* **2013**, *62*, 1914–1916. [CrossRef]

81. Benito, D.; Gómez, P.; Aspiroz, C.; Zarazaga, M.; Lozano, C.; Torres, C. Molecular characterization of *Staphylococcus aureus* isolated from humans related to a livestock farm in Spain, with detection of MRSA-CC130 carrying *mecC* gene: A zoonotic case? *Enferm. Infecc. Microbiol. Clin.* **2016**, *34*, 280–285. [CrossRef]

82. Worthing, K.A.; Coombs, G.W.; Pang, S.; Abraham, S.; Saputra, S.; Trott, D.J.; Jordan, D.; Wong, H.S.; Abraham, R.J.; Norris, J.M. Isolation of *mecC* MRSA in Australia. *J. Antimicrob. Chemother.* **2016**, *71*, 2348–2349. [CrossRef]

83. Aklilu, E.; Ying, C.H. First *mecC* and *mecA* positive livestock-associated Methicillin resistant *Staphylococcus aureus* (*mecC* MRSA/LA-MRSA) from dairy cattle in Malaysia. *Microorganisms.* **2020**, *8*, 147. [CrossRef] [PubMed]

84. Sakr, A.; Brégeon, F.; Mège, J.L.; Rolain, J.M.; Blin, O. *Staphylococcus aureus* nasal colonization: An update on mechanisms, epidemiology, risk factors, and subsequent infections. *Front. Microbiol.* **2018**, *9*, 2419. [CrossRef]

85. Brown, A.F.; Leech, J.M.; Rogers, T.R.; McLoughlin, R.M. *Staphylococcus aureus* colonization: Modulation of host immune response and impact on human vaccine design. *Front. Immunol.* **2014**, *4*, 507. [CrossRef] [PubMed]

86. Asadollahi, P.; Farahani, N.N.; Mirzaii, N.; Sajjad Khoramrooz, S.S.; van Belkum, A.; Asadollahi, K.; Dadashi, M.; Darban-Sarokhalil, D. Distribution of the most prevalent *spa* types among clinical isolates of Methicillin-Resistant and -Susceptible *Staphylococcus aureus* around the World: A Review. *Front. Microbiol.* **2018**, *9*, 163. [CrossRef] [PubMed]

87. Belmekki, M.; Mammeri, H.; Hamdad, F.; Rousseau, F.; Canarelli, B.; Biendo, M. Comparison of Xpert MRSA/SA Nasal and MRSA/SA ELITe MGB assays for detection of the *mecA* gene with susceptibility testing methods for determination of methicillin resistance in *Staphylococcus aureus* isolates. *J. Clin. Microbiol.* **2013**, *51*, 3183–3191. [CrossRef]

88. Stegger, M.; Lindsay, J.A.; Sørum, M.; Gould, K.A.; Skov, R. Genetic diversity in CC398 methicillin-resistant *Staphylococcus aureus* isolates of different geographical origin. *Clin. Microbiol. Infect.* **2010**, *16*, 1017–1019. [CrossRef]

89. Eriksson, J.; Espinosa-Gongora, C.; Stamphøj, I.; Rhod Larsen, A.; Guardabassi, L. Carriage frequency, diversity and methicillin resistance of *Staphylococcus aureus* in Danish small ruminants. *Vet. Microbiol.* **2013**, *163*, 110–115. [CrossRef]

90. Paterson, G.K.; Larsen, A.R.; Robb, A.; Edwards, G.E.; Pennycott, T.W.; Foster, G.; Mot, D.; Hermans, K.; Baert, K.; Peacock, S.J.; et al. The newly described *mecA* homologue, *mecA*(LGA251), is present in methicillin-resistant *Staphylococcus aureus* isolates from a diverse range of host species. *J. Antimicrob. Chemother.* **2012**, *67*, 2809–2813. [CrossRef]

91. Weinert, L.A.; Welch, J.J.; Suchard, M.A.; Lemey, P.; Rambaut, A.; Fitzgerald, J.R. Molecular dating of human-to-bovid host jumps by *Staphylococcus aureus* reveals an association with the spread of domestication. *Biol. Lett.* **2012**, *8*, 829–832. [CrossRef]

92. Ruiz-Ripa, L.; Gómez, P.; Alonso, C.A.; Camacho, M.C.; de la Puente, J.; Fernández-Fernández, R.; Ramiro, Y.; Quevedo, M.A.; Blanco, J.M.; Zarazaga, M.; et al. Detection of MRSA of Lineages CC130-*mecC* and CC398-*mecA* and *Staphylococcus delphini-lnu*(A) in Magpies and Cinereous Vultures in Spain. *Microb. Ecol.* **2019**, *78*, 409–415. [CrossRef] [PubMed]

93. Ford, A. *mecC*-harboring methicillin-resistant *Staphylococcus aureus*: Hiding in plain sight. *J. Clin. Microbiol.* **2017**, *56*, e01549-17. [CrossRef] [PubMed]

94. Haenni, M.; Châtre, P.; Dupieux, C.; Métayer, V.; Maillard, K.; Bes, M.; Madec, J.Y.; Laurent, F. *mecC*-positive MRSA in horses. *J. Antimicrob. Chemother.* **2015**, *70*, 3401–3402. [CrossRef]

95. Medhus, A.; Slettemeås, J.S.; Marstein, L.; Larssen, K.W.; Sunde, M. Methicillin-resistant *Staphylococcus aureus* with the novel *mecC* gene variant isolated from a cat suffering from chronic conjunctivitis. *J. Antimicrob. Chemother.* **2013**, *68*, 968–969. [CrossRef] [PubMed]

96. Haenni, M.; Châtre, P.; Tasse, J.; Nowak, N.; Bes, M.; Madec, J.Y.; Laurent, F. Geographical clustering of *mecC*-positive *Staphylococcus aureus* from bovine mastitis in France. *J. Antimicrob. Chemother.* **2014**, *69*, 2292–2293. [CrossRef]

97. Gindonis, V.; Taponen, S.; Myllyniemi, A.L.; Pyörälä, S.; Nykäsenoja, S.; Salmenlinna, S.; Lindholm, L.; Rantala, M. Occurrence and characterization of methicillin-resistant staphylococci from bovine mastitis milk samples in Finland. *Acta Vet. Scand.* **2013**, *55*, 61. [CrossRef]

Publisher's Note: MDPI stays neutral with regard to jurisdictional claims in published maps and institutional affiliations.

 *microorganisms*

Review

No Change, No Life? What We Know about Phase Variation in *Staphylococcus aureus*

Vishal Gor [1,*], Ryosuke L. Ohniwa [2] and Kazuya Morikawa [2,*]

1   Graduate School of Comprehensive Human Sciences, University of Tsukuba, Tsukuba, Ibaraki 305-8575, Japan
2   Faculty of Medicine, University of Tsukuba, Tsukuba, Ibaraki 305-8575, Japan; ohniwa@md.tsukuba.ac.jp
*   Correspondence: vishal.gor@outlook.com (V.G.); morikawa.kazuya.ga@u.tsukuba.ac.jp (K.M.)

Abstract: Phase variation (PV) is a well-known phenomenon of high-frequency reversible gene-expression switching. PV arises from genetic and epigenetic mechanisms and confers a range of benefits to bacteria, constituting both an innate immune strategy to infection from bacteriophages as well as an adaptation strategy within an infected host. PV has been well-characterized in numerous bacterial species; however, there is limited direct evidence of PV in the human opportunistic pathogen *Staphylococcus aureus*. This review provides an overview of the mechanisms that generate PV and focuses on earlier and recent findings of PV in *S. aureus*, with a brief look at the future of the field.

Keywords: *Staphylococcus aureus*; phase variation

1. Introduction

The Gram-positive human commensal *Staphylococcus aureus* is an opportunistic pathogen that imposes a major health and economic burden on a global scale [1]. *S. aureus* can colonize multiple sites of the human body, but the primary niche of commensal colonization is the anterior nares, with various other skin surfaces making up secondary niches. There are three main carrier-patterns of *S. aureus* amongst healthy individuals: persistent carriers (~20%), intermittent carriers (~30%), and non-carriers (~50%) [2]. Nasal carriage of *S. aureus* has been linked to a higher chance of contracting infection [2]. *S. aureus* is responsible for an astounding diversity of infections. It is the leading cause of infective endocarditis, osteoarticular infections, and surgical site infections and *S. aureus* bacteraemia is also prevalent [3,4]. *S. aureus* can also cause pneumonia and other respiratory infections, particularly in people living with cystic fibrosis [3]. Furthermore, *S. aureus* is supremely adept at colonizing alien surfaces within the body and is often responsible for infections associated with catheters, cannula, artificial heart valves, and prosthetic joints [3]. This diverse range of infections is enabled by a vast arsenal of virulence factors that are ready to be deployed in a variety of host environments [5,6]. Of particular concern is *S. aureus'* rapid development of antibiotic resistance. Methicillin Resistant *S. aureus* (MRSA) has broad-spectrum resistance against the β-lactam group of antibiotics and is a global danger with clones existing in both nosocomial and community settings [7]. MRSA is also a problem in the livestock sector, where it can co-infect both animals and humans [8]. The infamous development of antibiotic resistance, coupled with its worrying genetic plasticity, has earned *S. aureus* a place in the ESKAPE group of pathogens: a collection of bacteria that represent paradigms of acquisition, development, and transfer of antibiotic resistance [9]. Thus, to better combat this dangerous pathogen it is vitally important to study adaptation mechanisms of *S. aureus*.

Another particular trait of *S. aureus* that makes it notoriously difficult to combat in the clinical setting is phenotypic heterogeneity. An example of this is the phenomenon of persister cells, where sub-populations of *S. aureus* gain a resistance phenotype against antibiotic treatment resulting from arrested growth [10]. Persister cells may be generated in numerous ways, one of which is the formation of Small Colony Variants (SCVs) that are

Citation: Gor, V.; Ohniwa, R.L.; Morikawa, K. No Change, No Life? What We Know about Phase Variation in *Staphylococcus aureus*. *Microorganisms* **2021**, *9*, 244. https://doi.org/10.3390/microorganisms9020244

Academic Editor: Rajan P. Adhikari

Received: 31 December 2020
Accepted: 23 January 2021
Published: 25 January 2021

Publisher's Note: MDPI stays neutral with regard to jurisdictional claims in published maps and institutional affiliations.

characterized by auxotrophy for various compounds involved in the electron transport chain and slow growth, allowing them to escape the effects of many antibiotics [11,12]. Importantly, these populations do not acquire conventional resistance mechanisms against the antibiotics. This heterogenous phenomenon has severe clinical implications and is thought to be a significant cause of antibiotic treatment failure and chronic recurrent infections [13].

Heterogeneity is not limited to antibiotic resistance. As we discuss in this review, diverse traits, including pathogenicity factors, have recently been recognized as having sub-population patterns of expression. The scientific point-of-view has been increasingly focused on such heterogenous phenomena, yet progress is still in the relatively early stages and much work remains to be done. In this review, we summarize the information regarding bacterial Phase Variation (PV), a mechanism of high-frequency reversible gene expression switching (Section 2) and collate the known examples of *S. aureus* PV into one source (Section 3) to aid in future studies on heterogeneity in *S. aureus* (Section 4).

2. Bacterial Phase Variation

2.1. Background of Phase Variation

All living organisms are faced with the constant challenge of maintaining fitness in order to survive and reproduce, and this is no less true for bacterial species. Bacteria are under constant onslaught from fluctuations in their local environment, infection from bacteriophages, and (in the case of pathogenic bacteria) attack from their infected host. Although bacteria possess robust mechanisms of classical gene regulation that allow them to respond to extracellular changes (e.g., Bacterial Two-Component Systems), these alone may be unable to cope with the constant barrage of fluctuating pressures they face. These selective pressures are often focused on bacterial external proteins which form the first line of contact with the outside environment and this has led to development of what have been termed "contingency loci" [14,15]. Contingency loci are hypermutable genes that generate genetic and phenotypic variation allowing bacterial populations to survive unpredictable pressures. This hypermutability is conferred by the phenomena of Phase Variation (PV) and antigenic variation.

PV is a reversible gene expression switch that can alter expression between an ON and an OFF state and occurs through several genetic and epigenetic mechanisms [16]. It is characterized by high frequencies, usually exceeding 1×10^{-5} variants per total number of cells [17,18] which is orders of magnitude above the typical frequencies of spontaneous mutations (10^{-6} to 10^{-8} per cell per generation) [18]. Depending on the method of calculation, the frequency of PV may describe not only rate of the PV mechanisms but also the growth of the phase variants themselves. Antigenic variation is related to PV and occurs through similar mechanisms. However, rather than alternating between an ON and OFF state, antigenic variation mechanisms generate variations in the sequence of surface proteins resulting in the expression of different forms and structures of the antigenic proteins on the cell surface [17–19]. As such, due to the similar nature of the mechanisms involved, antigenic variation will not be separately addressed in this review.

As mentioned above, genes subject to PV often encode for cell-surface associated features such as adhesins, liposaccharide synthesis enzymes, and pili [20–22] but can also encode for virulence factors and secreted proteins such as iron acquisition machinery [23,24]. The collection of phase variable loci in a bacterial species is referred to as the "phasome" [16] and generally includes genes which are involved in bottlenecks experienced by the bacterial population. This is most clearly seen in pathogenic bacteria which undergo constant challenge from host immunity during the infection process. For example, PV mediated shutdown of liposaccharide synthesis genes in the invasive pathogen *Haemophilus influenzae* confers protection against neutrophil-mediated immune clearance but is detrimental in other environments [22,25,26]. In another example, PV in *Salmonella typhimurium* flagellae can modulate their antigenic properties and allow for evasion from host immunity [27].

It is likely that the original role of PV was as a mechanism of innate immunity against bacteria's greatest enemy: bacteriophages [28]. Although bacteriophages exist in exaggerated abundance relative to their bacterial hosts, their host range is often limited to just a few specific strains of a given bacterial species [29]. Thus, there is a constant cyclical arms race between bacteria and bacteriophages in order to stay one step ahead of each other [30], and PV plays an important role in both sides of this war. An example can once again be found in liposaccharide synthesis genes of *H. influenzae* in which PV can result in a switch from a sensitive to resistant phenotype against the HP1c1 phage [31]. On the other hand, PV in the *Escherichia coli* phage Mu causes a switch in expression between two sets of tail fibers resulting in modulation of the host specificity [32,33] with similar phenomena identified in other phages [34].

Considering the above information, it can be inferred that genetic loci susceptible to PV would be found in abundance amongst bacterial species that experience population bottlenecks. Typically, such bottlenecks often occur during the infectious process which imposes limits onto the bacterial population size. These bottlenecks reduce genetic diversity at a time when variation is most beneficial, and PV offers a solution to this hurdle and indeed, several pathogenic bacteria have been documented to undergo PV [18].

While PV is, by definition, a stochastic process, it occurs through several discrete mechanisms. Broadly speaking, mechanisms of generating PV can be discriminated into genetic and epigenetic mechanisms [16] both of which will be addressed in Sections 2.2 and 2.3 respectively.

2.2. Genetic Mechanisms of PV

There are three genetic mechanisms of PV which shall be discussed in the following chapters: Variation in length of DNA Short Sequence Repeats (SSRs) [35–37], DNA inversion [38], and DNA recombination [39,40].

2.2.1. Variation in Length of DNA Short Sequence Repeats (SSRs)

SSRs are homo- or hetero-nucleotide repeats in DNA that are highly prone to insertion/deletion (indel) errors due to Slipped-Strand Mispairings (SSMs) during DNA replication [35–37]. SSRs can be as complex as repeating units of tetranucleotides or as simple as a straight homonucleotide run. Indels in SSRs can result in frameshifts that largely have an ON↔OFF effect on protein function or gene expression (by resulting in abrupt termination of translation or inhibition of RNA polymerase binding, respectively Figure 1A) but can have an alternative gradation effect on gene expression as well. For example, alterations in the length of a dinucleotide TA10 tract in the promoter regions of the divergently transcribed *hifA* and *hifB* genes controlling fimbriae expression in *H. influenzae* can either significantly affect *hif* expression (TA10→TA9) or only moderately affect it (TA10→TA11) [41]. The evolution of the mutability of SSR tracts is largely driven by a combination of environmental and molecular drivers. The environmental drivers include factors such as the aforementioned population bottlenecks arising during infection processes. These bottleneck conditions exert a primary selective pressure for phenotypes that can survive them, e.g., a population that can shut down the expression of a surface protein that is targeted by host immunity. The necessity to survive this recurrent primary selection serves as a secondary layer of selection for plasticity of the gene itself.

The molecular factors are intrinsic to SSR tracts and include the DNA replication and the Mismatch Repair (MMR) [42]. The discriminating factors of SSRs can be broadly delineated into two groups: the composition of the repeating nucleotide unit (i.e., a homonucleotide or a heteronucleotide repeat) and the tract length. These in turn are differentially affected by the DNA replication and MMR machinery. Amongst these proteins are the DNA polymerase enzymes which include the polymerase responsible for the construction of new DNA strands (DNA polymerase III) as well as the polymerase responsible for DNA repair (DNA polymerase I). Studies have shown that these polymerases have an inherent frequency of generating addition/deletion errors when constructing new DNA strands [43,44].

Following DNA replication, any errors are corrected by the MMR machinery which is a suite of Mut proteins that target and fix errors in a strand specific manner. Inactivation of components from either of these suites of proteins results in a hypermutable phenotype and can lead to SSR alteration e.g., [45]. Additionally, the hypermutable phenotype that results from loss of the MMR machinery is directly responsible for genetic variability of bacteria and mutator phenotypes play an important role in bacterial adaptation [46]. For example, both *S. aureus* and *Pseudomonas aeruginosa* isolated from the lungs of people suffering from cystic fibrosis are commonly associated with antibiotic resistance caused by hypermutability [47–49]. Interestingly, while both the MMR machinery and DNA polymerases are involved in SSR evolution, they do not appear to be fully redundant. Several studies have shown that MMR is more responsible for variability of homonucleotide SSRs, especially for those which exceed eight nucleotides in length, whereas DNA polymerase I is exclusively responsible for mutations in heteronucleotide SSRs [50–52]. This could have evolutionary implications for the mechanisms of generating SSRs. For example, *H.influenza* is enriched with tetra-nucleotide SSRs [51] whose expansion/contraction is affected by DNA polymerase I. Furthermore, evidence suggests that the frequency of DNA polymerase I mediated errors differs between the leading and the lagging strands of newly synthesized DNA, implying that the direction of genes in the chromosome can also dictate the type of SSR that would evolve in them [53]. Lastly, an interesting study carried out by Lin et al. investigated the distribution of SSRs within the genomes of several bacterial species. They found that in many pathogenic species, SSRs were enriched towards the N-termini of protein coding sequences increasing the probability of frameshifts resulting in non-functional proteins [54,55]. This further suggests that bacteria have evolved SSRs in a manner to provide maximal PV.

2.2.2. DNA Inversion

DNA inversion was the first documented example of PV, though the mechanism was not known at the time the phenomenon was documented [38] (Figure 1B). It involves recognition of inverted repeat (IR) sequences by invertase enzymes and subsequent enzyme-mediated inversion of the DNA. An elaborate study was carried out by Jiang and colleagues who developed an algorithm to search published bacterial genome datasets for IR sequences that might be phase variable [56]. Not only did they identify that IR sequences were enriched in host-associates species (implying a benefit of PV during commensalism or infection) but they also discovered three antibiotic resistance genes regulated by invertible promoters: a macrolide resistance gene, a multidrug resistance cassette conferring resistance to macrolides and cephalosporins, and a cationic antibacterial peptide resistance operon [56]. The presence of antibiotics influenced the switch from an OFF to an ON state for these genes. Some of the invertible promoters seem to be located on genetic elements homologous to those conveyable by horizontal gene transfer mechanisms, raising the worrying possibility that these resistance gene switches can be transferred to other species [56].

2.2.3. DNA Recombination

Homologous recombination provides a pathway for DNA re-arrangement and subsequent PV. Events arising from recombination mechanisms are often due to DNA deletions, and thus tend to be in a one way ON→OFF direction. However, gene duplication or transfer events can often occur to balance out the accumulation of inactive variants in the population. A well characterized example of recombination mediated variation occurs in the *Neisseria gonorrhoea* pilus organelle, which is essential for full infectivity and natural transformation. *N. gonorrhoea* contains a *pilE* gene that encodes for a pilin protein that is the major component of the pilus, but also contains several silent *pilS* alleles several Kb away [39]. RecA-dependent recombination events can unidirectionally transfer large sections of the *pilS* allele into *pilE*, thus creating an OFF variant [40,57] (Figure 1C). The *N. gonorrhoea* pilus also undergoes PV by SSM-mediated variation in the length of a poly(G) tract in the *pilC* gene (which encodes for the adhesive tip of the pilus [58]) resulting in ON↔OFF switching [59,60].

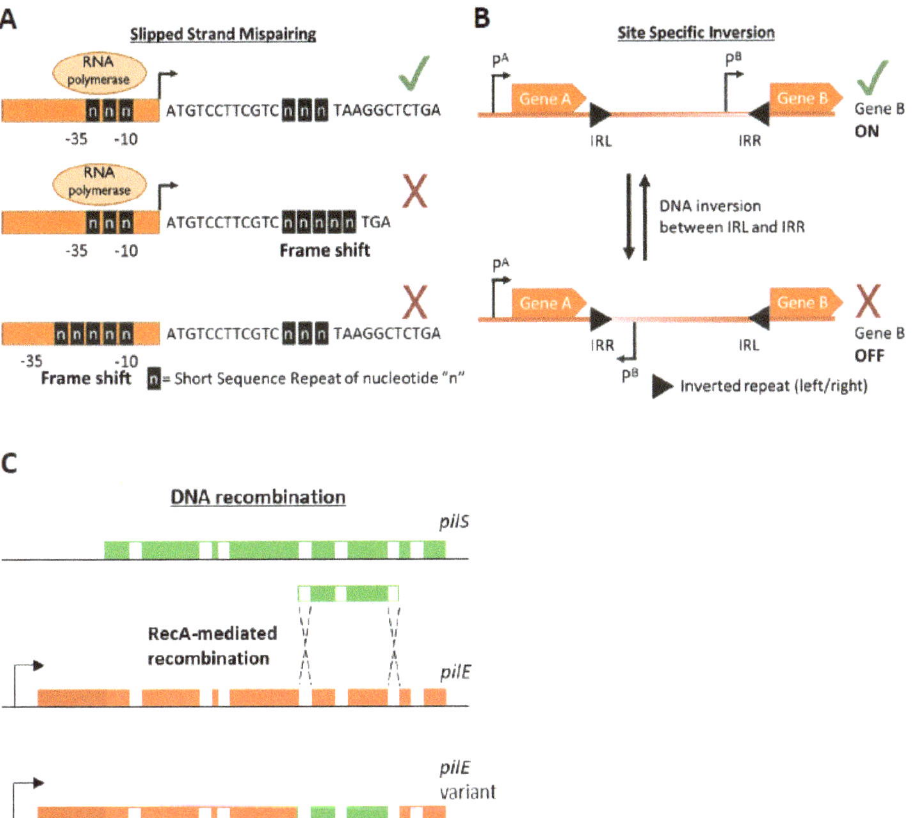

Figure 1. Genetic mechanisms of Phase Variation. A cartoon depicting the three main genetic mechanisms of Phase variation (PV). (**A**) Slipped-Strand Mispairing events within Short Sequence Repeats (SSR) result in expression (green tick mark) of truncated dysfunctional proteins (if SSR is in the CDS) or inhibition (red cross) of transcription by preventing RNA polymerase/transcription factor binding or by other mechanisms. For example, an interesting method of PV-mediated transcriptional control is shown by Danne et al. who demonstrate SSR alterations upstream of the *pilA* locus of *Streptococcus gallolyticus* can destabilize a premature transcription-terminating stem loop [61]. (**B**) Site-specific inversion is carried out by recombinases that recognize inverted repeat regions (Inverted Repeat Left/Right IRL/IRR) and flip the DNA sequence in between them. If a promoter region (e.g., p^B) lies within the sequence flanked by the inverted repeats this leads to shut down of gene expression. (**C**) RecA-mediated DNA recombination of *N. gonorrhoea pilS* into *pilE* results in the formation of new *pilE* variants. Both *pilS* and *pilE* contain variable regions (depicted in green and orange, respectively) interspersed with conserved regions (white) while *pilE* has a further 5′ conserved region (dark orange) and a promoter to initiate transcription.

2.3. Epigenetic Mechanisms of Phase Variation

An epigenetic trait has been defined as a heritable phenotype resulting from modified gene expression that is not due to any alterations in the DNA sequence of the chromosome [62,63]. In prokaryotes, DNA methylation occurs mainly at the nucleotide adenine although studies have shown that cytosine methylation can also occur [64–66]. DNA methylation usually occurs at specific target sites and is carried out either by methyltransferases that are part of dedicated Restriction–Modification (RM) systems or by orphan methyltransferases. A well-studied methylase responsible for bacterial epigenetic regulation of

PV is the DNA Adenine Methyltransferase (DAM) which is an orphan methyltransferase of the gammaproteobacterial family that is specific for GATC sites [64]. Methylation of DNA represses transcription, and thus PV can result if there are GATC sites within a gene promoter which also binds transcription factors, causing mutually exclusive binding competition between the transcription factor(s) and DAM. If there are numerous GATC sites within a promoter region then the mutually exclusive competition can result in differential methylation patterns of the promoter region resulting in switching between an ON and OFF state. A paradigm of this sort of PV is established by a series of intriguing reports studying the *pap* operon of *E. coli* and the *opvAB* operon of *Salmonella enterica* [21,67–69].

2.4. Combined Mechanisms of Phase Variation

There is growing evidence that shows that many bacterial species undertake a combined approach for PV to maximize the ability to generate rapid and diverse variation. This strategy involves generating PV through genetic mechanisms in genes of RM systems that can modify the transcriptome of the cell via epigenetic control. Such systems are referred to as "phasevarions" as they control phase-variable regulons [70] and are immensely powerful weapons in the arsenal of pathogens.

The earliest phasevarions identified are controlled by Type III RM systems. PV occurs in SSRs in the *mod* gene resulting in ON↔OFF variation and altered methylation states [71,72]. Strikingly, analyses of known Type III system sequences indicate that at least 20% of these systems contain SSRs and could potentially be phasevarions [73]. Furthermore, *mod* genes are highly conserved, with variation occurring mainly in the DNA recognition domain. This allows *mod* genes to exist within the species as multiple alleles, each of which controls distinct phasevarions [16].

There is some evidence of a Type II RM regulated phasevarion detected in *Campylobacter jejuni*, and gene expression patterns were detectably different upon RNAseq analysis, though no direct link to any altered phenotype was reported [74].

PV in Type I systems largely occurs through DNA inversion in the *hsdS* gene, creating multiple allelic variants of the specificity protein of the Type I system resulting in different gene targets upon PV [75]. An example of a Type I RM phasevarions can be seen in variable capsular expression controlling virulence in *Streptococcus pneumonia* [16,76].

In theory, phasevarions must be also seen by PV in other regulators of gene expression, such as transcription factors. Some examples of these are described in Section 3.

3. Known Examples of Phase Variation in *S. aureus*

S. aureus is an opportunistic pathogen that has claimed several distinct niches in the human body, thus being subject to a variety of different conditions and stresses against which it has accumulated diverse colonisation and pathogenicity factors. As such, it is primed to exploit the phenomenon of PV; however, there have been surprisingly few documented reports of PV examples, possibly due to a link between heterogenous phenotypes and PV not being made. This section (Section 3) will outline reports that have identified PV, or PV-like mechanisms, in *S. aureus*. Reports detailing heterogeneity arising from non-PV mechanisms will not be discussed as they have been detailed elsewhere [77].

Perhaps the best studied example of PV in *S. aureus* relates to its ability to form biofilms. A report from the early 1990s identified phase variation in the production of an extracellular polysaccharide coat (or "slime") whose production could be reversibly switched across generations of the same lineage with variants easily distinguishable by differential colony morphology on Congo Red Agar [78]. From there, the topic took on multiple approaches from various groups. Tormo and colleagues identified that expression of the Bap protein (a major surface component involved in biofilm formation that promotes primary attachment as well as intercellular adhesion) was phase variable [79]. However, although they confirmed that Bap-negative variants did not express the *bap* gene, they could not detect any sequence alterations, suggesting that the exact mechanism of PV was either indirect or occurred through epigenetic means. Investigating from another direction, Valle and colleagues discovered that *IS256* transposition can also result in biofilm PV by disrupting the *sarA* regulator and *icaC* [80]. The *icaADBC* operon encodes for genes involved in the synthesis of poly N-acetylglucosamine exopolysaccharide and its deacylated variant polysaccharide intercellular adhesion (PNAG/PIA) [81](other major extracellular components involved in biofilm formation that also have roles in immune evasion [82]). Interestingly, they further discovered that there is a connection between this variation and the global stress sigma factor σ^B, as a σ^B deletion mutant has significantly higher *IS256* copies and transposition frequencies [80]. However, there are yet more layers to this example. In 2003, Jefferson and colleagues discovered a 5-nucleotide SSR (TATTT) in the promoter region of the *ica* operon, whose expansion/contraction affected the binding of *ica* regulatory elements and shut down PNAG/PIA production [83]. In a subsequent study, Brooks and Jefferson discovered that there are further SSRs present in the operon in the form of tetranucleotide repeats within the *icaC* ORF (Open Reading Frame), and SSM events in those also reversibly control PNAG/PIA production [81]. Finally, an elaborate mechanism for the *icaC* SSM expansion has been proposed. The *icaC* tetranucleotide SSR can stably form a so called "mini dumbbell structure" by folding back on itself and making a small loop [84]. It has been proposed that if such a structure were to form during DNA replication it would increase the frequency of SSM events, resulting in expansion of the SSR [84].

Surface proteins are theoretically particularly prone to PV, and one of the first conclusively identified PV events occurs in the extracellular MapW protein, which may have functions in immune evasion based on its high degree of similarity with Major Histocompatibility Complex Class II molecules [85]. The *mapW* gene has a poly(A) tract that results in premature termination of translation. A change in the poly(A) tract can shift the reading frame and result in full-length protein being transcribed [85]. This example of PV varies the length of the protein product rather than switching between an ON↔OFF state, though it is yet to be confirmed if both the truncated and full-length protein have distinct functions.

A PV-like system was also found in the regulation of natural transformation just short of a decade ago [86]. The finding of staphylococcal natural transformation has implications regarding its rapid acquisition of antibiotic resistance genes, but intriguingly it was demonstrated that only a subset of the population can enter a competence state. The genes necessary for entering the competence state are controlled by the alternative sigma factor σ^H [87] and the transcription factor ComK that synergistically works with σ^H [88]. It was found that two independent mechanisms control σ^H expression in *S. aureus* [86]. The translation of the σ^H mRNA is likely repressed through the action of an inverted repeat loop in the 5′ UTR, and the still elusive de-repression mechanism allows σ^H expression in subpopulations. In addition, as a second genetic mechanism, at low frequencies ($\sim \leq 10^{-5}$) *sigH* undergoes a gene duplication event with downstream genes, effectively replacing the native 5′ UTR, and thus lifting repressive control [86]. This event is reversible and reverts to the native chromosomal structure at a frequency of 10^{-2}. Although the frequency of duplication is lower than that commonly associated with PV, the reversible nature of the mechanism coupled with the contingency-like nature of the *sigH* locus [89] allows for a justification of this phenomenon being discussed under the umbrella of PV.

A very recent study in our lab uncovered another example of PV in *S. aureus*, one with potentially far-reaching implications. Upon investigating the phenomenon of hemolytic heterogeneity commonly observed in *S. aureus*, we identified PV-controlled reversible shutdown of the central virulence regulatory system, the Accessory Gene Regulator (Agr) system [90]. PV occurred via two distinct mechanisms: the first was a duplication and inversion event within the ORF of *agrC* (which encodes for the sensor component of the Agr TCS) and the second involved alteration in the length of homonucleotide SSRs within the *agrA* ORF (which encodes for the TCS response regulator). The second mechanism was also identified in a single clinical isolate, and although we were unable to determine the clinical significance of this findings (owing to a minor frequency of clinical revertant strains), the results demonstrated that *S. aureus* has a phasevarion under control of the Agr system. Furthermore, a study that investigated *S. aureus* dermal colonization in children identified that chronic colonizers tended to have a mutationally inactivated Agr system. Importantly, we found that two of their samples (out of four) had frameshifts resulting from alterations in homonucleotide SSRs within the *agrC* and agrA ORFs [91]. The implications of this suggest that this phasevarion could come into play as a cryptic insurance strategy against host-mediated immune attack and may possibly even allow *S. aureus* to manipulate host phagocytic cells and use them as a Trojan horse to disseminate itself within the host (Figure 2). This is of particular consideration as under certain conditions bacterial infections can be established by a single surviving cell [92].

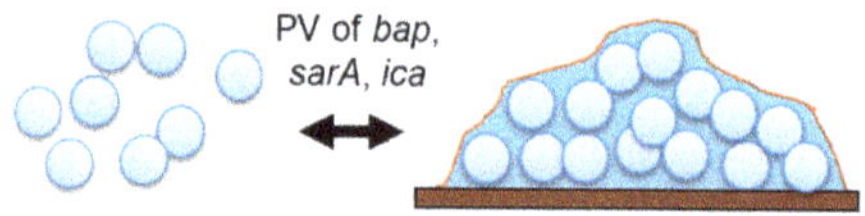

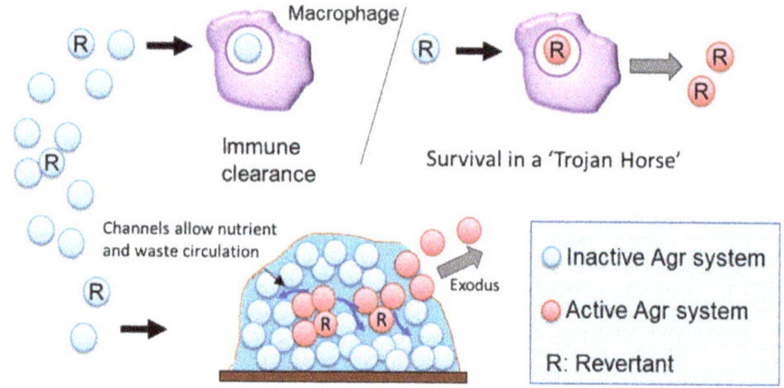

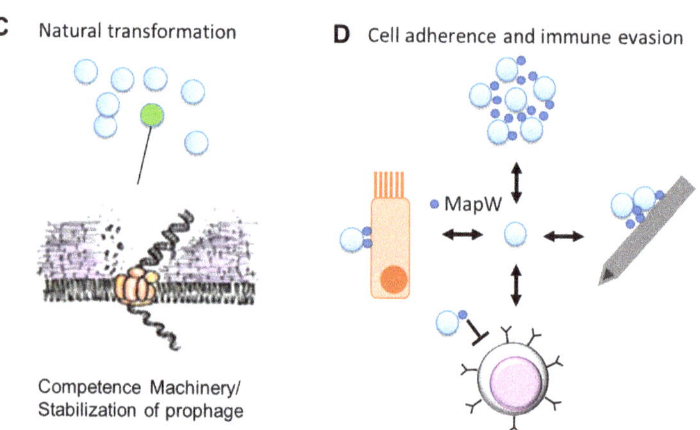

Figure 2. Phase Variation in *Staphylococcus aureus*. A cartoon depicting known PV in *S. aureus* and its roles. Double headed arrows indicate reversible PV events. (**A**) PV of Bap (Biofilm Associated Protein) (unclear mechanism) and of the *ica* (Intracellular Adhesion) operon (transposon insertion and SSR alteration) reversibly affects the biofilm-forming capability. (**B**) Phase variation of the Agr (Accessory Gene Regulator) system may have multiple possible roles [90]. It could serve as a cryptic insurance strategy against host immune attack, allowing phagocytosed revertant cells to activate

their reverted Agr system to survive within a "Trojan Horse" (Top). It may also aid in the proper structuring of biofilms, as revertant cells and their progeny can activate their Agr system in structured non-planktonic architecture and the resultant exoproteins form channels in the biofilm for circulation of waste and nutrients as well as facilitating the exodus of cells from the biofilm [93] (Bottom). (**C**) PV-like expression is one of two independent expression mechanisms known for the alternative sigma factor σ^H. Either mechanism allows for a subpopulation of cells to express the competence machinery and undergo natural transformation. (**D**) PV of the cell-wall associated MapW (MHC class II Analog Protein) may have multiple impacts [94]. MapW is involved in bacterial aggregation and may lead to biofilm formation; It is also implicated in the adherence to the host matrix and in-dwelling medical devices (e.g., cannulas); MapW has immunomodulatory effects and seems to suppress T-cells and their recruitment, though the exact mechanism remains elusive; MapW is important in adherence to, and internalization by, host non-phagocytic cells (e.g., epithelial cells). (Clockwise from top).

4. Future Perspectives

There is mounting evidence of PV being important in the evolution and adaptation of bacteria, with roles ranging from their arms race with phages to their pathogenic proficiency. *S. aureus* is an important global pathogen that can survive in a variety of different niches within the human body using its arsenal of virulence factors. Considering this, *S. aureus* should be primed to exploit PV in its lifestyle yet there remain few documented cases of clear PV phenomena. However, it is very possible that cases reminiscent of PV have been overlooked in the past and may be worth further investigating. For example, a study carried out by Aarestrup and colleagues as far back as 1999 [95] documented heterogenous expression of the alpha and beta hemolysins of *S. aureus* amongst strains that carried the hemolytic genes. This could be due to PV shutdown of these hemolysins in the non-expressing strains. We recently identified non-hemolytic clinical isolates that could revert hemolytic activity without any change in their Agr phenotype [96], which could support the idea that the heterogenous hemolytic phenotypes observed by Aarestrup arise from PV.

With modern developments in genetic and experimental technology, the way we can go about investigating phenomena such as PV has drastically changed. This is no more clearly demonstrated by Jiang et al. who mined genome databases for inverted repeat regions as a primary screen for potentially phase variable genes [56]. In an attempt to gain similar preliminary insight, we screened the genome of the highly virulent community-acquired Methicillin-Resistant strain MW2 for genes that contained homonucleotide SSRs of adenine and thymine that are 6 nucleotides or longer within the ORF and the putative promoter region of the gene. We focused on homonucleotide tracts as three cases of PV in *S. aureus* have been demonstrated to occur through homonucleotide tract-length alteration (*mapW* and 2 discrete SSRs in *agrA*, [85,90]). Our initial results were astonishing. More than 700 genes contained at least one Poly(A) or Poly(T) SSR, with a substantial number containing 3 to 4 SSRs, and one gene containing a stunning 26 SSRs (Figure S1). These initial data corroborate an extensive study carried out by Orsi et al. who identified that Poly(A)/Poly(T) tracts are overrepresented in numerous bacterial genomes [54]. Interestingly, we noticed a greater abundance of Poly(A) tracts compared to Poly(T) in coding sequences. This is corroborated by Orsi et al., who found a similar result when looking at tracts longer than 6 nucleotides in length, though the difference reduced with shorter tracts. Surprisingly however, although Orsi et al. identified that these SSRs are predominantly located near the 5' end of the CDS (coding sequence), we only observed this distribution pattern for Poly(T) SSRs, not Poly(A) (Figure 3). Taken together, PV in *S. aureus* may be severely under-reported, which is understandable as most PV may not be noticeable through conventional bulk analysis and elaborate experimental systems may be needed to observe the PV-mediated switch in gene expression. Intriguingly, approximately 13% of genes (from those with known function) that contained Poly(A) or Poly(T) SSRs were essential genes. If these genes are indeed subject to PV, it raises a question as to what circumstances or trade-offs could possibly merit the shutdown of essential genes as advantageous. Transcriptional slippage at Poly(A)/Poly(T) tracts can lead to a population of mRNAs with varying tract length, some of which may contain the correct number of nucleotides for the entire CDS to be in-frame [97]. This could enable low-level expression of essential genes that have been shut down by PV. Alternatively, the phenomenon of biological hysteresis [98], wherein functional proteins are inherited by daughter cells during splitting from the mother cell, could support daughter cells for a short period without further de novo synthesis, potentially giving chances of further changes to the SSR.

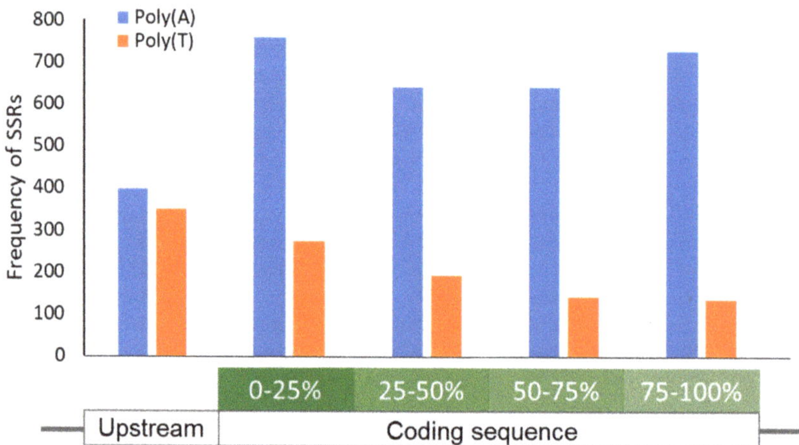

Figure 3. Location distribution of homonucleotide SSRs in *S. aureus*. A graph showing the location-dependent frequency of Poly(A)/(T) SSRs $\geq$6 nucleotides in length. A 100-nucleotide region upstream of all Open Reading Frames (ORFs) was also screened for SSRs. While Poly(T) SSRs are relatively enriched towards the 5′ of the ORF as reported [54], Poly(A) SSRs $\geq$6 nucleotides in length were found to be evenly distributed across the ORF. In contrast, the abundance of both SSR types is evenly matched in the region immediately upstream of the ORF.

The aim of this review is to stimulate interest in identifying PV in *S. aureus* and to increase attention to an area of study that warrants more investigation with modern technological approaches. Here, we have described a comprehensive understanding of the mechanisms of PV and the scope of the discoveries yet to be made in *S. aureus*. Further investigations focusing on PV in *S. aureus* are sure to lead to exciting new information, and the more we learn of the ingenious adaptation mechanisms this important pathogen employs during its infectious process, the better we will be equipped in dealing with it.

Supplementary Materials: The following are available online at https://www.mdpi.com/2076-2607/9/2/244/s1, Figure S1: Homonucleotide SSRs in S. aureus MW2.

Author Contributions: Conceptualization, V.G. and K.M.; sequence analysis, R.L.O.; writing—original draft preparation, V.G.; writing—review and editing, V.G. and K.M.; visualization, V.G., R.L.O. and K.M. All authors have read and agreed to the published version of the manuscript.

Funding: This work was partly supported by JSPS KAKENHI grant number 18H02652.

Conflicts of Interest: The authors declare no conflict of interest.

References

1. Cosgrove, S.E.; Qi, Y.; Kaye, K.S.; Harbarth, S.; Karchmer, A.W.; Carmeli, Y. The impact of methicillin resistance in *Staphylococcus aureus* bacteremia on patient outcomes: Mortality, length of stay, and hospital charges. *Infect. Control Hosp. Epidemiol.* **2005**, *26*, 166–174. [CrossRef] [PubMed]
2. Wertheim, H.F.L.; Melles, D.C.; Vos, M.C.; van Leeuwen, W.; van Belkum, A.; Verbrugh, H.A.; Nouwen, J.L. The role of nasal carriage in *Staphylococcus aureus* infections. *Lancet Infect. Dis.* **2005**, *5*, 751–762. [CrossRef]
3. Tong, S.Y.; Davis, J.S.; Eichenberger, E.; Holland, T.L.; Fowler, V.G., Jr. *Staphylococcus aureus* infections: Epidemiology, pathophysiology, clinical manifestations, and management. *Clin. Microbiol. Rev.* **2015**, *28*, 603–661. [CrossRef] [PubMed]

4.	Reddy, P.N.; Srirama, K.; Dirisala, V.R. An Update on Clinical Burden, Diagnostic Tools, and Therapeutic Options of *Staphylococcus aureus*. *Infect. Dis. Res. Treat.* **2017**, *10*, 1179916117703999. [CrossRef]

5.	Oliveira, D.; Borges, A.; Simões, M. *Staphylococcus aureus* Toxins and Their Molecular Activity in Infectious Diseases. *Toxins* **2018**, *10*, 252. [CrossRef] [PubMed]

6.	Tam, K.; Torres, V.J. *Staphylococcus aureus* Secreted Toxins and Extracellular Enzymes. *Microbiol. Spectr.* **2019**, *7*. [CrossRef]

7.	Gajdács, M. The Continuing Threat of Methicillin-Resistant *Staphylococcus aureus*. *Antibiotics* **2019**, *8*, 52. [CrossRef]

8.	Cuny, C.; Wieler, L.H.; Witte, W. Livestock-Associated MRSA: The Impact on Humans. *Antibiotics* **2015**, *4*, 521–543. [CrossRef]

9.	Rice, L.B. Federal Funding for the Study of Antimicrobial Resistance in Nosocomial Pathogens: No ESKAPE. *J. Infect. Dis.* **2008**, *197*, 1079–1081. [CrossRef]

10.	Chang, J.; Lee, R.-E.; Lee, W. A pursuit of *Staphylococcus aureus* continues: A role of persister cells. *Arch. Pharmacal. Res.* **2020**, *43*, 630–638. [CrossRef]

11.	Garcia, L.G.; Lemaire, S.; Kahl, B.C.; Becker, K.; Proctor, R.A.; Denis, O.; Tulkens, P.M.; Van Bambeke, F. Antibiotic activity against small-colony variants of *Staphylococcus aureus*: Review of in vitro, animal and clinical data. *J. Antimicrob. Chemother.* **2013**, *68*, 1455–1464. [CrossRef] [PubMed]

12.	Kahl, B.C. Small colony variants (SCVs) of *Staphylococcus aureus*—A bacterial survival strategy. *Infect. Genet. Evol.* **2014**, *21*, 515–522. [CrossRef] [PubMed]

13.	Fauvart, M.; De Groote, V.N.; Michiels, J. Role of persister cells in chronic infections: Clinical relevance and perspectives on anti-persister therapies. *J. Med. Microbiol.* **2011**, *60*, 699–709. [CrossRef] [PubMed]

14.	Moxon, R.; Bayliss, C.; Hood, D. Bacterial Contingency Loci: The Role of Simple Sequence DNA Repeats in Bacterial Adaptation. *Annu. Rev. Genet.* **2006**, *40*, 307–333. [CrossRef] [PubMed]

15.	Moxon, E.R.; Rainey, P.B.; Nowak, M.A.; Lenski, R.E. Adaptive evolution of highly mutable loci in pathogenic bacteria. *Curr. Biol.* **1994**, *4*, 24–33. [CrossRef]

16.	Phillips, Z.N.; Tram, G.; Seib, K.L.; Atack, J.M. Phase-variable bacterial loci: How bacteria gamble to maximise fitness in changing environments. *Biochem. Soc. Trans.* **2019**, *47*, 1131–1141. [CrossRef] [PubMed]

17.	Bayliss, C.D. Determinants of phase variation rate and the fitness implications of differing rates for bacterial pathogens and commensals. *FEMS Microbiol. Rev.* **2009**, *33*, 504–520. [CrossRef]

18.	Wisniewski-Dyé, F.; Vial, L. Phase and antigenic variation mediated by genome modifications. *Antonie Leeuwenhoek* **2008**, *94*, 493–515. [CrossRef]

19.	Van der Woude, M.W.; Baumler, A.J. Phase and antigenic variation in bacteria. *Clin. Microbiol Rev.* **2004**, *17*, 581–611. [CrossRef]

20.	Atack, J.M.; Winter, L.E.; Jurcisek, J.A.; Bakaletz, L.O.; Barenkamp, S.J.; Jennings, M.P. Selection and Counterselection of *Hia* Expression Reveals a Key Role for Phase-Variable Expression of Hia in Infection Caused by Nontypeable *Haemophilus influenzae*. *J. Infect. Dis.* **2015**, *212*, 645–653. [CrossRef]

21.	Blyn, L.B.; Braaten, B.A.; Low, D.A. Regulation of pap pilin phase variation by a mechanism involving differential dam methylation states. *EMBO J.* **1990**, *9*, 4045–4054. [CrossRef] [PubMed]

22.	Fox, K.L.; Atack, J.M.; Srikhanta, Y.N.; Eckert, A.; Novotny, L.A.; Bakaletz, L.O.; Jennings, M.P. Selection for phase variation of LOS biosynthetic genes frequently occurs in progression of non-typeable *Haemophilus influenzae* infection from the nasopharynx to the middle ear of human patients. *PLoS ONE* **2014**, *9*, e90505. [CrossRef] [PubMed]

23.	Richardson, A.R.; Stojiljkovic, I. HmbR, a Hemoglobin-Binding Outer Membrane Protein of *Neisseria meningitidis*, Undergoes Phase Variation. *J. Bacteriol.* **1999**, *181*, 2067–2074. [CrossRef] [PubMed]

24.	Ren, Z.; Jin, H.; Whitby, P.W.; Morton, D.J.; Stull, T.L. Role of CCAA Nucleotide Repeats in Regulation of Hemoglobin and Hemoglobin-Haptoglobin Binding Protein Genes of *Haemophilus influenzae*. *J. Bacteriol.* **1999**, *181*, 5865–5870. [CrossRef]

25.	Fox, K.L.; Yildirim, H.H.; Deadman, M.E.; Schweda, E.K.H.; Moxon, E.R.; Hood, D.W. Novel lipopolysaccharide biosynthetic genes containing tetranucleotide repeats in *Haemophilus influenzae*, identification of a gene for adding O-acetyl groups. *Mol. Microbiol.* **2005**, *58*, 207–216. [CrossRef]

26.	Langereis, J.D.; Weiser, J.N. Shielding of a Lipooligosaccharide IgM Epitope Allows Evasion of Neutrophil-Mediated Killing of an Invasive Strain of Nontypeable *Haemophilus influenzae*. *mBio* **2014**, *5*, e01478-14. [CrossRef]

27.	Ikeda, J.S.; Schmitt, C.K.; Darnell, S.C.; Watson, P.R.; Bispham, J.; Wallis, T.S.; Weinstein, D.L.; Metcalf, E.S.; Adams, P.; O'Connor, C.D.; et al. Flagellar Phase Variation of *Salmonella enterica* Serovar Typhimurium Contributes to Virulence in the Murine Typhoid Infection Model but Does Not Influence *Salmonella*-Induced Enteropathogenesis. *Infect. Immun.* **2001**, *69*, 3021–3030. [CrossRef]

28.	Bikard, D.; Marraffini, L.A. Innate and adaptive immunity in bacteria: Mechanisms of programmed genetic variation to fight bacteriophages. *Curr. Opin. Immunol.* **2012**, *24*, 15–20. [CrossRef]

29.	Hyman, P.; Abedon, S.T. Chapter 7—Bacteriophage Host Range and Bacterial Resistance. In *Advances in Applied Microbiology*; Academic Press: Cambridge, MA, USA, 2010; Volume 70, pp. 217–248.

30.	Stern, A.; Sorek, R. The phage-host arms race: Shaping the evolution of microbes. *Bioessays News Rev. Mol. Cell. Dev. Biol.* **2011**, *33*, 43–51. [CrossRef]

31.	Zaleski, P.; Wojciechowski, M.; Piekarowicz, A. The role of Dam methylation in phase variation of *Haemophilus influenzae* genes involved in defence against phage infection. *Microbiology* **2005**, *151*, 3361–3369. [CrossRef]

32. Kamp, D.; Kahmann, R.; Zipser, D.; Broker, T.R.; Chow, L.T. Inversion of the G DNA segment of phage Mu controls phage infectivity. *Nature* **1978**, *271*, 577–580. [CrossRef] [PubMed]
33. Van de Putte, P.; Cramer, S.; Giphart-Gassler, M. Invertible DNA determines host specificity of bacteriophage Mu. *Nature* **1980**, *286*, 218–222. [CrossRef]
34. Sandmeler, H. Acquisition and rearrangement of sequence motifs in the evolution of bacteriophage tail fibres. *Mol. Microbiol.* **1994**, *12*, 343–350. [CrossRef] [PubMed]
35. Levinson, G.; Gutman, G.A. Slipped-strand mispairing: A major mechanism for DNA sequence evolution. *Mol. Biol. Evol.* **1987**, *4*, 203–221. [CrossRef] [PubMed]
36. Streisinger, G.; Okada, Y.; Emrich, J.; Newton, J.; Tsugita, A.; Terzaghi, E.; Inouye, M. Frameshift mutations and the genetic code. *Cold Spring Harb. Symp. Quant. Biol.* **1966**, *31*, 77–84. [CrossRef] [PubMed]
37. Streisinger, G.; Owen, J.E. Mechanisms of spontaneous and induced frameshift mutation in bacteriophage T4. *Genetics* **1985**, *109*, 633–659. [CrossRef]
38. Andrewes, F.W. Studies in group-agglutination I. The salmonella group and its antigenic structure. *J. Pathol. Bacteriol.* **1922**, *25*, 505–521. [CrossRef]
39. Sechman, E.V.; Rohrer, M.S.; Seifert, H.S. A genetic screen identifies genes and sites involved in pilin antigenic variation in *Neisseria gonorrhoeae*. *Mol. Microbiol.* **2005**, *57*, 468–483. [CrossRef]
40. Mehr, I.J.; Seifert, H.S. Differential roles of homologous recombination pathways in *Neisseria gonorrhoeae* pilin antigenic variation, DNA transformation and DNA repair. *Mol. Microbiol.* **1998**, *30*, 697–710. [CrossRef]
41. Van Ham, S.M.; van Alphen, L.; Mooi, F.R.; van Putten, J.P.M. Phase variation of *H. influenzae* fimbriae: Transcriptional control of two divergent genes through a variable combined promoter region. *Cell* **1993**, *73*, 1187–1196. [CrossRef]
42. Bayliss, C.D.; Palmer, M.E. Evolution of simple sequence repeat–mediated phase variation in bacterial genomes. *Ann. N. Y. Acad. Sci.* **2012**, *1267*, 39–44. [CrossRef] [PubMed]
43. Thomas, A.; Kunkel, K.B. DNA Replication Fidelity. *Annu. Rev. Biochem.* **2000**, *69*, 497–529. [CrossRef]
44. Pham, P.T.; Olson, M.W.; McHenry, C.S.; Schaaper, R.M. The Base Substitution and Frameshift Fidelity of *Escherichia coli* DNA Polymerase III Holoenzyme in Vitro. *J. Biol. Chem.* **1998**, *273*, 23575–23584. [CrossRef]
45. Richardson, A.R.; Stojiljkovic, I. Mismatch repair and the regulation of phase variation in *Neisseria meningitidis*. *Mol. Microbiol.* **2001**, *40*, 645–655. [CrossRef] [PubMed]
46. Taddei, F.; Radman, M.; Maynard-Smith, J.; Toupance, B.; Gouyon, P.H.; Godelle, B. Role of mutator alleles in adaptive evolution. *Nature* **1997**, *387*, 700–702. [CrossRef]
47. Prunier, A.-L.; Malbruny, B.; Laurans, M.; Brouard, J.; Duhamel, J.-F.; Leclercq, R. High Rate of Macrolide Resistance in *Staphylococcus aureus* Strains from Patients with Cystic Fibrosis Reveals High Proportions of Hypermutable Strains. *J. Infect. Dis.* **2003**, *187*, 1709–1716. [CrossRef]
48. Prunier, A.-L.; Leclercq, R. Role of *mutS* and *mutL* Genes in Hypermutability and Recombination in *Staphylococcus aureus*. *J. Bacteriol.* **2005**, *187*, 3455–3464. [CrossRef]
49. Behzadi, P.; Baráth, Z.; Gajdács, M. It's Not Easy Being Green: A Narrative Review on the Microbiology, Virulence and Therapeutic Prospects of Multidrug-Resistant *Pseudomonas aeruginosa*. *Antibiotics* **2021**, *10*, 42. [CrossRef]
50. Martin, P.; Sun, L.; Hood, D.W.; Moxon, E.R. Involvement of genes of genome maintenance in the regulation of phase variation frequencies in *Neisseria meningitidis*. *Microbiology* **2004**, *150*, 3001–3012. [CrossRef]
51. Bayliss, C.D.; van de Ven, T.; Moxon, E.R. Mutations in poll but not mutSLH destabilize *Haemophilus influenzae* tetranucleotide repeats. *EMBO J.* **2002**, *21*, 1465–1476. [CrossRef]
52. Tran, H.T.; Keen, J.D.; Kricker, M.; Resnick, M.A.; Gordenin, D.A. Hypermutability of homonucleotide runs in mismatch repair and DNA polymerase proofreading yeast mutants. *Mol. Cell. Biol.* **1997**, *17*, 2859–2865. [CrossRef] [PubMed]
53. Gawel, D.; Jonczyk, P.; Bialoskorska, M.; Schaaper, R.M.; Fijalkowska, I.J. Asymmetry of frameshift mutagenesis during leading and lagging-strand replication in *Escherichia coli*. *Mutat. Res./Fundam. Mol. Mech. Mutagenesis* **2002**, *501*, 129–136. [CrossRef]
54. Orsi, R.H.; Bowen, B.M.; Wiedmann, M. Homopolymeric tracts represent a general regulatory mechanism in prokaryotes. *BMC Genom.* **2010**, *11*, 102. [CrossRef] [PubMed]
55. Lin, W.-H.; Kussell, E. Evolutionary pressures on simple sequence repeats in prokaryotic coding regions. *Nucleic Acids Res.* **2011**, *40*, 2399–2413. [CrossRef] [PubMed]
56. Jiang, X.; Hall, A.B.; Arthur, T.D.; Plichta, D.R.; Covington, C.T.; Poyet, M.; Crothers, J.; Moses, P.L.; Tolonen, A.C.; Vlamakis, H.; et al. Invertible promoters mediate bacterial phase variation, antibiotic resistance, and host adaptation in the gut. *Science* **2019**, *363*, 181–187. [CrossRef] [PubMed]
57. Henderson, I.R.; Owen, P.; Nataro, J.P. Molecular switches—The ON and OFF of bacterial phase variation. *Mol. Microbiol.* **1999**, *33*, 919–932. [CrossRef]
58. Rudel, T.; Scheuerpflug, I.; Meyer, T.F. *Neisseria* PilC protein identified as type-4 pilus tip-located adhesin. *Nature* **1995**, *373*, 357–359. [CrossRef]
59. Rytkönen, A.; Albiger, B.; Hansson-Palo, P.; Källström, H.; Olcén, P.; Fredlund, H.; Jonsson, A.-B. *Neisseria meningitidis* Undergoes PilC Phase Variation and PilE Sequence Variation during Invasive Disease. *J. Infect. Dis.* **2004**, *189*, 402–409. [CrossRef]
60. Jonsson, A.B.; Nyberg, G.; Normark, S. Phase variation of gonococcal pili by frameshift mutation in pilC, a novel gene for pilus assembly. *EMBO J.* **1991**, *10*, 477–488. [CrossRef]

61. Danne, C.; Dubrac, S.; Trieu-Cuot, P.; Dramsi, S. Single Cell Stochastic Regulation of Pilus Phase Variation by an Attenuation-like Mechanism. *PLoS Pathog.* **2014**, *10*, e1003860. [CrossRef]

62. Berger, S.L.; Kouzarides, T.; Shiekhattar, R.; Shilatifard, A. An operational definition of epigenetics. *Genes Dev.* **2009**, *23*, 781–783. [CrossRef]

63. Dupont, C.; Armant, D.R.; Brenner, C.A. Epigenetics: Definition, mechanisms and clinical perspective. *Semin. Reprod. Med.* **2009**, *27*, 351–357. [CrossRef]

64. Blow, M.J.; Clark, T.A.; Daum, C.G.; Deutschbauer, A.M.; Fomenkov, A.; Fries, R.; Froula, J.; Kang, D.D.; Malmstrom, R.R.; Morgan, R.D.; et al. The Epigenomic Landscape of Prokaryotes. *PLoS Genet.* **2016**, *12*, e1005854. [CrossRef]

65. Wion, D.; Casadesús, J. N6-methyl-adenine: An epigenetic signal for DNA–protein interactions. *Nat. Rev. Microbiol.* **2006**, *4*, 183–192. [CrossRef]

66. Estibariz, I.; Overmann, A.; Ailloud, F.; Krebes, J.; Josenhans, C.; Suerbaum, S. The core genome ^{m5}C methyltransferase JHP1050 (M.Hpy99III) plays an important role in orchestrating gene expression in *Helicobacter pylori*. *Nucleic Acids Res.* **2019**, *47*, 2336–2348. [CrossRef]

67. Van der Woude, M.; Braaten, B.; Low, D. Epigenetic phase variation of the *pap* operon in *Escherichia coli*. *Trends Microbiol.* **1996**, *4*, 5–9. [CrossRef]

68. Cota, I.; Sánchez-Romero, M.A.; Hernández, S.B.; Pucciarelli, M.G.; García-del Portillo, F.; Casadesús, J. Epigenetic Control of *Salmonella enterica* O-Antigen Chain Length: A Tradeoff between Virulence and Bacteriophage Resistance. *PLoS Genet.* **2015**, *11*, e1005667. [CrossRef]

69. Cota, I.; Blanc-Potard, A.B.; Casadesús, J. STM2209-STM2208 (opvAB): A Phase Variation Locus of *Salmonella enterica* Involved in Control of O-Antigen Chain Length. *PLoS ONE* **2012**, *7*, e36863. [CrossRef]

70. Srikhanta, Y.N.; Maguire, T.L.; Stacey, K.J.; Grimmond, S.M.; Jennings, M.P. The phasevarion: A genetic system controlling coordinated, random switching of expression of multiple genes. *Proc. Natl. Acad. Sci. USA* **2005**, *102*, 5547–5551. [CrossRef]

71. De Bolle, X.; Bayliss, C.D.; Field, D.; Van De Ven, T.; Saunders, N.J.; Hood, D.W.; Moxon, E.R. The length of a tetranucleotide repeat tract in *Haemophilus influenzae* determines the phase variation rate of a gene with homology to type III DNA methyltransferases. *Mol. Microbiol.* **2000**, *35*, 211–222. [CrossRef]

72. De Vries, N.; Duinsbergen, D.; Kuipers, E.J.; Pot, R.G.J.; Wiesenekker, P.; Penn, C.W.; van Vliet, A.H.M.; Vandenbroucke-Grauls, C.M.J.E.; Kusters, J.G. Transcriptional Phase Variation of a Type III Restriction-Modification System in *Helicobacter pylori*. *J. Bacteriol.* **2002**, *184*, 6615–6623. [CrossRef] [PubMed]

73. Atack, J.M.; Yang, Y.; Seib, K.L.; Zhou, Y.; Jennings, M.P. A survey of Type III restriction-modification systems reveals numerous, novel epigenetic regulators controlling phase-variable regulons; phasevarions. *Nucleic Acids Res.* **2018**, *46*, 3532–3542. [CrossRef] [PubMed]

74. Anjum, A.; Brathwaite, K.J.; Aidley, J.; Connerton, P.L.; Cummings, N.J.; Parkhill, J.; Connerton, I.; Bayliss, C.D. Phase variation of a Type IIG restriction-modification enzyme alters site-specific methylation patterns and gene expression in *Campylobacter jejuni* strain NCTC11168. *Nucleic Acids Res.* **2016**, *44*, 4581–4594. [CrossRef]

75. De Ste Croix, M.; Vacca, I.; Kwun, M.J.; Ralph, J.D.; Bentley, S.D.; Haigh, R.; Croucher, N.J.; Oggioni, M.R. Phase-variable methylation and epigenetic regulation by type I restriction-modification systems. *FEMS Microbiol. Rev.* **2017**, *41*, S3–S15. [CrossRef] [PubMed]

76. Sánchez-Romero, M.A.; Casadesús, J. The bacterial epigenome. *Nat. Rev. Microbiol.* **2020**, *18*, 7–20. [CrossRef]

77. Garcia-Betancur, J.C.; Goni-Moreno, A.; Horger, T.; Schott, M.; Sharan, M.; Eikmeier, J.; Wohlmuth, B.; Zernecke, A.; Ohlsen, K.; Kuttler, C.; et al. Cell differentiation defines acute and chronic infection cell types in *Staphylococcus aureus*. *eLife* **2017**, *6*, e28023. [CrossRef]

78. Baselga, R.; Albizu, I.; De La Cruz, M.; Del Cacho, E.; Barberan, M.; Amorena, B. Phase variation of slime production in *Staphylococcus aureus: Implications* in colonization and virulence. *Infect. Immun.* **1993**, *61*, 4857–4862. [CrossRef]

79. Tormo, M.Á.; Úbeda, C.; Martí, M.; Maiques, E.; Cucarella, C.; Valle, J.; Foster, T.J.; Lasa, Í.; Penadés, J.R. Phase-variable expression of the biofilm-associated protein (Bap) in Staphylococcus aureus. *Microbiology* **2007**, *153*, 1702–1710. [CrossRef]

80. Valle, J.; Vergara-Irigaray, M.; Merino, N.; Penadés, J.R.; Lasa, I. σ^B Regulates *IS256*-Mediated *Staphylococcus aureus* Biofilm Phenotypic Variation. *J. Bacteriol.* **2007**, *189*, 2886–2896. [CrossRef]

81. Brooks, J.L.; Jefferson, K.K. Phase Variation of Poly-N-Acetylglucosamine Expression in *Staphylococcus aureus*. *PLoS Pathog.* **2014**, *10*, e1004292. [CrossRef]

82. Cerca, N.; Jefferson, K.K.; Maira-Litrán, T.; Pier, D.B.; Kelly-Quintos, C.; Goldmann, D.A.; Azeredo, J.; Pier, G.B. Molecular Basis for Preferential Protective Efficacy of Antibodies Directed to the Poorly Acetylated Form of Staphylococcal Poly-N-Acetyl-β-(1-6)-Glucosamine. *Infect. Immun.* **2007**, *75*, 3406–3413. [CrossRef] [PubMed]

83. Jefferson, K.K.; Cramton, S.E.; Götz, F.; Pier, G.B. Identification of a 5-nucleotide sequence that controls expression of the *ica* locus in *Staphylococcus aureus* and characterization of the DNA-binding properties of IcaR. *Mol. Microbiol.* **2003**, *48*, 889–899. [CrossRef] [PubMed]

84. Guo, P.; Lam, S.L. Unusual structures of TTTA repeats in icaC gene of *Staphylococcus aureus*. *FEBS Lett.* **2015**, *589*, 1296–1300. [CrossRef] [PubMed]

85. Buckling, A.; Neilson, J.; Lindsay, J.; ffrench-Constant, R.; Enright, M.; Day, N.; Massey, R.C. Clonal Distribution and Phase-Variable Expression of a Major Histocompatibility Complex Analogue Protein in *Staphylococcus aureus*. *J. Bacteriol.* **2005**, *187*, 2917. [CrossRef]

86. Morikawa, K.; Takemura, A.J.; Inose, Y.; Tsai, M.; Nguyen Thi le, T.; Ohta, T.; Msadek, T. Expression of a cryptic secondary sigma factor gene unveils natural competence for DNA transformation in *Staphylococcus aureus*. *PLoS Pathog* **2012**, *8*, e1003003. [CrossRef]

87. Morikawa, K.; Inose, Y.; Okamura, H.; Maruyama, A.; Hayashi, H.; Takeyasu, K.; Ohta, T. A new staphylococcal sigma factor in the conserved gene cassette: Functional significance and implication for the evolutionary processes. *Genes Cells* **2003**, *8*, 699–712. [CrossRef]

88. Fagerlund, A.; Granum, P.E.; Håvarstein, L.S. *Staphylococcus aureus* competence genes: Mapping of the SigH, ComK1 and ComK2 regulons by transcriptome sequencing. *Mol. Microbiol.* **2014**, *94*, 557–579. [CrossRef]

89. Morikawa, K.; Ohniwa, R.L.; Kumano, M.; Okamura, H.; Saito, S.; Ohta, T. The sigH gene sequence can subspeciate staphylococci. *Diagn. Microbiol. Infect. Dis.* **2008**, *61*, 373–380. [CrossRef]

90. Gor, V.; Takemura, A.J.; Nishitani, M.; Higashide, M.; Medrano Romero, V.; Ohniwa, R.L.; Morikawa, K. Finding of Agr Phase Variants in *Staphylococcus aureus*. *mBio* **2019**, *10*, e00796-19. [CrossRef]

91. Nakamura, Y.; Takahashi, H.; Takaya, A.; Inoue, Y.; Katayama, Y.; Kusuya, Y.; Shoji, T.; Takada, S.; Nakagawa, S.; Oguma, R.; et al. Staphylococcus Agr virulence is critical for epidermal colonization and associates with atopic dermatitis development. *Sci. Transl. Med.* **2020**, *12*, eaay4068. [CrossRef]

92. Moxon, E.R.; Murphy, P.A. *Haemophilus influenzae* bacteremia and meningitis resulting from survival of a single organism. *Proc. Natl. Acad. Sci. USA* **1978**, *75*, 1534–1536. [CrossRef] [PubMed]

93. Periasamy, S.; Joo, H.-S.; Duong, A.C.; Bach, T.-H.L.; Tan, V.Y.; Chatterjee, S.S.; Cheung, G.Y.C.; Otto, M. How *Staphylococcus aureus* biofilms develop their characteristic structure. *Proc. Natl. Acad. Sci. USA* **2012**, *109*, 1281–1286. [CrossRef] [PubMed]

94. Harraghy, N.; Hussain, M.; Haggar, A.; Chavakis, T.; Sinha, B.; Herrmann, M.; Flock, J.-I. The adhesive and immunomodulating properties of the multifunctional *Staphylococcus aureus* protein Eap. *Microbiology* **2003**, *149*, 2701–2707. [CrossRef] [PubMed]

95. Aarestrup, F.M.; Larsen, H.D.; Eriksen, N.H.; Elsberg, C.S.; Jensen, N.E. Frequency of alpha- and beta-haemolysin in *Staphylococcus aureus* of bovine and human origin. A comparison between pheno- and genotype and variation in phenotypic expression. *APMIS Acta Pathol. Microbiol. Immunol. Scand.* **1999**, *107*, 425–430. [CrossRef]

96. Gor, V.; Hoshi, M.; Takemura, A.; Higashide, M.; Romero, V.; Ohniwa, R.; Morikawa, K. Prevalence of Agr phase variants in *Staphylococcus aureus*. *Access Microbiol.* **2020**, *2*, 2. [CrossRef]

97. Tamas, I.; Wernegreen, J.J.; Nystedt, B.; Kauppinen, S.N.; Darby, A.C.; Gomez-Valero, L.; Lundin, D.; Poole, A.M.; Andersson, S.G. Endosymbiont gene functions impaired and rescued by polymerase infidelity at poly(A) tracts. *Proc. Natl. Acad. Sci. USA* **2008**, *105*, 14934–14939. [CrossRef]

98. Veening, J.W.; Smits, W.K.; Kuipers, O.P. Bistability, epigenetics, and bet-hedging in bacteria. *Annu. Rev. Microbiol.* **2008**, *62*, 193–210. [CrossRef]

MDPI
St. Alban-Anlage 66
4052 Basel
Switzerland
Tel. +41 61 683 77 34
Fax +41 61 302 89 18
www.mdpi.com

Microorganisms Editorial Office
E-mail: microorganisms@mdpi.com
www.mdpi.com/journal/microorganisms